Advanced Fluid Dynamics and its Models

Advanced Fluid Dynamics and its Models

Edited by **Maria Forest**

NY RESEARCH PRESS

P R E S S

New York

Published by NY Research Press,
23 West, 55th Street, Suite 816,
New York, NY 10019, USA
www.nyresearchpress.com

Advanced Fluid Dynamics and its Models
Edited by Maria Forest

International Standard Book Number: 978-1-63238-015-9 (Hardback)

Printed in the United States of America.

Contents

Preface

This book aims to highlight the current researches and provides a platform to further the scope of innovations in this area. This book is a product of the combined efforts of many researchers and scientists, after going through thorough studies and analysis from different parts of the world. The objective of this book is to provide the readers with the latest information of the field.

Fluid dynamics is the sub-specialty of fluid mechanics dealing with the study of fluids in motion. This book demonstrates essential developments and applications in fluid dynamics modeling with emphasis on biomedical, bioengineering, chemical, civil and environmental engineering, aeronautics, astronautics, and automotive. This book will prove to be a valuable resource to scientists and engineers engaged in the study of fundamentals and applications of fluid dynamics.

I would like to express my sincere thanks to the authors for their dedicated efforts in the completion of this book. I acknowledge the efforts of the publisher for providing constant support. Lastly, I would like to thank my family for their support in all academic endeavors.

Editor

Surface Friction and Boundary Layer Thickening in Transitional Flow

Ping Lu, Manoj Thapa and Chaoqun Liu

Additional information is available at the end of the chapter

1. Introduction

Correct understanding of turbulence model, particularly boundary-layer turbulence model, has been a subject of significant investigation for over a century, but still is a great challenge for scientists[1]. Therefore, successful efforts to control the shear stress for turbulent boundary-layer flow would be much beneficial for significant savings in power requirements for the vehicle and aircraft, etc. Therefore, for many years scientists connected with the industry have been studying for finding some ways of controlling and reducing the skin-friction[2]. Experimentally, it has been shown that the surface friction coefficient for the turbulent boundary layer may be two to five times greater than that for laminar boundary layer[6]. By careful analysis of our new DNS results, we found that the skin-friction is immediately enlarged to three times greater during the transition from laminar to turbulent flow. We try to give the mechanism of this phenomenon by studying the flow transition over a flat plate, which may provide us an idea how to design a device and reduce shear stress.

Meanwhile, some of the current researches are focused on how to design a device that can artificially increase the thickness of the boundary layer in the wind tunnel. For instances, one way to increase is by using an array of varying diameter cross flow jets with the jet diameter reducing with distance downstream, and there are other methods like boundary layer fence, array of cylinders, or distributed drag method, etc. For detail information read [9]. However, there are few literatures which give the mechanism how the multi-level rings overlap and how boundary layer becomes thicker. By looking at the Figure 1 which is copied from the book of Schilichting, we can note that the boundary layer becomes thicker and thicker during the transition from laminar to turbulent flow. This phenomenon is also numerically proved by our DNS results by flow transition over a flat plate, which is shown in Figure 2 representing multiple level ring overlap. Moreover, we find that they never mix each other. More details will be given in the following sections.

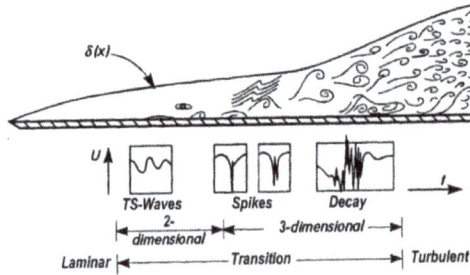

[16] (Copy of Figure 15.38, Page 474, Book of layer thickening Boundary Layer Theory by Schilichting et al, 2000)

Figure 1. Schematic of flow transition on a flat plate

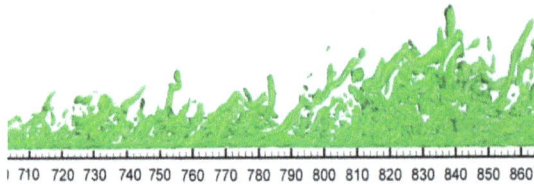

Figure 2. Vortex cycles overlapping and boundary

2. Case setup

The computational domain on a flat plate is displayed in Figure 3. The grid level is 1920x128x241, representing the number of grids in streamwise (x), spanwise (y), and wall normal (z) directions. The grid is stretched in the normal direction and uniform in the streamwise and spanwise directions. The length of the first grid interval in the normal direction at the entrance is found to be 0.43 in wall units ($z^+ = 0.43$) .

The parallel computation is accomplished through the Message Passing Interface (MPI) together with domain decomposition in the streamwise direction. The computational domain is partitioned into N equally-sized sub-domains along the streamwise direction. N is the number of processors used in the parallel computation. The flow parameters, including Mach number, Reynolds number, etc are listed in Table 1. Here, $x_{in} = 300.79\delta_{in}$ represents the distance between leading edge and inlet, $Lx = 798.03\delta_{in}$, $Ly = 22\delta_{in}$, $Lz_{in} = 40\delta_{in}$ are the lengths of the computational domain in x-, y-, and z-directions respectively and $T_w = 273.15K$ is the wall temperature.

M_∞	Re	x_{in}	Lx	Ly	Lz	T_w	T_∞
0.5	1000	300.79 δ_s	798.03 δ_{in}	22 δ_{in}	40 δ_{in}	273.15K	273.15K

Table 1. Flow parameters

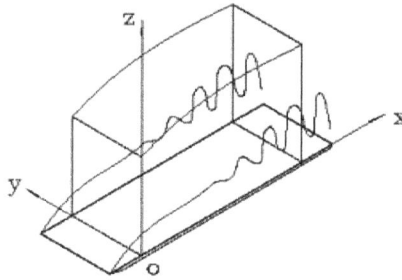

Figure 3. Computation domain

3. Code validation and DNS results visualization

To justify the DNS codes and DNS results, a number of verifications and validations have been conducted[5,12,13,14,15]

1. Comparison with Linear Theory

Figure 4(a) compares the velocity profile of the T-S wave given by our DNS results to linear theory. Figure 4(b) is a comparison of the perturbation amplification rate between DNS and LST. The agreement between linear theory and our numerical results is good.

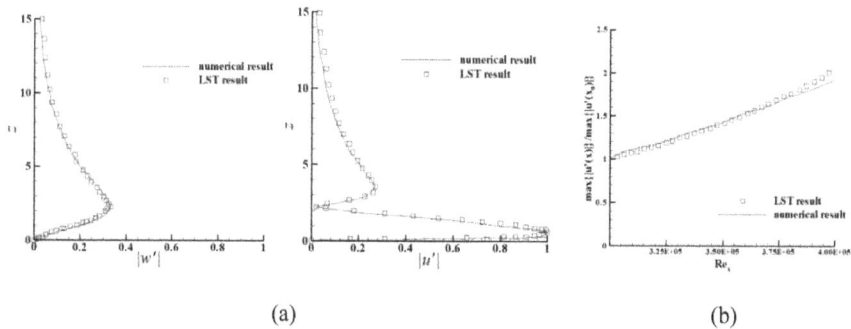

(a)   (b)

Figure 4. Comparison of the numerical and LST (a) velocity profiles at Rex=394300 (b) perturbation amplification rate

2. Skin friction and grid convergence

The skin friction coefficients calculated from the time-averaged and spanwise-averaged profiles on coarse and fine grids are displayed in Figure 5(a). The spatial evolution of skin friction coefficients of laminar flow is also plotted out for comparison. It is observed from these figures that the sharp growth of the skin-friction coefficient occurs after $x \approx 450\delta_{in}$, which is defined as the 'onset point'. The skin friction coefficient after transition is in good agreement with the flat-plate theory of turbulent boundary layer by Cousteix in 1989

(Ducros, 1996). The agreement between coarse and fine grid results also shows the grid convergence.

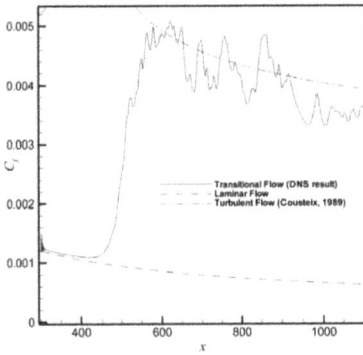

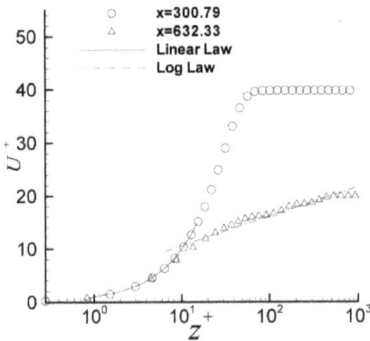

(I) Coarse Grids ($960 \times 64 \times 121$)

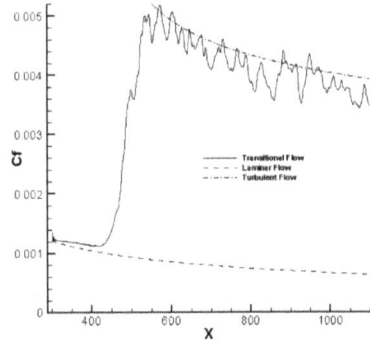

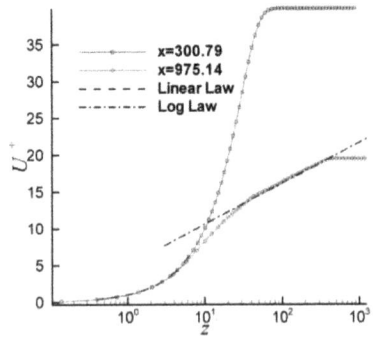

(II) Fine Grids (1920x128x241)

(I) Coarse Grids (960x64x121)

(II) Fine Grids (1920x128x241)

Figure 5. (a). Streamwise evolutions of the time-and spanwise-averaged skin-friction coefficient, (b). Log-linear plots of the time-and spanwise-averaged velocity profile in wall unit

3. Comparison with log law

Time-averaged and spanwise-averaged streamwise velocity profiles for various streamwise locations in two different grid levels are shown in Figure 5(b). The inflow velocity profiles at $x = 300.79\delta_{in}$ is a typical laminar flow velocity profile. At $x = 632.33\delta_{in}$, the mean velocity profile approaches a turbulent flow velocity profile (Log law). This comparison shows that the velocity profile from the DNS results is a turbulent flow velocity profile and the grid convergence has been realized. Figures 5(a) and 5(b) also show that grid convergence is obtained in the velocity profile.

4. Spectra and Reynolds stress (velocity) statistics

Figure 6 shows the spectra in x- and y- directions. The spectra are normalized by z at location of $Re_x = 1.07 \times 10^6$ and $y^+ = 100.250$. In general, the turbulent region is approximately defined by $y^+ > 100$ and $y / \delta < 0.15$. In our case, The location of $y / \delta = 0.15$ for $Re_x = 1.07 \times 10^6$ is corresponding to $y^+ \approx 350$, so the points at $y^+ = 100$ and 250 should be in the turbulent region. A straight line with slope of -5/3 is also shown for comparison. The spectra tend to tangent to the $k^{-\frac{5}{3}}$ law. The large oscillations of the spectra can be attributed to the inadequate samples in time when the average is computed.

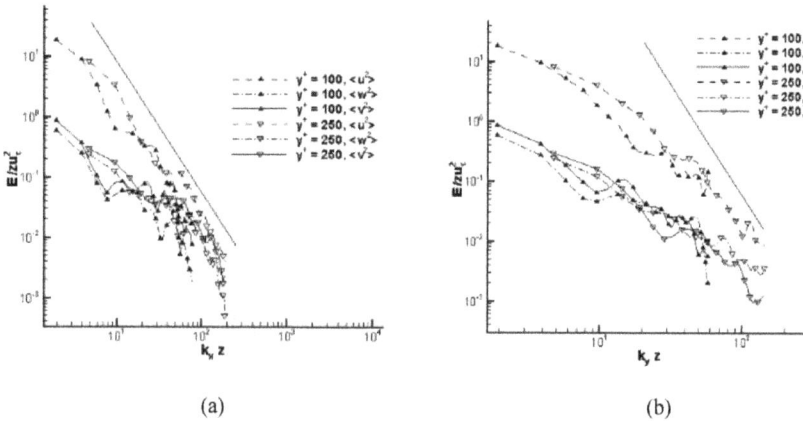

(a)   (b)

Figure 6. (a) Spectra in x direction; (b)Spectra in y direction

Figure 7 shows Reynolds shear stress profiles at various streamwise locations, normalized by square of wall shear velocity. There are 10 streamwise locations starting from leading edge to trailing edge are selected. As expected, close to the inlet where $Re_x = 326.79 \times 10^3$ where the flow is laminar, the values of the Reynolds stress is much smaller than those in the turbulent region. The peak value increases with the increase of x. At around $Re_x = 432.9 \times 10^3$, a big jump is observed, which indicates the flow is in transition. After $Re_x = 485.9 \times 10^3$, the Reynolds stress profile becomes close to each other in the turbulent region. So for this case, we can consider that the flow becomes turbulent after $Re_x = 490 \times 10^3$.

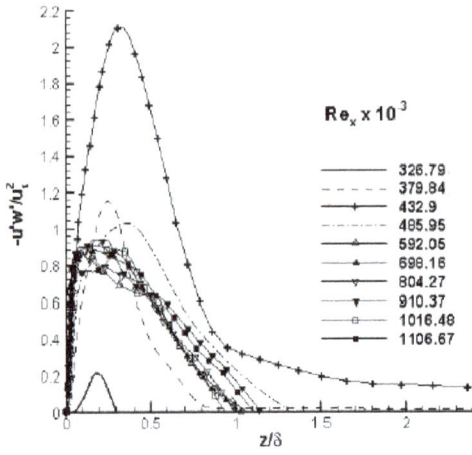

Figure 7. Reynolds stress

All these verifications and validations above show that our code is correct and our DNS results are reliable.

4. Small vortices generation and shape of positive spikes

A general scenario of formation and development of small vortices structures at the late stages of flow transition can be seen clearly by Figure 8.

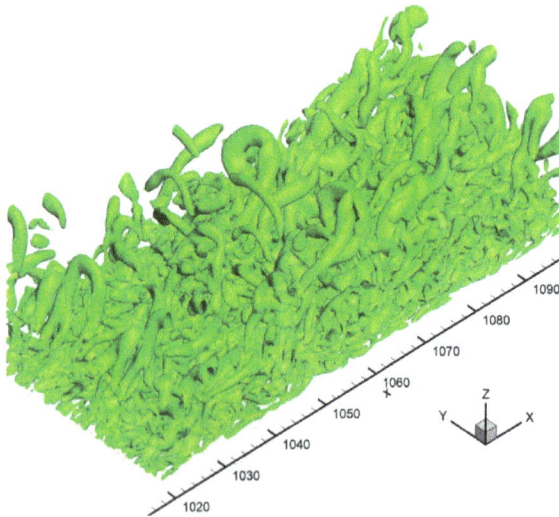

Figure 8. Visualization of flow transition at t=8.0T based on eigenvalue λ_2

Figure 9(a) is the visualization of λ_2 from the bottom view. Meanwhile, the shape of positive spikes along x-direction is shown in figure 9(b). We can see that from the top to bottom, originally the positive spike is generated by sweep motion, and then two spikes combine together to form a much stronger high speed area. Finally, two red regions (high speed areas) depart further under the ring-like vortex[5].

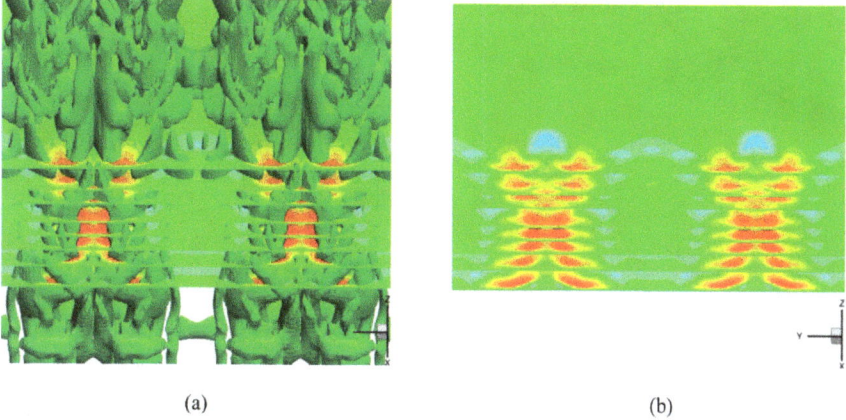

(a)   (b)

Figure 9. (a) bottom view of λ_2 structure; (b) visulation of shape of positive spikes along x-direction

In order to fully understand the relation between small length scale generation and increase of the skin friction, we will focus on one of two slices in more details first.

The streamwise location of the negative and positive spikes and their wall-normal positions with the co-existing small structures can be observed in this section. Figures 10(a) demonstrates that the small length scales (turbulence) are generated near the wall surface in the normal direction, and Figure 10(b) is the contour of velocity perturbation at an enlarged section x=508.633 in the streamwise direction. Red spot at the Figure 10(b) indicates the region of high shear layer generated around the spike. It shows that small vortices are all generated around the high speed region (positive spikes) due to instability of high shear layer, especially the one between the positive spikes and solid wall surface. For more references see[7,14,15].

5. Control of skin friction coefficient

The skin friction coefficient calculated from the time-averaged and spanwise-averaged profile is displayed in Figure 11. The spatial evolution of skin friction coefficients of laminar flow is also plotted out for comparison. It is observed from this figure that the sharp growth of the skin-friction coefficient occurs after $x \approx 450\delta_{in}$, which is defined as the 'onset point'.

The skin friction coefficient after transition is in good agreement with the flat-plate theory of turbulent boundary layer by Cousteix in 1989 (Ducros, 1996).

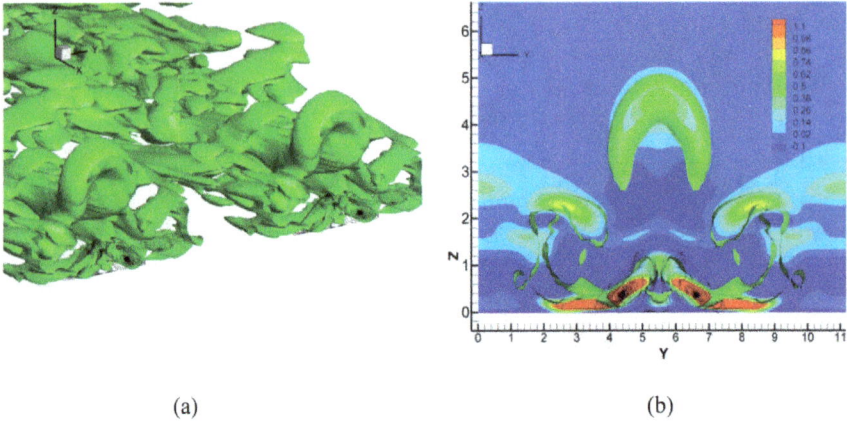

(a)   (b)

Figure 10. (a)Isosurface of λ_2 (b) Isosurface of λ_2 and streamtrace at x=508.633 velocity pertubation at x=508.633

The second sweep movement [5] induced by ring-like vortices combined with first sweep generated by primary vortex legs will lead to a huge energy and momentum transformation from high energy containing inviscid zones to low energy zones near the bottom of the boundary layers. We find that although it is still laminar flow at $x \approx 450\delta_{in}$ at this time step t=8.0T, the skin-friction is immediately enlarged at the exact location where small length scales are generated, which was mentioned in last section in Fig.10. Therefore, the generation of small length scales is the only reason why the value of skin-friction is suddenly increased, which has nothing to do with the viscosity change. In order to design a device to reduce the shear stress on the surface, we should eliminate or postpone the positive spike generation, which will be discussed in more details next.

Figure12 shows the four ring-like vortices at time step t=8.0T from the side view. We concentrated on examination of relationship between the downdraft motions and small length scale vortex generation and found out the physics of the following important phenomena. When the primary vortex ring is perpendicular and perfectly circular, it will generate a strong second sweep which brings a lot of energy from the inviscid area to the bottom of the boundary layer and makes that area very active. However, when the heading primary ring is skewed and sloped but no longer perfectly circular and perpendicular, the second sweep immediately becomes weak. This phenomenon can be verified from the Figure 13 that the sweep motion is getting weak as long as the vortex rings do not keep perfectly circular and perpendicular. By looking at Figure 14 around the region of x=508, we note that there is a high speed area (red color region) under the ring-like vortex, which is caused by the strong sweep motion. However, for the ring located at x=536, we can see there is no high speed region below the first ring located at x=536 due to the weakness of the sweep motion. In addition, we can see that the structure around the ring is quite clean. This is because the small length scale structures are rapidly damped. That gives us an idea that we can try to change the gesture and shape of the vortex rings in order to reduce the intensity of positive spikes. Eventually, the skin friction can be reduced consequently.

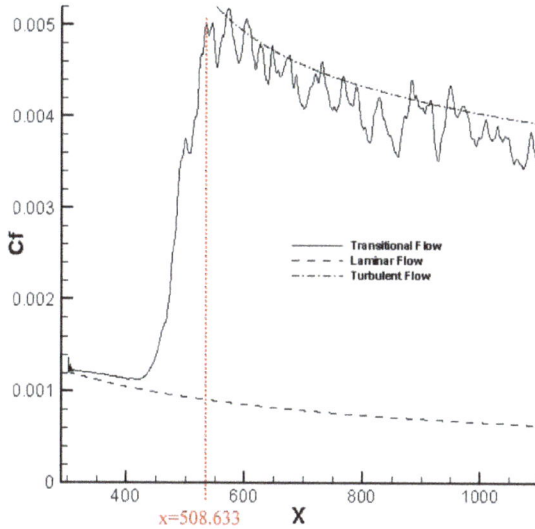

Figure 11. Streamwise evolutions of the time-and spanwise-averaged skin-friction coefficient

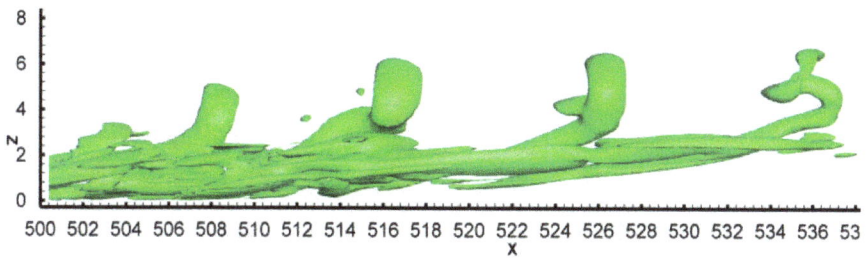

Figure 12. Side view for multiple rings at t=8.0T

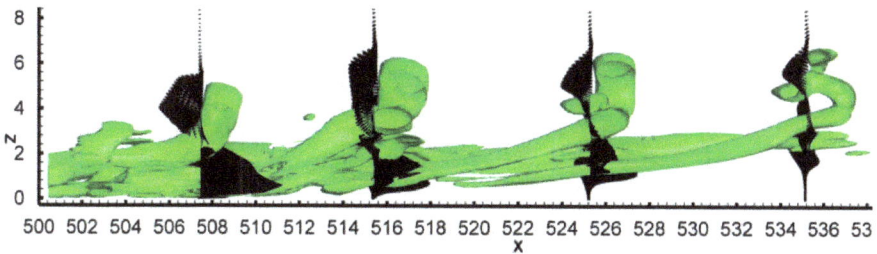

Figure 13. Side view for multiple rings with vector distribution at t=8.0T- sweep motion is weaker

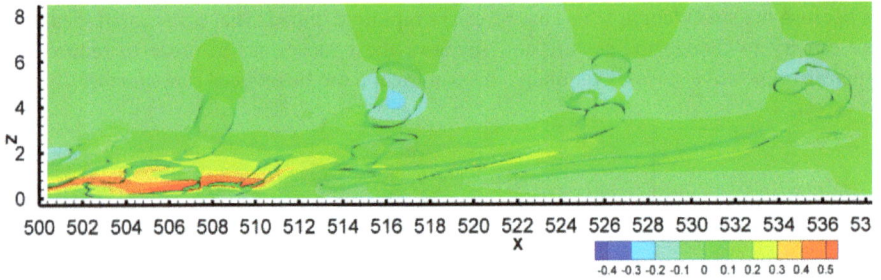

Figure 14. Side view for multiple rings with velocity perturbation at t=8.0T

6. Universal structure of turbulent flow

This section illustrates a uniform structure around each ring-like vortex existing in the flow field (Figure 15). From the λ_2 contour map and streamtrace at the section of x= 530.348 δ_{in} shown in Figure 16, we have found that the prime streamwise vorticity creates counter-rotated secondary streamwise vorticity because of the effect of the solid wall. The secondary streamwise vorticity is strengthened and the vortex detaches from the solid wall gradually. When the secondary vortex detaches from the wall, it induces new streamwise vorticity by the interaction of the secondary vortex and the solid wall, which is finally formed a tertiary streamwise vortex. The tertiary vortex is called the U-shaped vortex, which has been found by experiment and DNS. For detailed mechanisms read [10,14].

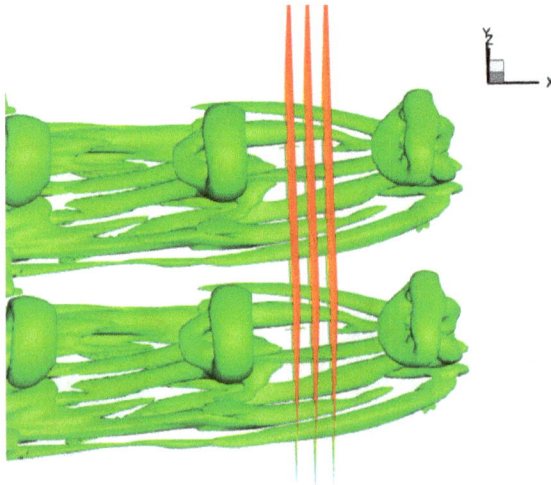

Figure 15. Top view with three cross-sections

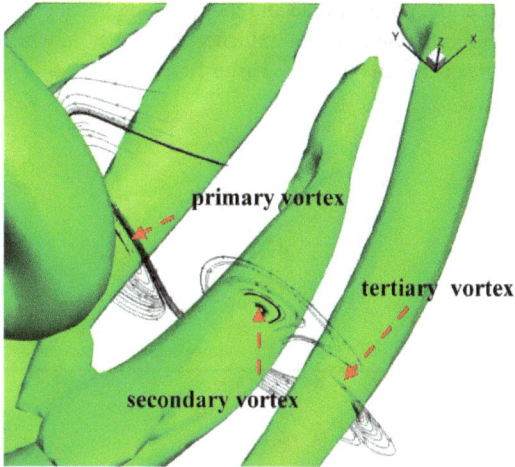

Figure 16. Structure around ring-like vortex in streamwise direction

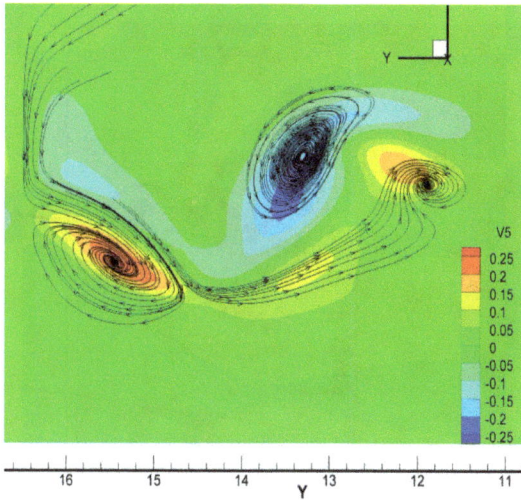

Figure 17. Stream traces velocity vector around ringlike vortex

7. Multi-level rings overlap

A side view of *isosurface of* λ_2 [8] with a cross-section at x=590 at time step t=9.2T is given in the Figure 18 which clearly illustrates that there are more than one ring-like vortex cycle overlapped together and the thickness of boundary layer becomes much thicker than before. Next, Figure 19 was obtained from the same time step and shows that there are two ring cycles which are located at the purple frame and the red frame. This phenomenon confirms that the growth of the second cycle does not influence the first cycle which is because there is a counter-rotating vortex between those two vortex rings[14].

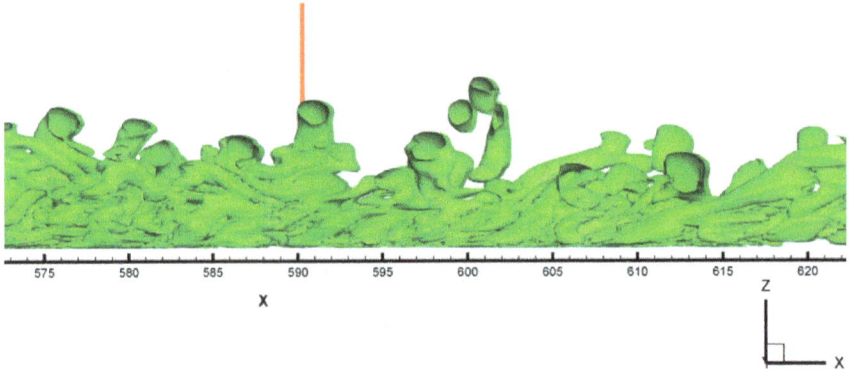

Figure 18. Side view of isosurface λ_2 with cross section

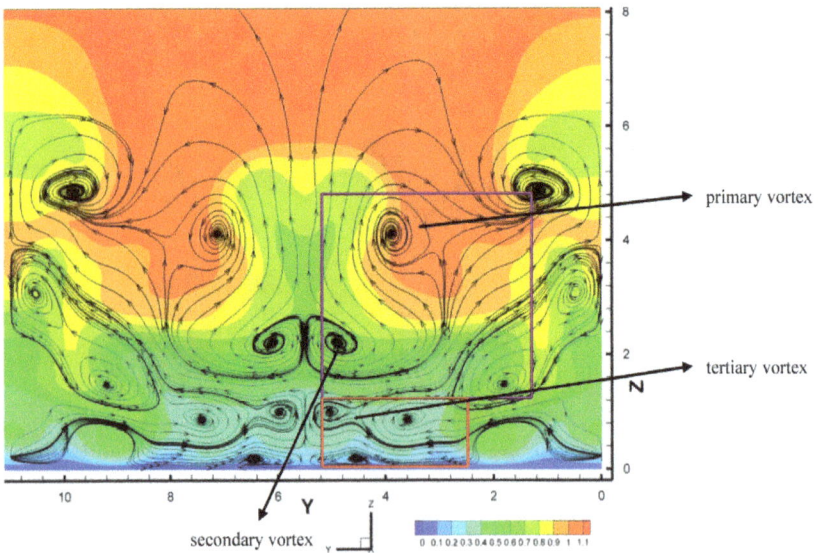

Figure 19. Cross-section of velocity distribution and streamtrace

8. Conclusion

Although flow becomes increasingly complex at the late stages of flow transition, some common patterns still can be observed which are beneficial for understanding the mechanism that how to control the skin friction and why the boundary layer becomes thicker. Based on our new DNS study, the following conclusions can be made.

1. The skin-friction is quickly enlarged when the small length scales are generated during the transition process. It clearly illustrates that the shear stress is only related to velocity gradient rather than viscosity change.

2. If the ring is deformed and/or the standing position is inclined, the second sweep and then the positive spikes will be weakened. The consequence is that small length scales quickly damp. This is a clear clue that we should mainly consider the sharp velocity gradients for turbulence modeling instead of only considering the change of viscous coefficients in the near wall region.

3. Because the ring head moves faster than the ring legs does and more small vortices are generated near the wall region, the consequence is that the multi-level ring cycles will overlap.

4. Multiple ring cycles overlapping will lead to the thickening of the transitional boundary layer. However, they never mix each other. That is because the two different level cycles are separated by a vortex trees which has a different sign with the bottom vortex cycle.

Nomenclature

M_∞ = Mach number
Re = Reynolds number
δ_{in} = inflow displacement thickness
T_w = wall temperature
T_∞ = free stream temperature
Lz_{in} = height at inflow boundary
Lz_{out} = height at outflow boundary
Lx = length of computational domain along x direction
Ly = length of computational domain along y direction
x_{in} = distance between leading edge of flat plate and upstream boundary of computational domain
A_{2d} = amplitude of 2D inlet disturbance
A_{3d} = amplitude of 3D inlet disturbance
ω = frequency of inlet disturbance
α_{2d}, α_{3d} = two and three dimensional streamwise wave number of inlet disturbance
β = spanwise wave number of inlet disturbance
R = ideal gas constant
γ = ratio of specific heats
μ_∞ = viscosity

Author details

Ping Lu, Manoj Thapa and Chaoqun Liu

University of Texas at Arlington, Arlington, TX, USA

9. References

[1] A.Cheskidov. Theoretical skin-friction law in a turbulent boundary layer. Physics Letters A. Issues 5-6, 27 June 2005, Pages 487-494

[2] A.A.Townsend. Turbulent friction on a flat plate., Emmanuel College, Cambridge. Skinsmodelltankens meddelelse nr. 32, Mars 1954.

[3] Bake S, Meyer D, Rist U. Turbulence mechanism in Klebanoff transition:a quantitative comparison of experiment and direct numerial simulation. J.Fluid Mech, 2002 , 459:217-243

[4] Boroduln V I, Gaponenko V R, Kachanov Y S, et al. Late-stage transition boundary-Layer structure: direct numerical simulation and exerperiment. Theoret.Comput.Fluid Dynamics, 2002,15:317-337.

[5] Chen, L and Liu, C., Numerical Study on Mechanisms of Second Sweep and Positive Spikes in Transitional Flow on a Flat Plate, Journal of Computers and Fluids, Vol 40, p28-41, 2010

[6] GROSS, J. F. Skin friction and stability of a laminar binary boundary layer on a flat plate. Research memo. JAN 1963

[7] Guo, Ha; Borodulin, V.I.; Kachanov, Y.s.; Pan, C; Wang, J.J.; Lian, X.Q.; Wang, S.F., Nature of sweep and ejection events in transitional and turbulent boundary layers, J of Turbulence, August, 2010

[8] Jeong J., Hussain F. On the identification of a vortex, J. Fluid Mech. 1995, 285:69-94

[9] J.E. Sargison, et al. Experimental review of devices to artificially thicken wind tunnel boundary layers. 15th Australian Fluid Mechanics Conference.

[10] Kachanov, Y.S. On a universal mechanism of turbulence production in wall shear flows. In: Notes on Numerical Fluid Mechanics and Multidisciplinary Design. Vol. 86. Recent Results in Laminar-Turbulent Transition. — Berlin: Springer, 2003, pp. 1–12.

[11] Lee C B., Li R Q. A dominant structure in turbulent production of boundary layer transition. Journal of Turbulence, 2007, Volume 55, CHIN. PHYS. LETT. Vol. 27, No. 2 (2010) 024706, 2010b

[12] Liu, X., Chen, L., Liu, C. , Study of Mechanism of Ring-Like Vortex Formation in Late Flow Transition AIAA Paper Number 2010-1456, 2010, 2010c

[13] Liu, C. and Chen, L., Parallel DNS for Vortex Structure of Late Stages of Flow Transition, Journal of Computers and Fluids, 2010d

[14] P. Lu and C. Liu. DNS Study on Mechanism of Small Length Scale Generation in Late Boundary Layer Transition. Physica D: Nonlinear Phenomena, Volume 241, Issue 1, 1 January 2012, Pages 11-24

[15] P. Lu, Z. Wang, L. Chen and C. Liu. Numerical Study on U-shaped Vortex Formation in Late Boundary Layer Transition. Journal of Computer & Fluids, CAF-D-11-00081, 2011

[16] Schlichting, H. and Gersten, K., Boundary Layer Theory, Springer, 8th revised edition, 2000

Chapter 2

Physical Modeling of Gas Pollutant Motion in the Atmosphere

Ondrej Zavila

Additional information is available at the end of the chapter

1. Introduction

Air pollution is becoming an increasingly serious global issue. Factories produce large amounts of pollutants that damage the environment and harm human health. From this point of view, problems of motion and dispersion of pollutants in the atmosphere relate not only to environmental studies but also to other disciplines, such as safety engineering.

An understanding of the physical principles of pollutants' motion and dispersion is important in order to determine the impact of air pollution on the environment and humans. This study only deals with these physical principles of pollutants' motion and dispersion. Possible chemical reactions in the atmosphere are not covered.

For the purpose of the study, a simple model of a typical real situation was defined. Physical parameters of the model were gradually modified to achieve visible changes in results so that general principles could be defined. The above-mentioned demonstration model represents a chimney situated in a simple flat terrain. Gas pollutant is discharged from the chimney and carried by flowing air. Gas pollutant plume is detected and visualized with a numerical model as iso-surfaces or contours of pollutant concentrations in two-dimensional cut planes of three-dimensional geometry.

The dependence of the pollutant plume shape, size and inclination on modification of three physical parameters was investigated. The selected parameters included pollutant density, air flow velocity and model scale.

The results are presented in the form of text, commented figures and tables. ANSYS Fluent 13.0 CFD (Computational Fluid Dynamics) code was used to demonstrate and visualize all problem variants (see [1],[2]). The numerical model of the pollutant plume motion created in this software had been verified by an experiment conducted in the low-speed wind tunnel in the Aerodynamic Laboratory of the Academy of Sciences of the Czech Republic in Novy

Knin (see [3],[4],[5]). One of the aims of the study is also to demonstrate that physical modeling of pollutant plume motion and dispersion with severely downscaled models has its limitations that should be known and considered to avoid obtaining false results.

2. Physical modeling in wind tunnels

Wind tunnels are facilities where specific air flow regime can be set up with a certain level of precision as required by investigators. They are primarily used to study aerodynamic characteristics of bodies. The history of wind tunnels dates back to 1751 and is connected with the name of Francis H. Wenham, a Council Member of the Aeronautical Society of Great Britain. In the beginning, wind tunnels were used mainly in aviation development. Nowadays, they are used for many applications in, for instance, automotive industry, building industry, environmental and safety engineering.

Wind tunnels can be classified, for example, by the velocity of air flow (see [6]) as:

a. *Low-speed wind tunnels* – air flow velocity in the measuring section of the tunnel is low enough to consider the air as an incompressible medium;
b. *High-speed wind tunnels* – air flow velocity in the measuring section of the tunnel is high enough to consider the air as a compressible medium;
c. *Subsonic wind tunnels* - air flow velocity in the measuring section of the tunnel is high enough (subsonic) to consider the air as a compressible medium;
d. *Supersonic wind tunnels* - air flow velocity in the measuring section of the tunnel is high enough (supersonic) to consider the air as a compressible medium.

Except for applications using only clean air, there are many applications in which various pollutants are applied. Pollutants are introduced to visualize the air flow field or to simulate the motion and dispersion of pollution in the real atmosphere. These applications can be found, for example, in environmental and safety engineering. Results of measurements using downscaled models in wind tunnels can be used as base data for local emergency planning studies.

2.1. Criteria of physical similarity

Two phenomena can be considered to be similar (despite different geometrical scales) if three types of similarity match: geometric, kinematic and dynamic. Criteria of geometric similarity require that the ratios of main corresponding dimensions on the model and the original pattern be constant. Also, main corresponding angles on the original pattern and the model must be of the same value. Criteria of kinematic similarity require that the ratios of velocities at corresponding points be the same for both the original pattern and the model. Criteria of dynamic similarity require that the ratios of the main forces at corresponding points be the same for both the original pattern and the model.

Forces can be divided into two groups: areal forces and volume (weight) forces. Areal forces include friction forces, compression forces, and capillary (surface) forces. Volume (weight) forces include inertial forces, gravity forces and impulse forces (resulting from the change in momentum). According to the type of phenomena, these forces can be put into mathematical relation and criteria (numbers) of similarity can be established. In fluid mechanics, the Reynolds number, Euler number, Newton number, Froude number, Weber number, and Mach number are the most widely known criteria. Each of them expresses ratio between two different forces. In practice, it is not possible to achieve correspondence between the original pattern and the model in all criteria. Therefore, it is always up to the investigators who must use their knowledge and experience to choose the right and most important criterion (or criteria) for the investigated phenomenon. As a result, investigators usually work with one or two dominant criteria of similarity [7],[8],[9].

2.2. Applying froude number

The Froude number expresses the ratio between gravity forces and inertial forces. Gravity forces cause vertical movements of the plume (climbing or descending) and inertial forces cause horizontal movements of the plume. The Froude number can be therefore considered as a criterion of dynamic similarity, which should be of the same value for both the scaled model and the real pattern (see [7],[8],[9]).

The Froude number can be defined as

$$Fr \approx \frac{F_{I-air}}{F_{G-pollutant}} = \frac{\rho_{air} \cdot S \cdot v_{air}^2}{\rho_{pollutant} \cdot g \cdot V} = \frac{\rho_{air} \cdot l_2 \cdot l_3 \cdot v_{air}^2}{\rho_{pollutant} \cdot g \cdot l_1 \cdot l_2 \cdot l_3} = \frac{\rho_{air} \cdot v_{air}^2}{\rho_{pollutant} \cdot g \cdot l_1} \tag{1}$$

where F_{I-air} is the inertial force due to the air acting on pollutant element [N], $F_{G-pollutant}$ is the gravity force acting on pollutant element [N], ρ_{air} is the air density [kg/m³], $\rho_{pollutant}$ is the pollutant density [kg/m³], S is the surface of pollutant element acted on by the flowing air [m²], v_{air} is the air flow velocity [m/s], g is the gravity acceleration constant [m/s²], V is the volume of pollutant element released from the pollutant source per 1 second [m³], l_1 is the 1st characteristic dimension of the pollutant source (length of pollutant cubic element) [m], l_2 is the 2nd characteristic dimension of the pollutant source (width of pollutant cubic element) [m] and l_3 is the 3rd characteristic dimension of the pollutant source (height of pollutant cubic element) [m]. l_3 can be replaced by $v_{pollutant}$ that represents the velocity of the pollutant released from the source in vertical direction [m/s]. Fr is a dimensionless constant [-] whose value determines whether the inertial force or the gravity force will dominate in the specific pollutant plume motion scenario.

All important characteristics are illustrated in Figure 1. Pollutant element was simplified into a rectangular cuboid with dimensions of l_1, l_2 and l_3 to make practical calculations easier. Of course, the spout of a real chimney can be of a different shape, most commonly circular or elliptical. In this case it is advisable to calculate the surface of the spout and transform the shape into a square or a rectangle with dimensions of l_1 and l_2. The value of l_3 remains the same (despite the shape of the spout) and is replaced by the velocity with which the pollutant leaves the source in vertical direction $v_{pollutant}$.

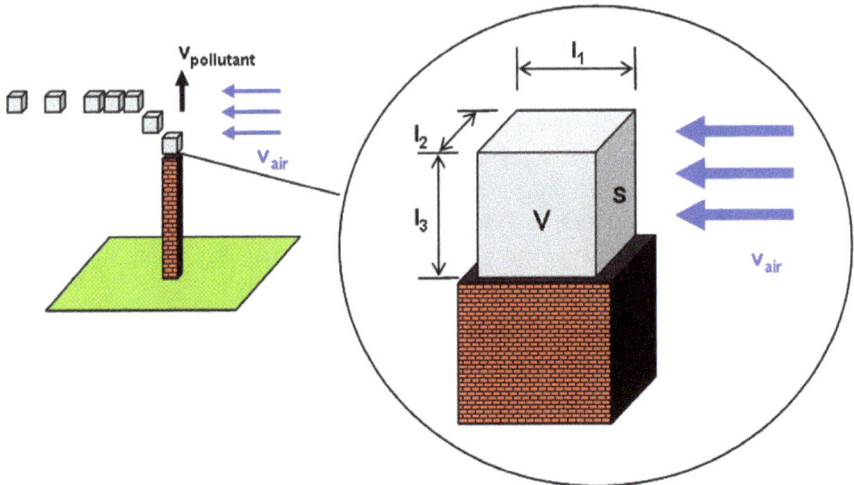

Figure 1. Air flow action on gas pollutant element leaking from the nozzle (chimney)

If $Fr < 1$, gravity forces are assumed greater than inertial forces. Hence, vertical motions (climbing or descending) of the gas pollutant plume can be expected due to different densities of the pollutant and the air. Plumes of light gas pollutants will tend to climb, whereas plumes of heavy gas pollutants will tend to descend.

If $Fr = 1$, gravity forces are assumed equal to inertial forces. Hence, gas pollutant plumes are carried by flowing air along with manifesting partial vertical motions.

If $Fr > 1$, inertial forces are assumed greater than gravity forces. Hence, vertical motions of the gas pollutant plume are limited or nonexistent. The gas pollutant plume is carried by strong flowing air, regardless of the pollutant—air density difference or weight of the pollutant.

This third scenario causes common difficulties when planning gas pollutant plume motion and dispersion experiments with downscaled models in low-speed wind tunnels. At small dimensions of measuring sections of common wind tunnels and, thus, low scales of models, the air flow may be too great to allow vertical motions of gas pollutant plumes. Proper conditions often cannot be assured in such cases.

2.3. Influence of air velocity

According to Equation (1) in Section 1.2, the air flow velocity v_{air} influences the inertial force F_{I-air} that causes gas pollutant horizontal motion. If all the other physical characteristics are constant, the following principles can be formulated: The greater the air flow velocity v_{air} is, the greater the inertial force F_{I-air} is. The greater the inertial force is, the more limited the pollutant plume vertical motions (climbing or descending) are.

The air flow velocity $v_{air,\,Fr=1}$ for $Fr = 1$ (i.e., inertial and gravity forces are equal) can be deduced from Equation (1) mentioned in Section 1.2:

$$1 = \frac{F_{I-air}}{F_{G-pollu\tan t}} = \frac{\rho_{air} \cdot S \cdot v_{air,\,Fr=1}{}^2}{\rho_{pollu\tan t} \cdot g \cdot V} = \frac{\rho_{air} \cdot l_2 \cdot l_3 \cdot v_{air,\,Fr=1}{}^2}{\rho_{pollu\tan t} \cdot g \cdot l_1 \cdot l_2 \cdot l_3} = \frac{\rho_{air} \cdot v_{air,\,Fr=1}{}^2}{\rho_{pollu\tan t} \cdot g \cdot l_1} \tag{2}$$

Thus, the air flow velocity $v_{air,\,Fr=1}$ is

$$v_{air,\,Fr=1} = \sqrt{\frac{\rho_{pollu\tan t} \cdot g \cdot l_1}{\rho_{air}}} = \sqrt{\frac{\rho_{pollu\tan t} \cdot g \cdot l_1 \cdot l_2 \cdot l_3}{\rho_{air} \cdot l_2 \cdot l_3}} = \sqrt{\frac{\rho_{pollu\tan t} \cdot g \cdot V}{\rho_{air} \cdot S}} \tag{3}$$

However, one must realize that the change in air flow velocity influences also air flow field turbulent characteristics. For example, turbulent intensity is influenced when the air flows around solid objects or in a complex terrain. Investigators must consider whether these changes have a serious impact on accuracy of the experiment or mathematical model. This is very important for modeling gas pollutant motion and dispersion in a complex geometry (complex terrain) where the real model of turbulent flow field is the key element of the simulation. If the air turbulent flow field is seriously influenced by the change in the air flow velocity, the results of the analysis can be misleading. This approach is therefore not suitable for such cases and a different parameter of the model must be changed (see Sections 1.4 and 1.5).

2.4. Influence of pollutant density

According to Equation (1) in Section 1.2, the pollutant density $\rho_{pollu\tan t}$ influences the gravity force $F_{G-pollu\tan t}$ that causes gas pollutant vertical motions (climbing or descending). If all the other physical characteristics are constant, the following principles can be formulated: If the pollutant density $\rho_{pollu\tan t}$ is greater than the air density ρ_{air}, the pollutant tends to descend (i.e., the gas pollutant plume descends). If the pollutant density $\rho_{pollu\tan t}$ is lower than the air density ρ_{air}, the pollutant tends to climb (i.e., the gas pollutant plume climbs). The greater the gravity forces $F_{G-pollu\tan t}$ are, the more significant the gas pollutant plume vertical movements (climbing or descending) are.

The pollutant density $\rho_{pollu\tan t,\,Fr=1}$ for $Fr = 1$ (i.e., inertial and gravity forces are equal) can be deduced from Equation (1) in Section 1.2:

$$1 = \frac{F_{I-air}}{F_{G-pollu\tan t}} = \frac{\rho_{air} \cdot S \cdot v_{air}{}^2}{\rho_{pollu\tan t,\,Fr=1} \cdot g \cdot V} = \frac{\rho_{air} \cdot l_2 \cdot l_3 \cdot v_{air}{}^2}{\rho_{pollu\tan t,\,Fr=1} \cdot g \cdot l_1 \cdot l_2 \cdot l_3} = \frac{\rho_{air} \cdot v_{air}{}^2}{\rho_{pollu\tan t,\,Fr=1} \cdot g \cdot l_1} \tag{4}$$

Thus, the pollutant density $\rho_{pollu\tan t,\,Fr=1}$ is

$$\rho_{pollu\tan t,\,Fr=1} = \frac{\rho_{air} \cdot v_{air}{}^2}{g \cdot l_1} = \frac{\rho_{air} \cdot v_{air}{}^2 \cdot l_2 \cdot l_3}{g \cdot l_1 \cdot l_2 \cdot l_3} = \frac{\rho_{air} \cdot v_{air}{}^2 \cdot S}{g \cdot V} \tag{5}$$

A change in pollutant density in order to achieve the optimum ratio between inertial and gravity forces would be often the ideal solution. However, there is a problem. The densities of pollutants range within a narrow interval—approximately of one of order of magnitude—which is usually is not enough to compensate the Froude number differences resulting from, e.g., a substantial change of the model scale.

A typical example can be the physical modeling of gas pollutant plumes in low-speed wind tunnels where the scale of the model is around 1:1000. In such a case, there is a need to change the pollutant density 100, or even 1000 times, which is impossible. This is why change in pollutant density can be used to achieve only a small change in the Froude number. These small changes, however, may not be sufficient for a successful execution of the experiment or numerical modeling.

The change of the flowing gas density could be an alternative solution. For example, air could be replaced by a different gas with a different value of density. However, this change influences turbulent flow field characteristics, which may be undesirable (see Section 1.3).

2.5. Influence of model scale

According to Equation (1) in Section 1.2, the model scale can be expressed by using the value l_1 that represents the 1st characteristic dimension of the pollutant source (i.e., length of pollutant cubic element) (see Figure 1). The model scale influences both inertial and gravity forces and changes their ratio. If all the other physical characteristics are constant, the following principles can be formulated: The greater the model scale is, the greater the influence of gravity forces is. The smaller the model scale is, the greater the influence of inertial forces is. Gravity forces are, e.g. , greater in a model scaled at 1:4 than in that scaled at 1:1000 (see Section 2.6).

The value of the 1st characteristic dimension of the pollutant source (length of pollutant cubic element) $l_{1, Fr=1}$ for $Fr = 1$ (inertial and gravity forces are equal) can be deduced from Equation (1) in Section 1.2:

$$1 = \frac{F_{I-air}}{F_{G-pollutant}} = \frac{\rho_{air} \cdot S_{Fr=1} \cdot v_{air}^2}{\rho_{pollutant} \cdot g \cdot V_{Fr=1}} = \frac{\rho_{air} \cdot l_{2, Fr=1} \cdot l_{3, Fr=1} \cdot v_{air}^2}{\rho_{pollutant} \cdot g \cdot l_{1, Fr=1} \cdot l_{2, Fr=1} \cdot l_{3, Fr=1}} \tag{6}$$

or

$$1 = \frac{F_{I-air}}{F_{G-pollutant}} = \frac{\rho_{air} \cdot v_{air}^2}{\rho_{pollutant, Fr=1} \cdot g \cdot l_{1, Fr=1}} \tag{7}$$

Thus, $l_{1, Fr=1}$ is

$$l_{1, Fr=1} = \frac{\rho_{air} \cdot v_{air}^2}{\rho_{pollutant} \cdot g} = \frac{\rho_{air} \cdot v_{air}^2 \cdot l_{2, Fr=1} \cdot l_{3, Fr=1}}{\rho_{pollutant} \cdot g \cdot l_{2, Fr=1} \cdot l_{3, Fr=1}} = \frac{\rho_{air} \cdot v_{air}^2 \cdot S_{Fr=1}}{\rho_{pollutant} \cdot g \cdot S_{Fr=1}} \tag{8}$$

T model scale $M_{Fr=1}$ [-] for $Fr = 1$ (inertial and gravity forces are equal) is given by

$$M_{Fr=1} = \frac{1}{X} \tag{9}$$

where

$$X = \frac{l_1}{l_{1,\,Fr=1}} \tag{10}$$

The value l_1 represents the 1st characteristic dimension of the pollutant source (length of pollutant cubic element) in the original model. The value $l_{1,\,Fr=1}$ is the 1st characteristic dimension of the pollutant source in the model where $Fr = 1$ (i.e., inertial and gravity forces are equal).

If the model scale is changed, the change of the inertial force is given by

$$F_{I1} = X^2 \cdot F_{I2} \tag{11}$$

and the change of the gravity force is given by

$$F_{G1} = X^3 \cdot F_{G2} \tag{12}$$

The value X is a model scale factor [-]. If $X > 1$, the model is smaller than its original pattern (i.e., the model is downscaled). If $X < 1$, the model is larger than its original pattern (i.e., the model is enlarged). Equations (11) and (12) are deduced from Equation (1) for the Froude number. The situation is illustrated in Figure 2.

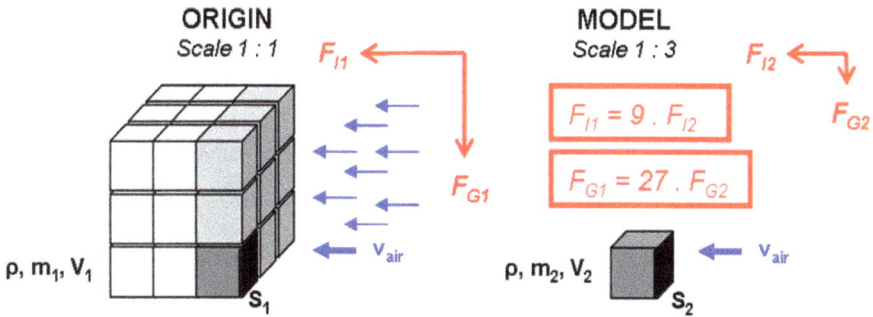

Figure 2. Inertial and gravity forces acting on gas pollutant element in two different model scales

A change in model scale always causes a change in the ratio of inertial and gravity forces. Therefore, some problems cannot be reliably modeled at other than approximately original scales. The range of deviation from the original depends on the discretion of investigators. Investigators must decide whether the tolerance of the results is acceptable.

Modeling of gas pollutant plume motions with severely downscaled models is a typical example of this problem. Downscaled pollutant plume models will not correspond to the originally scaled patterns without modifying some key physical characteristics (air flow velocity, turbulent characteristics, etc.).

The problem can be solved by using a numerical mathematical model verified by a clearly defined experiment of the same type of physical phenomenon. Once the numerical model is verified, it can be used for numerical simulation of any problem of the same physical principles, whatever the model scale is. Sometimes it is impossible to do the same with a physical experiment.

It can be concluded that severely downscaled experiments are not suitable for modeling gas pollutant plume motions because of possible absence of vertical movements. It is more advisable to use physical experiment data only for verification of the numerical mathematical model (code, software).

3. CFD numerical modeling of gas pollutant motion

CFD (Computational Fluid Dynamics) is an abbreviation for a category of sophisticated mathematical codes for fluid mechanics computations. These codes offer a wide range of applications including modeling of turbulent fluid flows, heat transfers, and species motions with or without chemical reactions. It is an effective tool for prediction and reconstruction of many physical phenomena, especially in mechanical engineering, civil engineering, safety engineering, power engineering, environmental engineering, and in many other disciplines.

3.1. CFD numerical modeling basics

CFD (Computational Fluid Dynamic) codes are based on the numerical solution of systems of partial differential equations that express the law of conservation of mass (continuity equation), the law of conservation of momentum (Navier-Stokes equations) and the law of conservation of energy (energy equation). This basic set of equations can be supplemented by additional equations that express heat transfer (heat transfer equations – convection, conduction or radiation), or species transport (species transport equations – gas, liquid or solid). The system of equations is then solved with an appropriate numerical method.

At the start of the modeling process, the geometrical shape of a two- or three-dimensional geometry is created that represents the object of modeling and its close surroundings. The size and shape of the geometry must enable the investigated phenomena to be captured.

In the second step, the geometry is covered with a grid. The grid divides the geometry into a finite number of two- or three-dimensional elements (cells) of a certain shape in which relevant physical quantities will be calculated. The quantities are calculated at the middle of the cells. Values in intermediate spaces are interpolated or extrapolated. The type and quality of grid influence the calculation time and the quality of results. A quality grid is smooth enough to capture the modeled phenomenon and its cells have as little shape deformation as possible compared to the ideal original shape.

In the third step, investigators choose the most suitable mathematical model and sub-models for the calculation of the problem. For gas pollutant plume motion and dispersion, the model of turbulence for the air flow field calculation is chosen first, than the species transport model is chosen for species transport modeling.

In the fourth step, boundary and initial conditions are defined. Boundary conditions are physical constants or variables that characterize the physical conditions at the boundaries of the geometry, i.e., at its inlet, outlet or walls. Boundary conditions do not change during the course of the calculation. They can be defined as a constant, a spatial-dependent function or a time-dependent function (polynomial, linear or periodic). Initial conditions are physical constants or variables that characterize the initial state of the system. Unlike boundary conditions, the initial conditions change in the course of the calculation after each iteration. If data for setting initial conditions are missing, at least approximate values should be entered. Initial conditions influence the calculation run-time, i.e., the time required to reach convergence.

In the fifth step, investigators set mathematical parameters of the calculation, e.g., solution schemes, under-relaxation parameters, and criteria of convergence. If required, as in case of calculations of time-dependent problems, special post-processing functions are set in this step, too.

Afterwards, the calculation is started and its progress monitored. Lastly, results are evaluated with available standard software tools or with preset post-processing functions (animations, time-dependent progress of physical variables, etc.) (see [1],[2]).

3.2. Basic equations

The following equations are used for basic air flow field calculations:

Continuity equation for compressible fluid flow

The continuity equation expresses the law of conservation of mass. For unsteady (time-dependent) compressible fluid flows, it can be written as a differential equation in form of

$$\frac{\partial \rho}{\partial t} + \frac{\partial \left(\rho \cdot \bar{u}_j \right)}{\partial x_j} = 0 \tag{13}$$

where ρ is the fluid density [kg/m³], t is time [s], $\bar{u}_j$ is the time-averaged j-coordinate of the fluid flow velocity [m/s], and x_j is a coordinate of the Cartesian coordinate system [-].

Equations for compressible fluid flow

These equations—known as Navier-Stokes equations—express the law of conservation of momentum. For calculating turbulent flows, time-averaged values must be applied. The substitution of the time-averaged values into the Navier-Stokes equations gives the Reynolds equations. The equation of conservation of momentum for compressible fluids can be written in the form corresponding to that in differential form as

$$\frac{\partial(\rho \cdot \bar{u}_i)}{\partial t} + \frac{\partial(\rho \bar{u}_i \cdot \bar{u}_j)}{\partial x_j} = -\frac{\partial \bar{p}}{\partial x_i} + \frac{\partial}{\partial x_j} \cdot \tau_{ji} + \rho \cdot \delta_{i3} \cdot g + \rho \cdot f_c \cdot \varepsilon_{ij3} \cdot \bar{u}_j + \rho \cdot f_j \qquad (14)$$

where ρ is the fluid density [kg/m³], t is time [s], $\bar{u}_j$ is the time-averaged j-coordinate of the fluid flow velocity [m/s], x_j is a coordinate of the Cartesian coordinates system [-], $\bar{p}$ is the time-averaged value of pressure [Pa], τ_{ji} is the tensor of viscous stress [Pa], δ_{i3} is the Kronecker delta [-], ε_{ij3} is the unit tensor for centrifugal forces [-], f_j is the j-coordinate of force [N], and g is the gravity acceleration [m/s²] (if gravity forces are included).

The meaning of terms in Equation (14):

1st term on the left	....	Unsteady term (acceleration)
2nd term on the left	....	Convective term
1st term on the right	....	Influence of pressure forces
2nd term on the right	....	Influence of viscous stress
3rd term on the right	....	Influence of gravity forces (buoyancy)
4th term on the right	....	Influence of Earth rotation (Coriolis force)
5th term on the right	....	Influence of volumetric forces

The equation of turbulent kinetic energy k and the equation of turbulent dissipation rate ε are used to express the turbulent flow field variables. The exact equation for k can be deduced from the Navier-Stokes equations and written as

$$\frac{\partial k}{\partial t} + \frac{\partial \overline{u}_j \cdot k}{\partial x_j} = -\frac{\partial}{\partial x_j}\left[\overline{u'_j \cdot \left(\frac{u'_l \cdot u'_l}{2} + \delta_{jl} \cdot \frac{p'}{\rho}\right)}\right] + v_t \cdot \frac{\partial^2 k}{\partial x_j^2} - \overline{u'_l \cdot u'_j} \cdot \frac{\partial \overline{u}_l}{\partial x_j} - v \cdot \overline{\frac{\partial u'_l}{\partial x_j} \cdot \frac{\partial u'_l}{\partial x_j}} \qquad (15)$$

where k is the turbulent kinetic energy [m²/s²], t is time [s], $\bar{u}_j$ is the time-averaged j-coordinate of the fluid flow velocity [m/s], x_j is a coordinate of the Cartesian coordinate system [-], ρ is the fluid density [kg/m³], p' is a component of the pressure fluctuation [Pa], and v_t is the turbulent kinematic viscosity [Pa.s].

The meaning of terms in Equation (15):

1st term on the left	....	Turbulent energy rate
2nd term on the left	....	Convective transport of turbulent energy
1st term on the right	....	Turbulent diffusion due to pressure and velocity fluctuation
2nd term on the right	....	Molecular diffusion
3rd term on the right	....	Production of turbulent kinetic energy due to sliding friction
4th term on the right	....	Viscous dissipation

The turbulent kinetic energy k from Equation (15) is

$$k = \frac{1}{2} \cdot \left(\overline{u_1'^2} + \overline{u_2'^2} + \overline{u_3'^2} \right) = \frac{1}{2} \cdot \overline{u_j'^2} \tag{16}$$

where $\overline{u}_j$ represents time-averaged flow velocity components [m/s].

The exact equation for ε can be deduced from the Navier-Stokes equations and written as

$$\frac{\partial \varepsilon}{\partial t} + \frac{\partial \overline{u}_j \cdot \varepsilon}{\partial x_j} = \frac{\partial}{\partial x_j} \cdot \left(\frac{v_t}{\sigma_\varepsilon} \cdot \frac{\partial \varepsilon}{\partial x_j} \right) + C_{1\varepsilon} \cdot v_t \cdot \left(\frac{\partial \overline{u}_j}{\partial x_l} + \frac{\partial \overline{u}_l}{\partial x_j} \right) \cdot \frac{\partial \overline{u}_l}{\partial x_j} - C_{2\varepsilon} \cdot \frac{\varepsilon^2}{k} \tag{17}$$

where ε is the turbulent dissipation rate [m²/s³], t is time [s], $\overline{u}_j$ is the time-averaged j-coordinate of the fluid flow velocity [m/s], x_j is a coordinate of the Cartesian coordinate system [-], v_t is the turbulent kinematic viscosity [m²/s], σ_ε, $C_{1\varepsilon}$ and $C_{2\varepsilon}$ are empirical constants [-], and k is the turbulent kinetic energy [m²/s²].

The turbulent kinetic viscosity v_t is

$$v_t = C_v \cdot \frac{k^2}{\varepsilon} \tag{18}$$

where C_v is an empirical constant [-].

Energy equation

The energy equation expresses the law of conservation of energy. According to this law, change in the total fluid energy $\overline{E}$ in the volume V is determined by the change of the inner energy and kinetic energy, and by the flux of both energies through surface S [m²] that surrounds volume V [m³]. The final equation can be written as

$$\frac{\partial}{\partial t} \cdot \left[\rho \cdot \overline{E} \right] + \frac{\partial}{\partial x_j} \cdot \left[\rho \cdot \overline{u}_j \cdot \overline{E} \right] = \rho \cdot \overline{u}_j \cdot f_j - \frac{\partial \left(p \cdot \overline{u}_j \right)}{\partial x_j} + \frac{\partial \left(\tau_{ji} \cdot \overline{u}_j \right)}{\partial x_l} - \frac{\partial \overline{q}_j}{\partial x_j} \tag{19}$$

where t is time [s], ρ is the fluid density [kg/m³], $\overline{E}$ is the time-averaged value of energy [J/kg], $\overline{u}_j$ is the time-averaged j-coordinate of the flow field velocity [m/s], x_j is a coordinate of the Cartesian coordinate system [-], p is the pressure [Pa], τ_{ji} is the tensor of viscous stress [Pa], and $\overline{q}_j$ is the time-averaged j-coordinate of heat flux [J/(m².s)].

The meaning of terms in Equation (19):

1st term on the left	Energy rate
2nd term on the left	Convective transport of energy
1st term on the right	Work of outer volumetric forces (gravity force)
2nd term on the right	Thermodynamic reversible energy flux into the volume V and its change caused by pressure forces
3rd term on the right	Irreversible growth of energy due to dissipation caused by viscosity
4th term on the right	Vector of heat flux according to the Fourier law

The following equation is used for the calculations of species transport:

Species transport equation

In the model, time-averaged values of the local species mass fraction $\overline{Y}_{i'}$ are calculated. These values are described by a balance equation similar to the energy equation (see Equation 19). The energy equation includes both convective and diffuse components of the transport. It can be written in conservative form as

$$\frac{\partial}{\partial t}\left(\rho \cdot \overline{Y}_{i'}\right) + \frac{\partial}{\partial x_j}\cdot\left(\rho \cdot \overline{u}_j \cdot \overline{Y}_{i'}\right) = -\frac{\partial}{\partial x_i}J_{j,i'} + R_{i'} + S_{i'} \tag{20}$$

where $\overline{u}_j$ is the time-averaged j-coordinate of the flow field velocity [m/s], $R_{i'}$ is the production rate of species i' due to chemical reaction [kg/(m³.s)], and $S_{i'}$ is the growth production rate from distributed species [kg/(m³.s)]. The Equation (20) is valid for $N-1$ species, where N is the total number of components presented in the mathematical model [-]. Species distribution can be performed under various conditions, for which it can generally be divided into laminar and turbulent flow distribution. The value $J_{j,i'}$ represents the j-coordinate of i'-species diffuse flux [kg/(m³.s)]. To express the i'-species diffuse flux in the turbulent flow regime, the model employs

$$J_{i'} = -\left(\frac{\mu_t}{Sc_t}\right)\cdot\frac{\partial \overline{Y}_{i'}}{\partial x_j} \tag{21}$$

where μ_t is the turbulent dynamic viscosity [Pa.s], $\overline{Y}_{i'}$ is the time-averaged species i' mass fraction [-], and Sc_t is the Schmidt turbulent number [-] (set at the default value of 0.7) (see [1],[2]).

3.3. Example of numerical simulation – Gas pollutant motion in wind tunnel

ANSYS Fluent 13.0, one of the world's most sophisticated CFD codes, was chosen for the numerical simulation of the gas pollutant plume motion and dispersion. The gauging section of the low-speed wind tunnel with a small nozzle representing a chimney in a flat, simple terrain was the object of the numerical simulation. Gas pollutant enters the gauging section through the top of the nozzle (chimney) and is carried by flowing air (see Figure 3 to Figure 6).

Three-dimensional cubic geometry in five different model scales was designed for the purpose of this analysis (see Table 1).

The geometry grid at all modeled scales consisted of 569 490 three-dimensional cells. The problem was defined as steady (time-independent) with accuracy set at the value of 0.0001 (criterion of convergence).

RANS (Reynolds-averaged Navier-Stokes equations) approach was used for turbulent characteristics definition. RNG $k-\varepsilon$ model of turbulence was used for the air flow field basic calculation. The Boussinesq hypothesis of swirl turbulent viscosity was applied for the

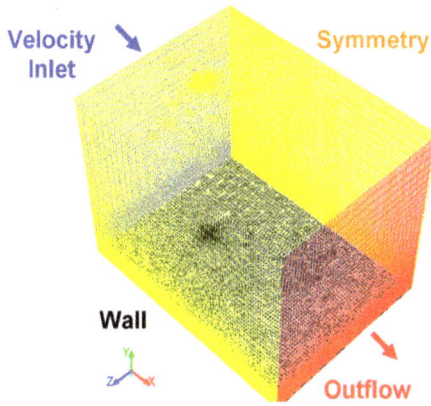

Figure 3. Geometry and grid (isometric view)

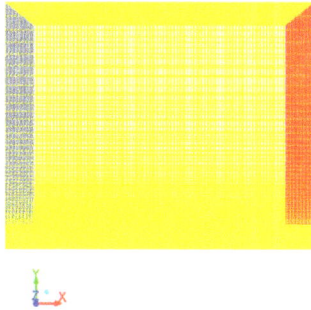

Figure 4. Geometry and grid (right side view)

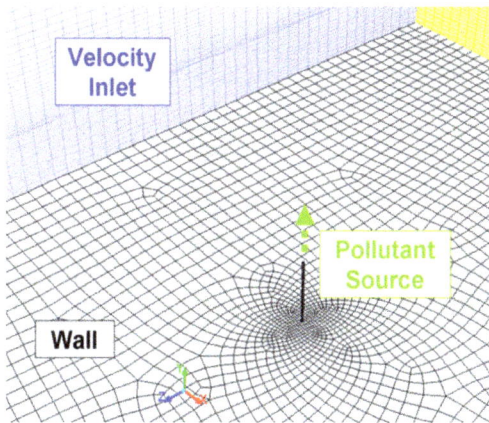

Figure 5. Pollutant source

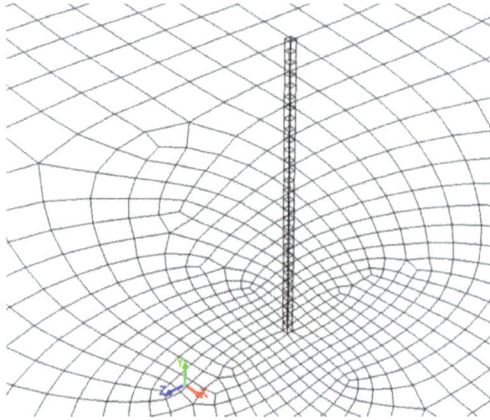

Figure 6. Pollutant source (detail)

Model Scale [-]	Geometry Dimensions [m]	Nozzle (Chimney) Dimensions [m]	Nozzle (Chimney) Spout Surface [m]	Coordinates of the Nozzle (Chimney) [m]	Note
1:1	X = 2000 Y = 1500 Z = 1500	X = 3.5 Y = 200 Z = 3.5	X = 3.5 Z = 3.5	X = 500 Y = 0 Z = 750	Real Pattern
1:4.04	X = 495.04 Y = 371.28 Z = 371.28	X = 0.866 Y = 49.5 Z = 0.866	X = 0.866 Z = 0.866	X = 123.76 Y = 0 Z = 185.64	-
1:35.51	X = 56.32 Y = 42.24 Z = 42.24	X = 0.098 Y = 5.632 Z = 0.098	X = 0.098 Z = 0.098	X = 14.08 Y = 0 Z = 21.12	-
1:101.82	X = 19.64 Y = 14.73 Z = 14.73	X = 0.034 Y = 1.964 Z = 0.034	X = 0.034 Z = 0.034	X = 4.91 Y = 0 Z = 7.36	-
1:1000	X = 2.0 Y = 1.5 Z = 1.5	X = 0.0035 Y = 0.2 Z = 0.0035	X = 0.0035 Z = 0.0035	X = 0.5 Y = 0 Z = 0.75	Wind Tunnel

Table 1. Geometry dimensions and pollutant source coordinates by model scale

turbulent viscosity calculation (see [1],[2]). Species transport model was used for the species motion calculation. Both models worked simultaneously. No additional gas pollutant dispersion model was applied. The operating pressure was set at 101 325 [Pa], the operating temperature was 300 [K], and the gravity acceleration -9.81 [m/s^2] in the geometry. Considering the pollutant source close surroundings, the ranges of the Reynolds number

Re were 250–1250 [-] (model scale 1:1000, referential air flow velocity 1–5 [m/s], and nozzle spout diameter 0.0035 [m]) and 250000–1250000 [-] (model scale 1:1, referential air flow velocity 1–5 [m/s], and chimney spout diameter 3.5 [m]).

Boundary conditions were set to *Velocity Inlet* at the inlet, *Outflow* at the outlet, *Wall* for the floor, *Symmetry* for the walls, *Wall* for pollutant walls, and *Velocity Inlet* for the nozzle (spout of the chimney).

Profiles of the flow field physical characteristics were determined at the inlet of the geometry (see Table 2) based on experimental data from a low-speed wind tunnel for a 1:1000-scale model (see [3],[4],[5],[11]). For other model scales, the profiles were modified to keep the trend of curves.

Title	Equation
Vertical profile of air flow velocity (X-direction)	$v_x = 0.2371 \cdot \ln(Y + 0.00327) + 1.3571$
Vertical profile of air turbulent intensity	$I = -0.0673 \cdot \ln(Y + 0.00327) + 0.1405$
Vertical profile of turbulent kinetic energy	$k = 1.5 \cdot (v_x \cdot I)^2$
Vertical profile of turbulent dissipation rate	$\varepsilon = \dfrac{1.225 \cdot 0.09 \cdot (k^3)}{1.4}$

Table 2. Air flow velocity profile and turbulent characteristics profiles in geometry

In Table 2 the parameter v_x represents the air flow velocity in the direction of X-axis [m/s], I is the intensity of turbulence [%], Y is the vertical coordinate of the geometry [m], k is the turbulent kinetic energy [m²/s²], and ε is the turbulent dissipation rate [m²/s³].

The pollutant source was designed as a nozzle (chimney) of a cuboid shape. Its dimensions and spout surfaces are shown in Table 1. For all model scales, the pollutant velocity $v_{polutant}$ was set at 0.5 [m/s], the intensity of turbulence in the pollutant source at 10 [%], the pollutant mass fraction in the pollutant source at 0.95 [-], and the air mass fraction in the pollutant source at 0.05 [-]. The hydraulic diameter of the pollutant source was set at 3.5 [m] for 1:1-scale model, 0.866 [m] for 1:4.04-scale model, 0.0986 [m] for 1:35.51-scale model, 0.344 [m] for 1:101.821-scale model, and 0.0035 [m] for 1:1000-scale model scale 1:1000.

Three different pollutants were chosen to be tested: helium, methanol and 1,2-dichlorethane. Helium (ρ = 0.1625 [kg/m³]) has a lower density than air, i.e., it is lighter than air (ρ = 1.225 [kg/m³]). Methanol (ρ = 1.43 [kg/m³]) has approximately the same density as air, i.e., it is approximately of the same weight as air. 1,2-dichlorethane (ρ = 4.1855 [kg/m³]) has a greater density than air, i.e., it is heavier than air. Plumes of pollutants lighter than air tend to climb, whereas those heavier than air tend to descend. However, this is not always the case. Pollutant plume vertical movements can be influenced by several other physical factors as demonstrated in the analysis (see Sections 2.4, 2.5 and 2.6).

3.4. Analysis of results by air flow velocity

The aim of this analysis is to compare gas pollutant plume shapes and motions for three different gas pollutants (helium, methanol and 1,2-dichlorethane) at different values of the air flow velocity v_{air}. The demonstration of the problem was performed with a 1:1-scale three-dimensional geometry representing the real pattern of a simple terrain with a chimney (see Table 1). The referential air flow velocities v_{air} at the level of the chimney spout (pollutant source) were 1 [m/s], 3 [m/s] and 5 [m/s].

The resulting values of the Froude number Fr for each pollutants and air flow velocities are shown in Table 3. Results were calculated using the ANSYS Fluent 13.0 software and were visualized as iso-surfaces of pollutant concentrations or as contours of pollutant concentration fields (see Figure 7 to Figure 12). The contours were plotted in two-dimensional planes of the geometry, sc., the central vertical longitudinal plane, the floor (ground) plane, and the outlet plane.

Pollutant	Chemical Symbol/Formula	Model Scale [-]	Air Flow Velocity [m/s]	Froude number [-]	Note
Helium (gas)	He	1 : 1	1	0.248	Fr < 1
		1 : 1	3	2.230	Fr > 1
		1 : 1	5	6.195	Fr > 1
Methanol (gas)	CH₃OH	1 : 1	1	0.028	Fr ≪ 1
		1 : 1	3	0.253	Fr < 1
		1 : 1	5	0.704	Fr < 1
1,2-dichlorethane (gas)	CH₂ClCH₂Cl	1 : 1	1	0.010	Fr ≪ 1
		1 : 1	3	0.088	Fr ≪ 1
		1 : 1	5	0.246	Fr < 1

Table 3. Values of the Froude number for one model scale and three different air flow velocities

The following figures show that with increasing air flow velocity v_{air} the pollutant plume vertical movements are reduced. The pollutant plume inclines horizontally at the level of the chimney spout (pollutant source) showing no tendency to climb or descend. This is because the inertial force F_{I-air} increases as the air flow velocity v_{air} increases. Hence, the pollutant plume vertical movements are reduced or totally eliminated.

The air flow velocity also influences the size and shape of the pollutant plume. With increasing air flow velocity v_{air} the pollutant plume tends to be narrower and longer. However, a further increase in the air flow velocity makes the pollutant plume shorter because of greater rate of the pollutant dispersion. The plume range at certain concentration of the pollutant therefore decreases with increasing air flow velocity.

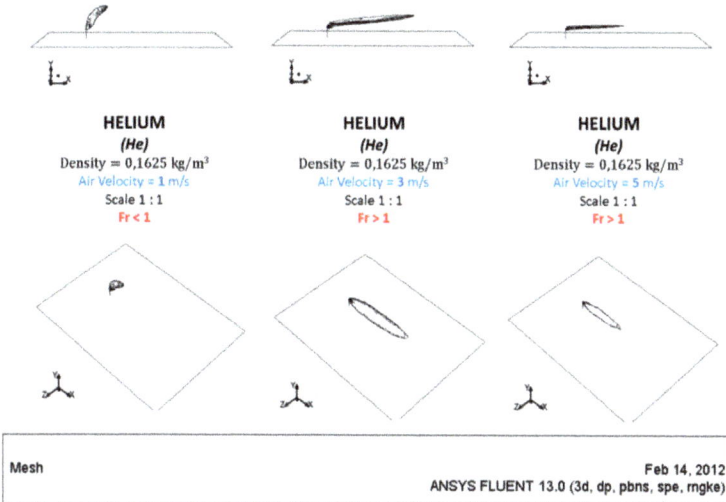

Figure 7. Gas pollutant plume motion analysis by air flow velocity (iso-surfaces of gaseous helium concentration with mass fraction of 0.0001 [-])

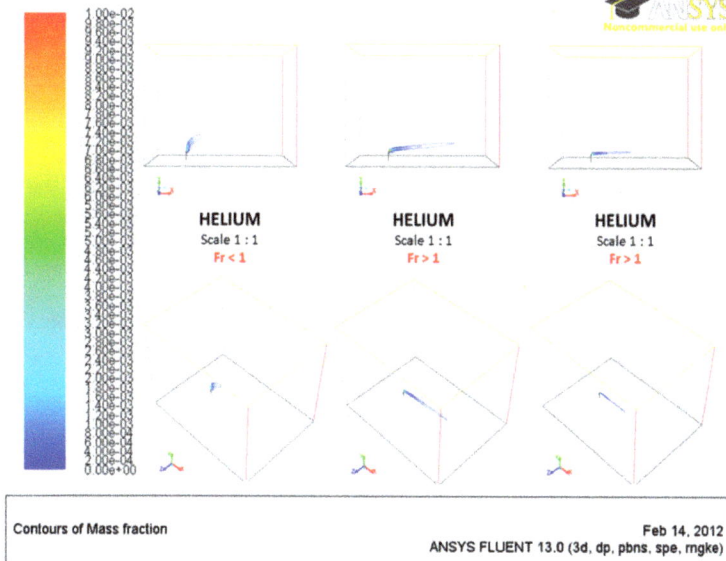

Figure 8. Gas pollutant plume motion analysis by air flow velocity (contours of gaseous helium concentration with mass fraction range of 0–0.01 [-])

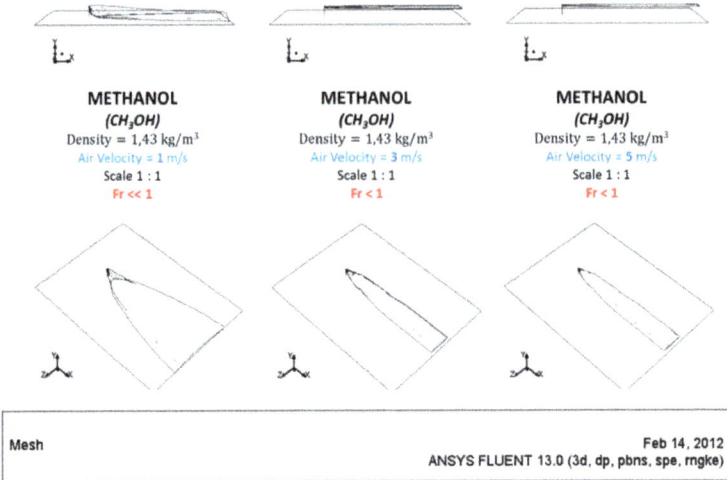

Figure 9. Gas pollutant plume motion analysis by air flow velocity (iso-surfaces of gaseous methanol concentration with mass fraction of 0.0001 [-])

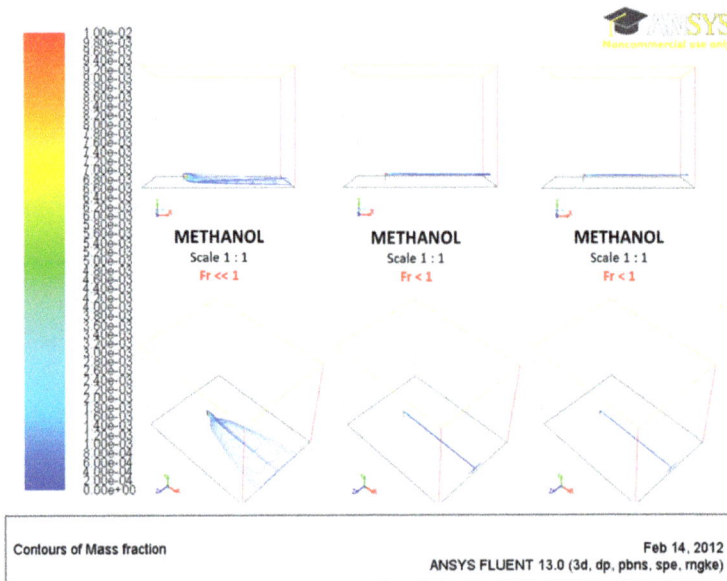

Figure 10. Gas pollutant plume motion analysis by air flow velocity (contours of gaseous methanol concentration with mass fraction range of 0–0.01 [-])

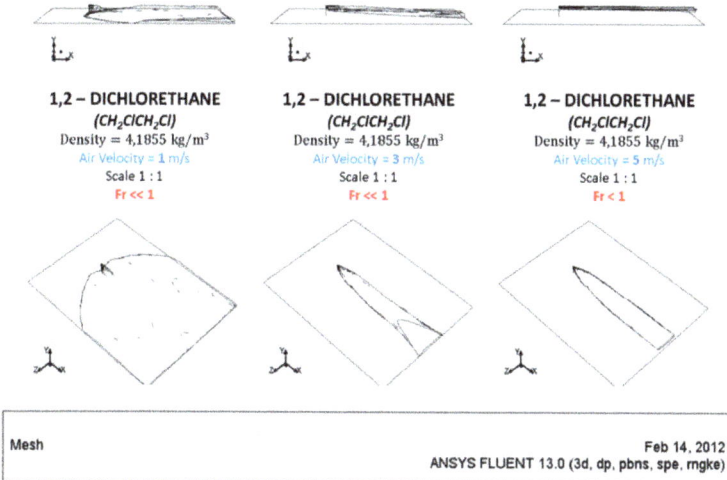

Figure 11. Gas pollutant plume motion analysis by air flow velocity (iso-surfaces of gaseous 1,2-dichlorethane concentration with mass fraction of 0.0001 [-])

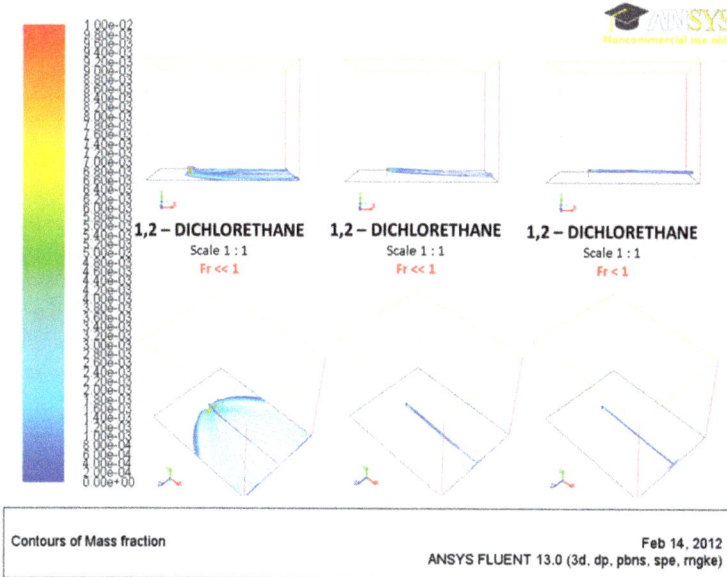

Figure 12. Gas pollutant plume motion analysis by air flow velocity (contours of gaseous 1,2-dichlorethane concentration with mass fraction range of 0–0.01 [-])

3.5. Analysis of results by pollutant density

The aim of this analysis is to compare gas pollutant plume shapes and motions for three different gas pollutants (helium, methanol and 1,2-dichlorethane) at different values of their density. Demonstration of the problem was performed with a 1:1-scale three-dimensional geometry representing the real pattern of a simple terrain with a chimney (see Table 1). The referential air flow velocity v_{air} at the level of the chimney spout (pollutant source) was 1 [m/s].

The resulting values of the Froude number Fr for each individual pollutant are shown in Table 4. Results were calculated using the ANSYS Fluent 13.0 software and were visualized as iso-surfaces of pollutant concentrations or as contours of pollutant concentration fields (see Figure 13 and Figure 14). The contours were plotted in two-dimensional planes of the geometry, sc., the central vertical longitudinal plane, the floor (ground) plane, and the outlet plane.

Pollutant	Chemical Symbol/Formula	Model Scale [-]	Air Flow Velocity [m/s]	Froude number [-]	Note
Helium (gas)	He	1 : 1	1	0.248	Fr < 1
Methanol (gas)	CH₃OH	1 : 1	1	0.028	Fr << 1
1,2-dichlorethane (gas)	CH₂ClCH₂Cl	1 : 1	1	0.010	Fr << 1

Table 4. Values of the Froude number for one model scale and one air flow velocity

The figures show that: If the pollutant density $\rho_{pollutant}$ is lower than the air density ρ_{air} the gas pollutant plume tends to climb (for helium see Figure 13 and Figure 14). If the pollutant density $\rho_{pollutant}$ is approximately the same as the air density ρ_{air} the gas pollutant plume neither climbs nor descends (for methanol see Figure 13 and Figure 14). If the pollutant density $\rho_{pollutant}$ is greater than the air density ρ_{air} the gas pollutant plume tends to descend (for 1,2-dichlorethane see Figure 13 and Figure 14). The range of vertical movements is determined by gravity force $F_{G-pollutant}$ that influences pollutant plume at given conditions. The greater the gravity force $F_{G-pollutant}$ is compared to the inertial force F_{I-air}, the more significant vertical movement of the plume is, i.e., light pollutant plume climbs and heavy pollutant plume descends.

The pollutant density $\rho_{pollutant}$ also influences the pollutant plume dispersion. The greater the pollutant density is, the longer the range of the plume is. At given air flow velocity v_{air}, the plume dispersion of pollutants with a low density is faster and easier than that of pollutants with a greater density.

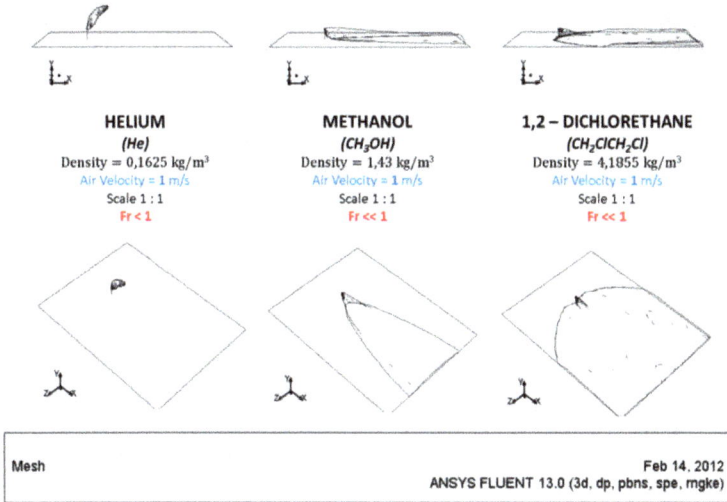

Figure 13. Gas pollutant plume motion analysis by pollutant density (iso-surfaces of gaseous helium, methanol, and 1,2-dichlorethane concentrations with mass fraction of 0.0001 [-]; 1:1-scale model)

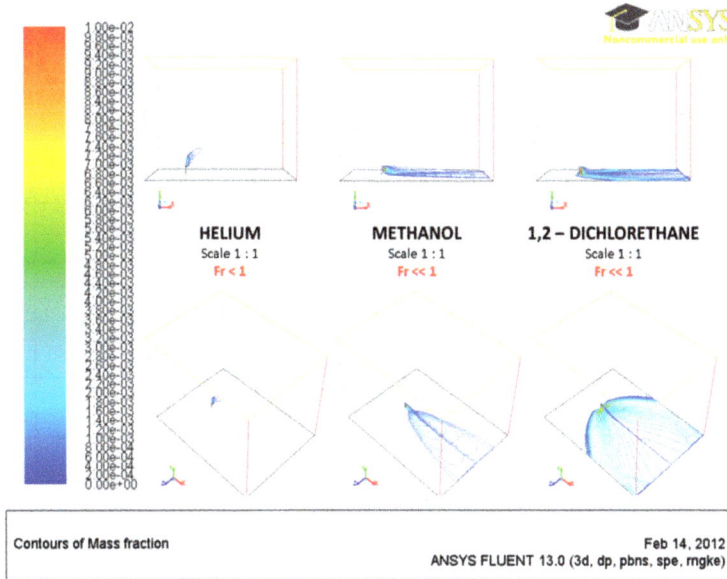

Figure 14. Gas pollutant plume motion analysis by pollutant density (contours of gaseous helium, methanol, and 1,2-dichlorethane concentrations with mass fraction range of 0–0.01 [-]; 1:1-scale model)

3.6. Analysis of results by model scale

The aim of this analysis is to compare gas pollutant plume shapes and motions for three different gas pollutants (helium, methanol, and 1,2-dichlorethane) at different model scales. The demonstration of the problem was performed with a three-dimensional geometry at three selected scales (see Table 1) for each of the three pollutants. The first 1:1-scale model represents the real pattern of a simple terrain with a chimney (pollutant source) where $Fr <$ 1, i.e., the gravity force $F_{G-pollutant}$ is greater than the inertial force F_{I-air}. The second model (scaled at 1:4.04, 1:35.51, and 101.82, respectively) represents the state when $Fr = 1$, i.e., the gravity force $F_{G-pollutant}$ equals the inertial force F_{I-air}. The third 1:1000-scale model represents the gauging section of a low-speed wind tunnel with a nozzle (pollutant source) on the floor where $Fr > 1$, i.e., the inertial force F_{I-air} is greater than the gravity force $F_{G-pollutant}$. The referential air flow velocity v_{air} at the level of the chimney spout (pollutant source) was 1 [m/s].

The resulting values of the Froude number Fr for each pollutant and model scale are shown in Table 5. Results were calculated using the ANSYS Fluent 13.0 software and were visualized as iso-surfaces of pollutant concentration or as contours of pollutant concentration fields (see Figure 15 to Figure 20). Contours were plotted in two-dimensional planes of the geometry, sc. the central vertical longitudinal plane, the floor (ground) plane, and the outlet plane.

Pollutant	Chemical Symbol/Formula	Model Scale [-]	Air Flow Velocity [m/s]	Froude number [-]	Note
Helium (gas)	He	1 : 1	1	0.248	Fr < 1
		1 : 4.04	1	1	Fr = 1
		1 : 1000	1	247.806	Fr >> 1
Methanol (gas)	CH₃OH	1 : 1	1	0.028	Fr << 1
		1 : 35.51	1	1	Fr = 1
		1 : 1000	1	28.160	Fr >> 1
1,2-dichlorethane (gas)	CH₂ClCH₂Cl	1 : 1	1	0.010	Fr << 1
		1 : 101.82	1	1	Fr = 1
		1 : 1000	1	9.822	Fr > 1

Table 5. Values of the Froude number for different model scales and one air flow velocity

The figures (see Figure 15 to Figure 20) show that: If the model scale changes and all other characteristics remain unchanged, the inertial and gravity forces and their ratio change too. Therefore, the size, shape and inclination of the pollutant plume change.

According to Equation (11) in Section 1.5, the inertial force is proportional to the square of the model scale. According to Equation (12) in Section 1.5, the gravity force is proportional to the third power of the model scale. Therefore, the change in the gravity force due to the change of the model scale is considerably greater than the change in the inertial force. The lower the model scale is, the greater the dominance of inertial forces is compared to gravity forces), and vice versa.

Also, the greater the pollutant density, the lower the model scale is if $Fr = 1$, i.e., the inertial and gravity forces are equal.

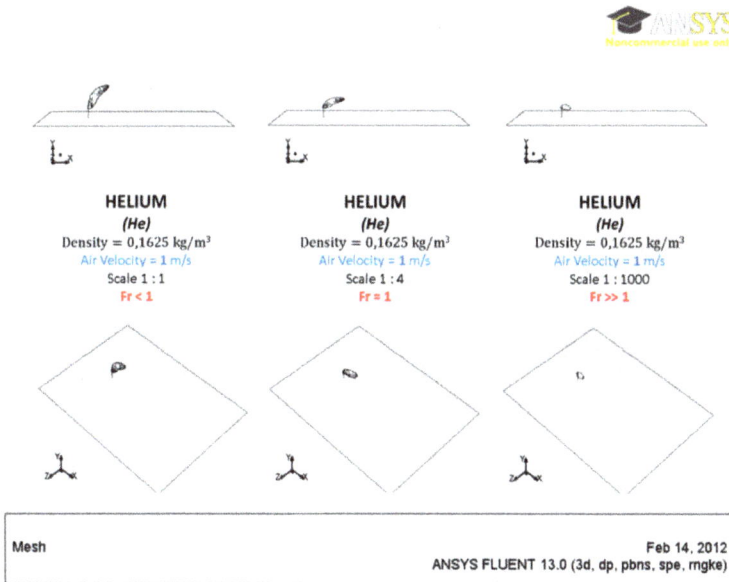

Figure 15. Gas pollutant plume motion analysis by model scale (iso-surfaces of gaseous helium concentration with mass fraction of 0.0001 [-] for three different model scales)

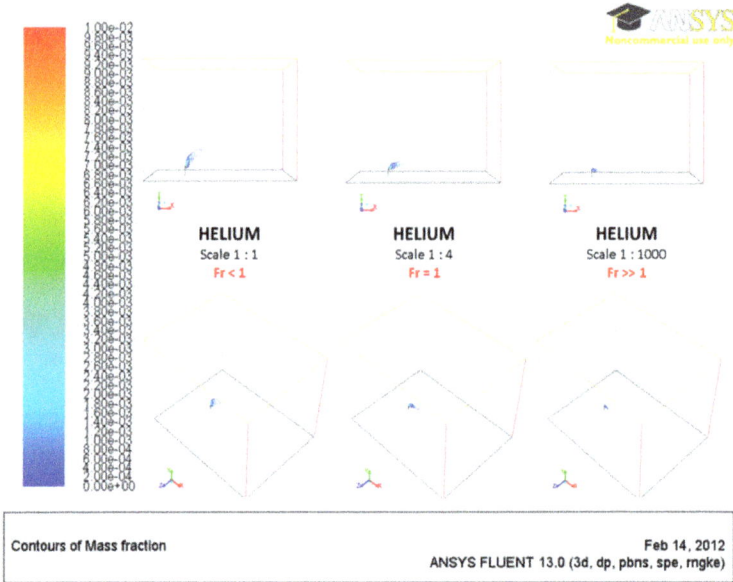

Figure 16. Gas pollutant plume motion analysis by model scale (contours of gaseous helium concentration with mass fraction range of 0–0.01 [-] for three different model scales)

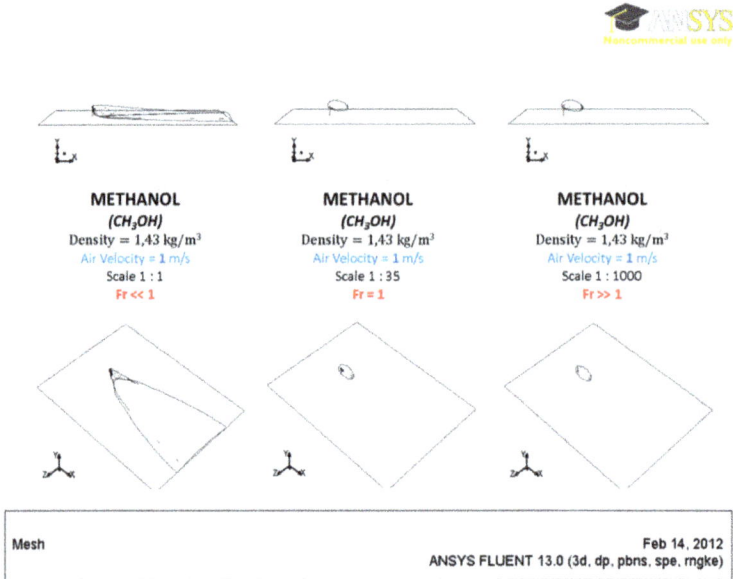

Figure 17. Gas pollutant plume motion analysis by model scale (iso-surfaces of gaseous methanol concentration with mass fraction of 0.0001 [-] for three different model scales)

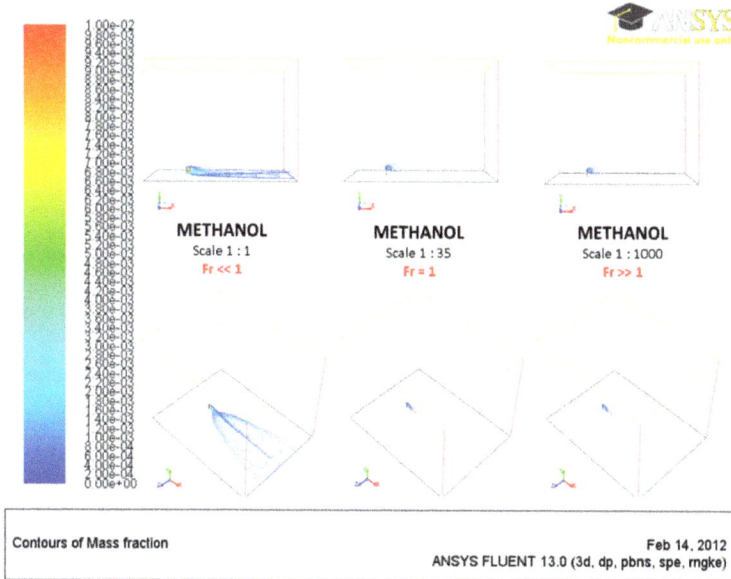

Figure 18. Gas pollutant plume motion analysis by model scale (contours of gaseous methanol concentration with mass fraction range of 0–0.01 [-] for three different model scales)

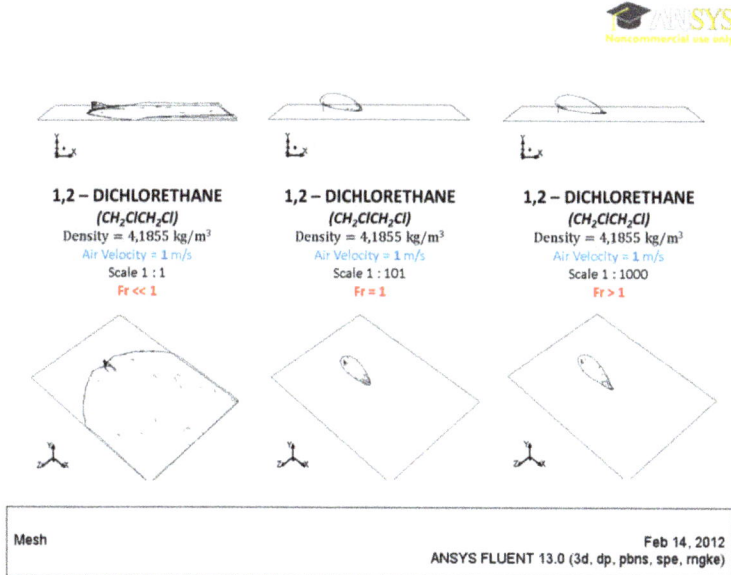

Figure 19. Gas pollutant plume motion analysis by model scale (iso-surfaces of gaseous 1,2-dichlorethane concentration with mass fraction of 0.0001 [-] for three different model scales)

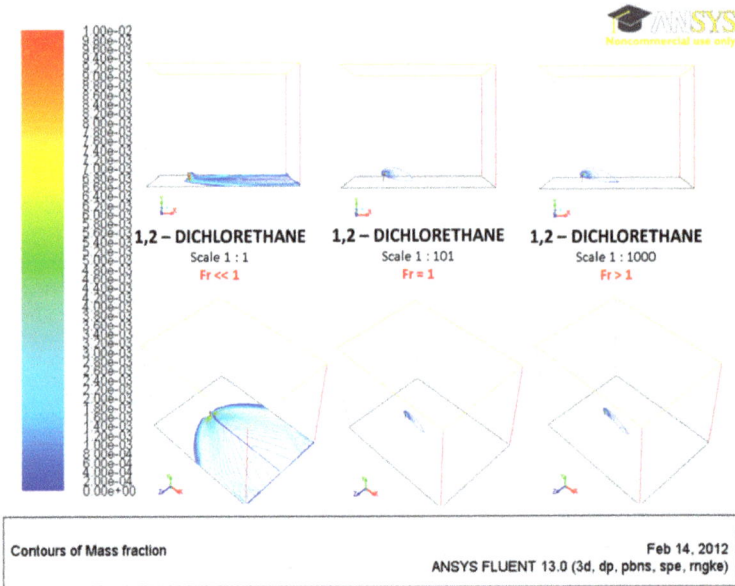

Figure 20. Gas pollutant plume motion analysis by model scale (contours of gaseous 1,2-dichlorethane concentration with mass fraction range 0–0.01 [-] for three different model scales)

4. Conclusion

The aim of the analyses was to lay down principles for physical and mathematical modeling of gas pollutant plume motion and dispersion in real atmospheric conditions. The influences of the air flow velocity, pollutant's density, and model scale on pollutant plume size, shape, and inclination were investigated.

The Froude number was chosen as a criterion of physical similarity for the pollutant plume behavior in the atmosphere. Basic mathematical rules and principles (see Chapter 1) were formulated upon study of available fluid mechanics literature (see [7],[8],[9],[10],[11]). All mathematical and physical assumptions were next verified by numerical simulation using the ANSYS Fluent 13.0 software. Air flow field was modeled using the RNG $k - \varepsilon$ model of turbulence, the gas pollutant motion was modeled using the Species Transport Model, both in the same three-dimensional geometry consisting of 569 490 grid cells. Turbulent characteristics were defined using RANS approach. No additional dispersion model was applied (see Chapter 2).

Object of modeling was gauging section of the low speed wind tunnel (for model scale of 1:1000) or big real terrain (for model scale of 1:1) with a pollutant source in form of nozzle (or chimney, respectively) situated on the section floor (ground). The gauging section with the nozzle represented a chimney in a simple, flat terrain. The chimney was considered to be a pollutant source for three different gas pollutants (helium, methanol, and 1,2-

dichlorethane). The numerical simulation was performed for five model scales, three gas pollutants with different densities, and three different air flow velocities. The simulations were steady (time-independent) with the accuracy of 0.0001 (criterion of convergence). Final results were visualized as iso-surfaces of pollutant concentrations and contours of pollutant concentration fields with the concentration limit value of 0,001. The contours were plotted in two-dimensional planes of the geometry, sc. the central vertical longitudinal plane, the floor (ground) plane and the outlet plane. The numerical model had been verified by an experiment performed in a low-speed wind tunnel (see [3],[4],[5]).

The following principles based on the results of the Froude number analysis of pollutant plume motion and dispersion in real atmosphere can be defined:

1. The greater the air flow velocity is, the greater the inertial forces are. These forces influence pollutant plume and reduce its vertical motions (inclination). With increasing air flow velocity the pollutant plume inclines horizontally at the level of the chimney spout (pollutant source), but with further increase in the air flow velocity it becomes narrower and shorter (see Section 2.4).
2. The greater the difference between pollutant density and air density is, the more significant the tendency towards vertical movements (climbing or descending) of the plume is. The plume of pollutant with lower density than air tends to climb, whereas the plume of pollutant with greater density than air tends to descend. The density of the pollutant also influences the pollutant plume dispersion. The greater the pollutant density is, the longer the range of the plume is. At given air flow velocity v_{air}, the plume dispersion of pollutants with a low density is faster and easier than that of pollutants with a greater density (see Section 2.5).
3. If the model scale changes and all other characteristics remain unchanged, the inertial and gravity forces and their ratio change too. Therefore, the size, shape and inclination of the pollutant plume change. The inertial force is proportional to the square of the model scale, whereas the gravity force is proportional to the third power of the model scale (see Section 1.5). The change in the gravity force due to the change of model scale is considerably greater than the change in the inertial force. The lower the model scale is, the greater the dominance of inertial forces is compared to gravity forces. Also, the greater the pollutant density is, the lower the model scale is if $Fr = 1$, i.e., the inertial and gravity forces are equal (see Section 2.6).

From the above it follows that if investigators want to respect and follow the basics of physical phenomena, they must consider criteria of physical similarity very carefully, in particular criteria of dynamic similarity (see Section 1.1). Some physical phenomena, however, cannot be modeled in any model scale but the original one without changing the basis of the phenomena (see Sections 1.5 and 2.6).

This analysis is intended for those who are interested in gas pollutant plume motion in the atmosphere and in theory of physical similarity. The conclusions of the analysis can be used for further experiment design works or for checking results of mathematical modeling.

Author details

Ondrej Zavila

Vysoka Skola Banska, Technical University of Ostrava, Faculty of Safety Engineering, Ostrava, Vyskovice, Czech Republic

Acknowledgement

The author acknowledges the financial support of the SPII 1a10 45/07 project of the Ministry of the Environment of the Czech Republic.

5. References

[1]   Kozubkova M (2008) Modeling of Fluid Flow (in Czech), FLUENT, CFX. Ostrava: VSB - Technical University of Ostrava. 153 p.

[2]   Bojko M (2008) Guide for Training of Flow Modeling – FLUENT (in Czech). Ostrava: VSB - Technical University of Ostrava. 141 p.

[3]   Civis S, Zelinger Z, Strizik M, Janour Z (2001) Simulation of Air Pollution in a Wind Tunnel. Spectroscopy from Space. Dordrecht: Kluwer Academic. pp. 275-299

[4]   Zelinger Z, Strizik M, Kubat P, Janour Z, Berger P, Cerny A, Engst P (2004) Laser Remote Sensing and Photoacoustic Spectrometry Applied in Air Pollution Investigation. Opt. Lasers Eng. 42. pp. 403-412

[5]   Zelinger Z, Strizik M, Kubat P, Civis S, Grigorova E, Janeckova R, Zavila O, Nevrly V, Herecova L, Bailleux S, Horka V, Ferus M, Skrinsky J, Kozubkova M, Drabkova S, Janour Z (2009) Dispersion of Light and Heavy Pollutants in Urban Scale Models: CO_2 Laser Photoacoustic Studies. Applied spectroscopy. Society for Applied Spectroscopy. pp. 430-436.

[6]   Benson, Tom (2010) Wind Tunnel Index. Available: http://www.grc.nasa.gov/WWW/k-12/airplane/shortt.html. Accessed 2012 March 4.

[7]   Carnogurska M, Prihoda M (2011) Application of Three-dimensional Analysis for Modelling Phenomena in the field of Power Engineering (in Slovak). Kosice: Technical University of Kosice. 214 p.

[8]   Incropera F.P, Dewitt D.P, Bergman T.L, Lavine A.S (2007) Fundamentals of Heat and Mass Transfer. New York: John Wiley & Sons. 997 p.

[9]   Shaughnessy E.J, Katz M.I, Schaffer J.P (2005) Introduction to Fluid Mechanics. New York: Oxford University Press. 1118 p.

[10] Drabkova S, Platos P (2003) Numerical Simulation as a Tool for the Solution and Understanding of Practical Air Pollution Problems. Proc. of the Conference on Modelling Fluid Flow (CMFF'03). Budapest: Budapest University of Technology and Economics. pp. 501-506.

[11] Stull B.R (1994) An Introduction to Boundary Layer Meteorology. Dordrecht: Kluwer Academic Publisher. 666 p.

Chapter 3

Numerical Investigation for Steady and Unsteady Cavitating Flows

Hatem Kanfoudi, Hedi Lamloumi and Ridha Zgolli

Additional information is available at the end of the chapter

1. Introduction

Cavitation is a particular two-phase flow with phase transition (vaporization/condensation) driven by pressure change without any heating. It can be interpreted as the rupture of the liquid continuum due to excessive stresses. Modeling of cavitating flows as a multi-fluid is a complex problem especially when the 3D consideration is adopted. However most recently research works are based on the mixture consideration of homogeneous fluid composed by two phases of liquid and vapor, which is also used by our model....

Minimizing the nuisance of cavitation is a great challenge in the design phase of a marine propeller. For efficiency reasons, the propeller usually needs to be operated in cavitating conditions but one still needs to avoid the effects of vibrations, noise and erosion. However, cavitation is a complex phenomenon not yet neither reliably assessable nor fully understood. Experimental observations can only give a part of the answer due to the obvious limitations in the measurement techniques; one example is measuring reentrant jets and internal flow, where flow features are hidden for optical measurement techniques by the cavity itself. Standard simulation tools used in design typically include potential flow solvers, lifting surface or boundary element approaches, with strict theoretical limits on cavitation modeling that only in the hands of an experienced designer may give satisfactory propeller designs. Adding to the challenge is a lack of theoretical knowledge of the physical mechanisms leading to harmful cavitation and thus how to modify a design if some form of nuisance is detected.

The transport equation models of cavitation suggested by Alajbegovic, Grogger et Philipp [1] and Yuan, et al.[2], use the simplistic Rayleigh model. This model describes the limiting case of inertia-controlled growth of a spherical bubble in a liquid under a step variation in pressure of the surrounding liquid. However, this model cannot accurately describe bubble collapse and neglects a number of effects, which determine the behavior of cavitation bubbles.

Recently, progress has been made in the development of numerical models for calculation of cavitation flows. Though the models may differ in terms of realization (using the single-fluid or multi-fluid frame-work, the Eulerian- Eulerian or Eulerian- Lagrangian approaches), all of them are empirical to a certain level.

Modeling of cavitation flow as a multi-fluid is a complex problem which does not lead to satisfying results especially when the cavitation is modeled as 3D. Therefore in the engineering practice cavitation flow is often modeled as a single-fluid, where the cavitation area is handled as an area with the pressure lower then the vapour pressure. This approach always leads to the result, and the requirement of computer time is many times lower in comparison with multi-phase flow models. Moreover the steady solution of multiphase flow model may not be found at all due to the unsteady nature of cavitation flow.

Significant progress has been achieved recently in the development of homogeneous-mixture (single-fluid) models for the simulation of three-dimensional transient cavitating flows (Chen and Heister [3] [4]; Kunz, Boger and Stinebring [5]; Ahuja, Hosangadi et Arunajatesan [6]; Yuan, et al. [2]; Singhal, et al.[7]; Kubota, Kato et Yamaguchi [8]). These models allow single-fluid solvers to be applied to the conservation equations for the mixture, without increase in computational cost due to the increase in the number of conservation equations when applying the multi-fluid flow concept.

The present work is an investigation to develop a relevant physical model to simulate the cavitating flow. The main goal is the development of computational methodologies which can provide detailed description of the numerical set up for modeling and simulation the cavitation with the CFD code.

2. Mathematical formulation

To simulate cavitating flows, the two phases, liquid and vapour, need to be represented in the problem, as well as the phase transition mechanism between the two. Here, we consider a one fluid, single-fluid (mixture), introduced through the local vapour volume fraction and having the spatial and temporal variation of the vapour fraction described by a transport equation including source terms for the mass transfer rate between the phases. The numerical model solves the Reynolds averaged Navier-Stockes equations, coupled with a localized vapour transport model for predicting cavitation.

The fundamental equations governing the flow are taken with incompressible fluid case as given by the Navier-Stockes equations, which control the transport of momentum within the fluid (2) in addition to the mass conservation constraint (1):

$$\frac{\partial \rho_m}{\partial t} + \frac{\partial \left(\rho_m u_j \right)}{\partial x_j} = 0 \qquad (1)$$

$$\frac{\partial \left(\rho_m u_i \right)}{\partial t} + \frac{\partial \left(\rho_m u_j u_i \right)}{\partial x_j} = -\frac{\partial p}{\partial x_i} + \frac{\partial}{\partial x_j}\left[\left(\mu_m + \mu_t \right)\left(\frac{\partial u_i}{\partial x_j} + \frac{\partial u_j}{\partial x_i} \right) \right] \tag{2}$$

The effective density and viscosity of the mixture are given (3) and (4) respectively :

$$\rho_m = \rho_v \alpha + \left(1 - \alpha \right)\rho_l \tag{3}$$

$$\mu_m = \mu_v \alpha + \left(1 - \alpha \right)\mu_l \tag{4}$$

Where α is the vapor fraction ($\alpha \rightarrow 1$: vapor and for $\alpha \rightarrow 0$: liquid). The density and the viscosity of liquid and vapor are assumed to be constant. To compute the volume fraction we need a closure model, the distribution of values α was obtained on each cell of the computational domain will guide the attendance rate of the vapour.

Vapour fraction transport equation

The transport equation for the vapour scalar fraction α is given by:

$$\frac{\partial \rho_v \alpha}{\partial t} + \frac{\partial \left(\rho_v \alpha u_j \right)}{\partial x_j} = \dot{m}^+ + \dot{m}^- \tag{5}$$

This transport equation of volume fraction of vapour, with appropriate source terms to regulate the mass transfer between phases (liquid/vapor), is solved.

The proposed model formulation

a. This work deals with a numerical simulation of cavitation process around a hydrofoil. The numerical approach is based on predicting of the collapse process of pocket of vapor due to liquid compressibility. The model based is the Rayleigh Plesset (dynamics of spherical bubbles) integrated in the term source. The interface velocity investigated during the collapse process is given by:

$$\dot{R} = -\sqrt{\left(\frac{2}{3} \frac{p - p_v}{\rho_l} \right)\left(\frac{R_0^3}{R^3} - 1 \right)}.$$

b. finally, the proposed model can be express :

$$\dot{m}^- = C_c \alpha^{\frac{1}{6}}\left(1 - \alpha \right)^2 \frac{\rho_v \rho_l}{\rho_m}\sqrt{\frac{2}{3}\frac{p - p_v}{\rho_l}}; \quad \dot{m}^+ = C_e \alpha^{\frac{2}{3}}\left(1 - \alpha \right)^{\frac{4}{3}} \frac{\rho_v \rho_l}{\rho_m}\sqrt{\frac{2}{3}\frac{p_v - p}{\rho_l}} \tag{6}$$

with :

$$C_c = -10 R_0^{\frac{3}{2}} n_0^{\frac{5}{6}}; \quad C_e = 5\sqrt[3]{n_0} \tag{7}$$

The value of n_0 is a calibrate parameter with experimental results; it is define the bubble density.

3. Numerical result

3.1. Numerical study

To validate the proposed model, a confrontation to experimental measures and to the numerical models (Yuan et al., Schnerr and Sauer and EOS). The application is a NACA0009 hydrofoil, truncated at 90% of the original chord length. It has the final dimensions of 100 mm of chord length. The 3D test section is modeled by a quasi 2D domain, with three rows of cells in spanwise direction for the numerical domain. The same mesh and numerical setup is used for the computations with the two models.

3.1.1. Geometry

The domain is 9 blocks (See Fig. 1). The boundary conditions are set using a velocity inlet and a pressure at the outlet (the parameter which fixes the cavitation number).

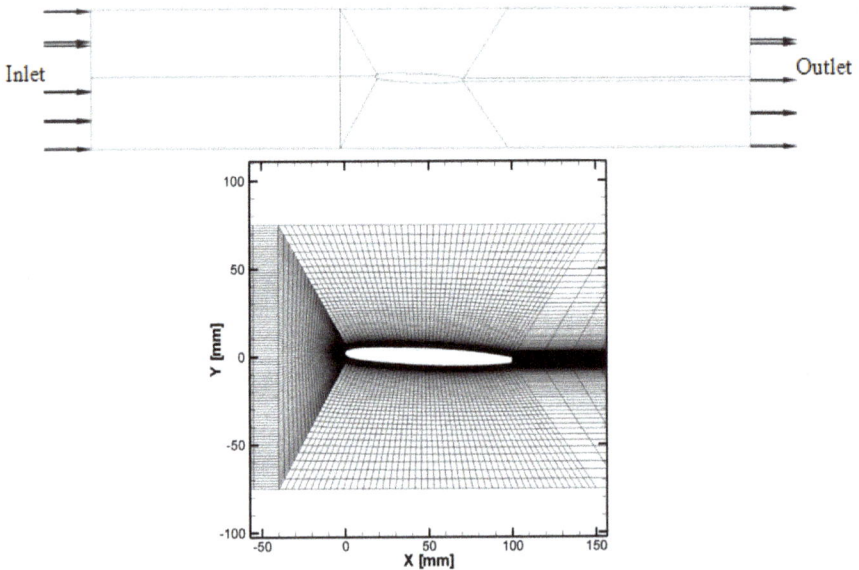

Figure 1. NACA0009 domain grid, y+=10, O-grid structure, with 9 blocks;

The equations of the non-truncated hydrofoil are:

$$0 \leq \frac{y}{c} \leq 0.45 \quad \frac{y}{c} = a_0 \left(\frac{x}{c}\right)^{\frac{1}{2}} + a_1 \left(\frac{x}{c}\right) + a_2 \left(\frac{x}{c}\right)^2 + a_3 \left(\frac{x}{c}\right)^3$$

$$0.45 \leq \frac{y}{c} \leq 1 \quad \frac{y}{c} = b_1 \left(1-\frac{x}{c}\right) + b_2 \left(1-\frac{x}{c}\right)^2 + b_3 \left(1-\frac{x}{c}\right)^3$$

$$\begin{cases} a_0 = +0.173688 \\ a_1 = -0.244183 \quad b_1 = +0.1737 \\ a_2 = +0.313481 \quad b_2 = -0.185355 \\ a_3 = -0.275571 \quad b_3 = +0.33268 \end{cases} \qquad (8)$$

3.1.2. Mesh quality

Numerical solutions of fluid flow and heat transfer problems are only approximate solutions. In addition to the errors that might be introduced in the course of the development of the solution algorithm, in programming or setting up the boundary conditions… The discretization approximations introduce errors which decrease as the grid is refined, and that the order of the approximation is a measure of accuracy. However, on a given grid, methods of the same order may produce solution errors which differ by as much as an order of magnitude. Theoretically, the errors in the solution related to the grid must disappear for an increasingly fine mesh [HYPERLINK \l "Fer96" 9].

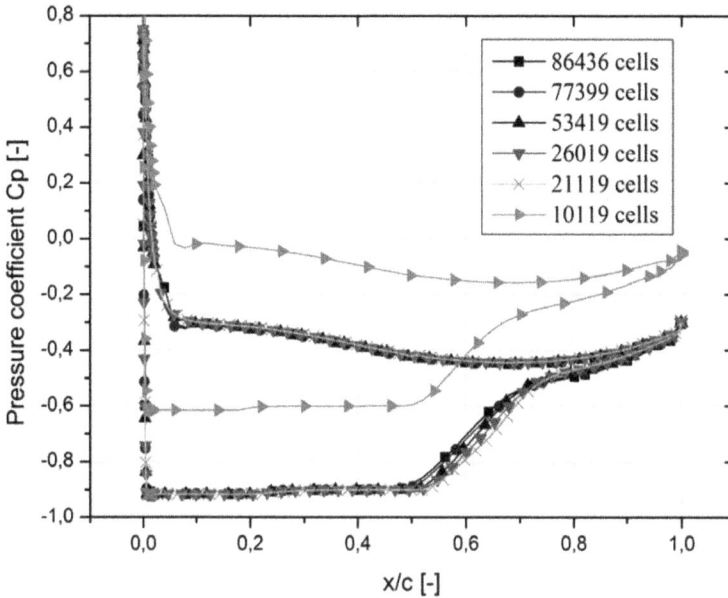

Figure 2. Impact of the quality of the mesh in numerical result, σ=0.4

The pressure coefficient at σ=0.4 was taken as the parameter to evaluate six grids (See Fig. 2) and determine the influence of the mesh size on the solution. The selected convergence criteria were a maximum residual of 10^{-4}. According to this figure, the grid with 53419 cells is considered to be sufficiently reliable to ensure mesh independence.

Constants	Mark	Value
Reference velocity@ Inlet	C_{ref}	20 [m s^{-1}]
Length of chord		0.11 [m]
Saturation pressure		3164 [Pa]
Liquid density		997 [kg m^{-3}]
Vapour density		0.023 [kg m^{-3}]
Reference time		0.0055 [s]
Pressure @ Outlet	P_{out} : calculate as a function of the value of	

Table 1. Numerical parameters

Mass, momentum, turbulence and scalar transport equations were solved with Anys CFX which uses a coupled solver, which solves the hydrodynamic equations (for u, v, w, p) as a single system. This solution approach uses a fully implicit discretisation of the equations at any given time step. For steady state problems the time-step behaves like an 'acceleration parameter', to guide the approximate solutions in a physically based manner to a steady-state solution. This reduces the number of iterations required for convergence to a steady state, or to calculate the solution for each time step in a time dependent analysis. For the advection discretization, the High Resolution scheme is employed and the second order backward Euler scheme for the transient term, to assure accurate solution and to reduce the numerical diffusion for the solution [10].

In these calculations turbulence effects were considered using turbulence models, as the k-ε RNG models, with the modification of the turbulent viscosity. To model the flow close to the wall, standard wall-function approach was used. For this model, the used numerical scheme of the flow equations was the segregated implicit solver.

3.2. Validation steady flow

3.2.1. Turbulence model discussion

For steady state flows, there is a priori no difference between the two equations formulations. The use of SST in the case of 2D steady hydrofoil computations instead of k-ω or k-ε is justified in this way.

SST model works by solving a turbulence/frequency-based model (k–ω) at the wall and k-ε in the bulk flow. A blending function ensures a smooth transition between the two models. The SST model performance has been studied in a large number of cases. In a NASA Technical Memorandum, SST was rated the most accurate model for aerodynamic applications [11].

3.2.2. Pressure coefficient validation

A validation of the proposed model is described in this section. Firstly, we proceed with a distribution of the pressure coefficient around the surface of the hydrofoil and then we adjust with the profile velocity.

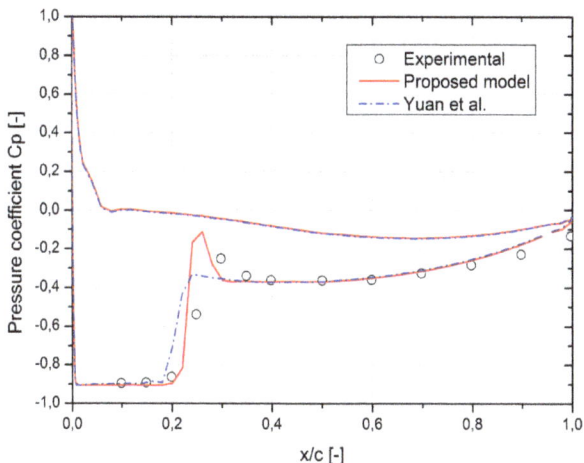

Figure 3. Pressure distribution comparison between proposed model and Yuan et al. model with experimental (measurement reported from [12], on NACA0009, i=2.5°, σ=0.9, = 30 m/s.

Fig. 3 presents the pressure coefficient distribution around the NACA 0009 for the proposed model and the Yuan et al. model with the experimental measurements. The proposed model shows a peak at the closing-pocket (condensation), this is due to the collapse velocity; it tends to compress the vapor pocket (collapse). The results from the proposed model can be satisfied compared to the experimental measurement.

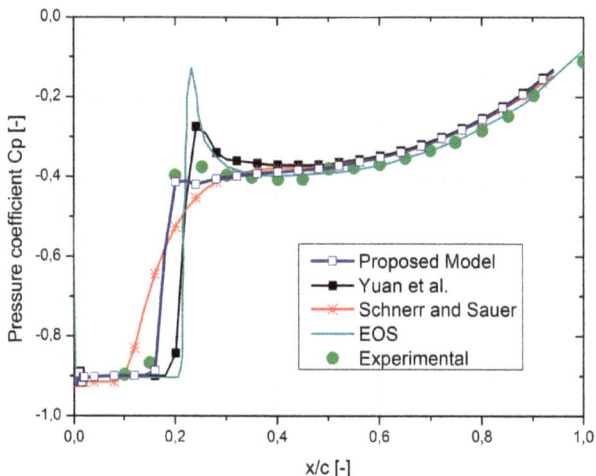

Figure 4. Pressure distribution comparison between proposed model, Yuan et al. model, Schnerr and Sauer model and EOS with experimental (measurement reported from [13]), on NACA0009, i=2.5°, σ=0.9, = 20 m/s.

We remark the good agreement of the numerical values compare to experimental data for $R_0=10^{-6}$ m and $n_0=10^{14}$ bubbles/m^3. These constants are fixed for all this study.

Experimental data concerning the NACA0009 hydrofoil are reported for different cavitation numbers, and compared with the results of the computations in Fig 4.

These figures show satisfactory results of the proposed models in predicting the pressure distribution on the hydrofoil. As expected, the cavity becomes larger with decreasing cavitation number. However the models exhibit very different flow behavior at the cavitation detachment and closure regions. This is the result of dynamic bubble which is implementing in the proposed model. Clearly the model can perfectly predict the cavity of the vapor.

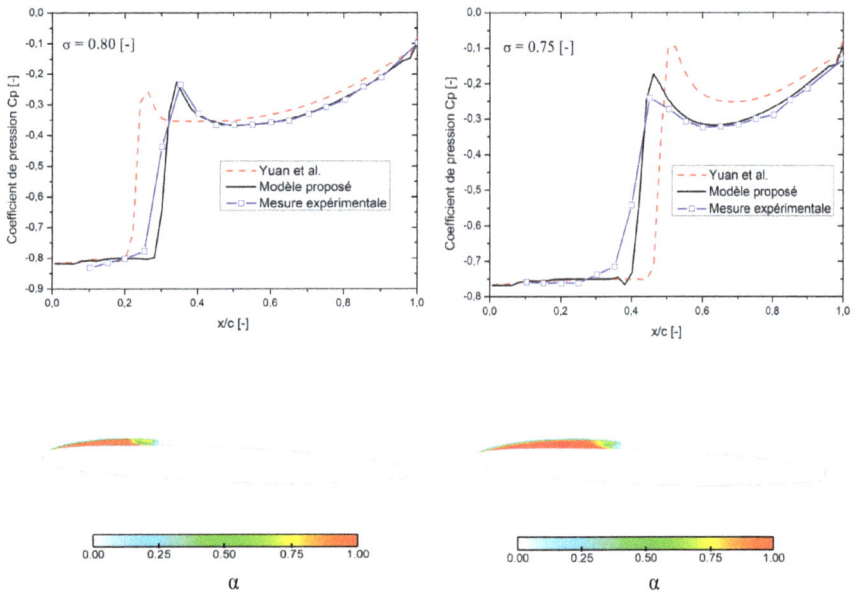

Figure 5. Pressure distribution confrontation between proposed model, Yuan et al. [2] and with experimental (measurement reported from [13]), on NACA0009, U_{ref}= 20 m/s.

Figure 5 show satisfactory results of the proposed model in the pressure distribution on the hydrofoil. As expected, the cavity becomes larger with decreasing cavitation number. However the model exhibit very different flow behavior at the cavitation detachment and closure regions. This is the result of dynamic bubble which is taken account in the proposed model (interface velocity). Clearly the model can perfectly predict the cavity of the vapour.

3.2.3. Velocity distribution validation

Figure 6 show the predicted velocity profile of the main flow far from the wall is in good agreement with measurements. In the near-wall and in the wake region, the proposed model tends to reproduce the flow re-entrant jet.

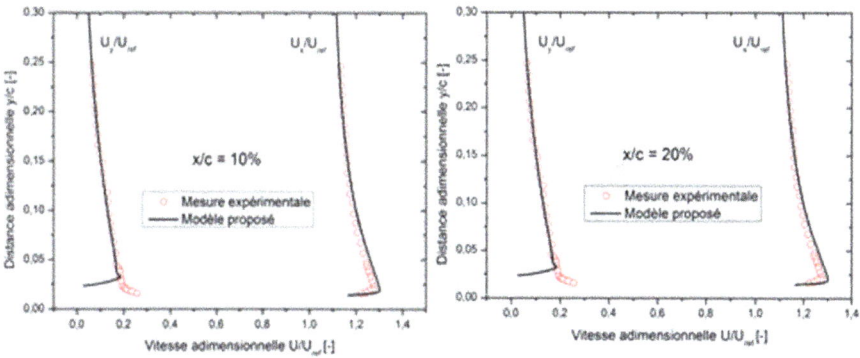

Figure 6. Comparison between computed and measured averaged dimensionless velocity profiles at 10% and 20% of the chord, (chord=100[mm]) (measurement reported from [12]).

3.3. Validation unsteady flow

3.3.1. 2D configuration(NACA0009)

The vapor cavity is characterized by a thick main cavity and an important shed cavity volume in a disorganized way, such as the cavity can be divided into many small cavities. For one period of cavity creation and collapse, one can remark two distinct life cycles highlighted by the lift and drag signals. Firstly, starting from the maximum cavity length, the closure region is in the small adverse pressure gradient and the reentrant jet is too thin, such as the cavity closure region is continuously broken into small vapor volumes. Secondly, as the reentrant jet reaches the region of the high pressure gradient, the reentrant jet is more important and the whole cavity is extracted and shed downstream. We consider the non-truncated variant of the 2D NACA0009 hydrofoil at high incidence angles. The flow parameters are $i=5°$ and $σ=1.2$, and reflected by a high pressure gradient over the hydrofoil leading to an unsteady state flow behavior and shedding of large transient cavities. The domain is taken larger than the experimental test section to avoid numerical problem mainly due to reflections on the boundaries.

Cavitating flow are highly sensitive to turbulent fluctuations present in the flow [14] [15]. The effects of modeling turbulence quantities have an enormous impact on the cavitation dynamics and the overall flow structure. The k-ε RNG turbulence model and LES model are adopted in this study.

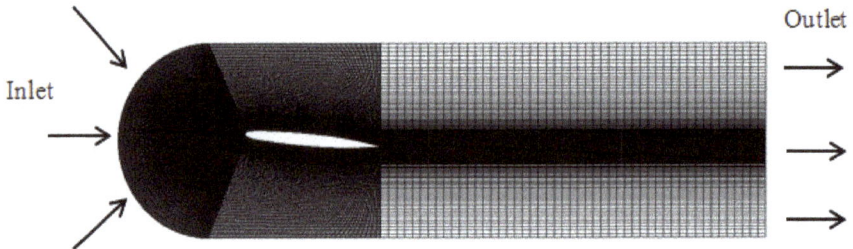

Figure 7. Domain mesh, C-type; $y^+=1$; $i=5°$

3.3.2. Result and discussion

In this section we present the behavior of the proposed model with LES and the modified turbulence model. Most of characteristics of the unsteady flow is defined as a function of lift and drag coefficients, this coefficient can be expressed as:

$$C_l = \frac{L}{\frac{1}{2}\rho_l C_{ref}^2 A}; \quad C_d = \frac{D}{\frac{1}{2}\rho_l C_{ref}^2 A} \qquad (9)$$

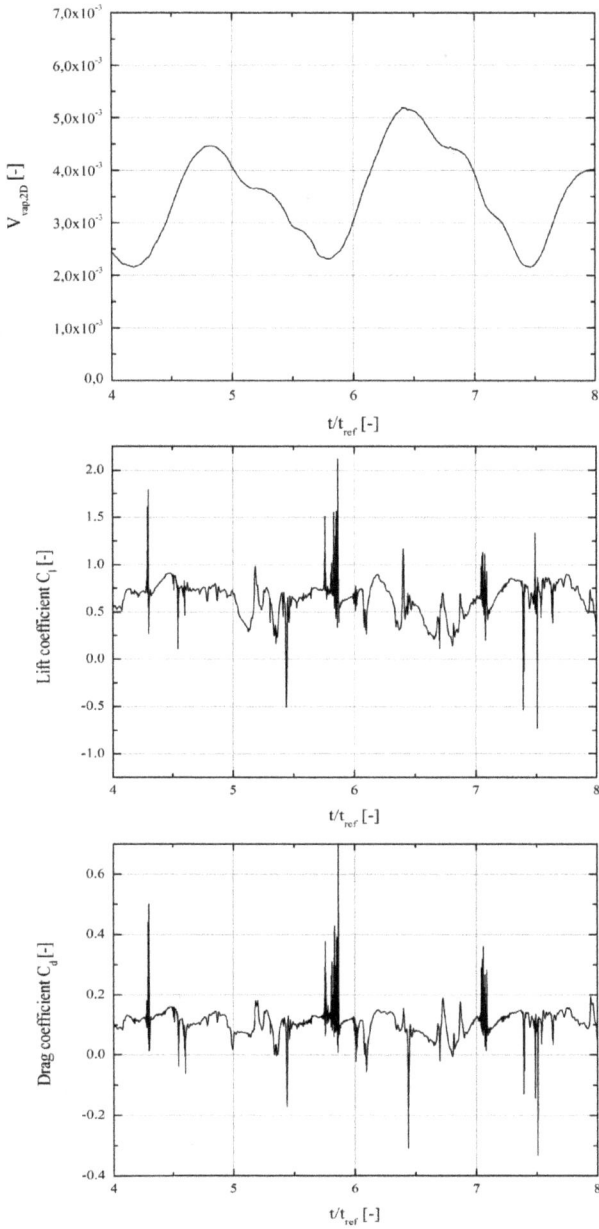

Figure 8. The time history of the lift, the drag coefficient, and total vapour volume for both cavitation numbers σ =1.2, T=0.008 s (the proposed model).

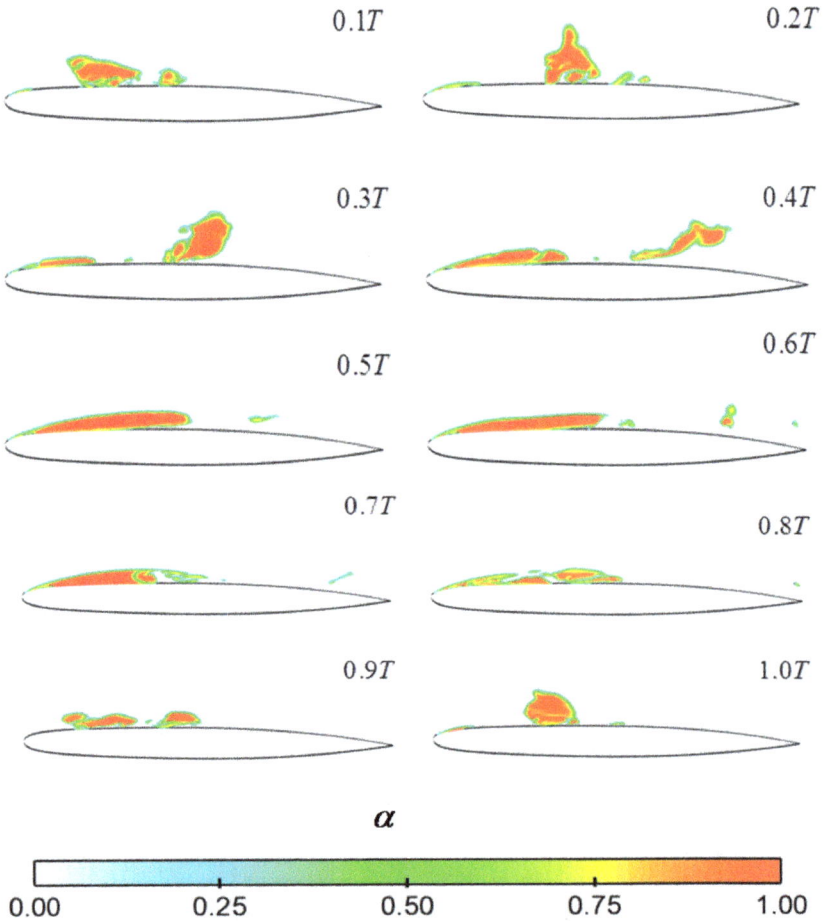

Figure 9. Volume fraction of vapor for σ=1.2, for life cycle (period T=0.01s).

For flows about 2D hydrofoils the dimensionless total vapour volume $V_{vap,2D}$ is defined as :

$$V_{vap,2D} = \frac{1}{c^2}\sum_{i=1}^{N}\alpha_i V_i \tag{10}$$

where N is the total number of control volumes and α is the volume fraction of vapour in each control volume, with the volume Vi of the fluid in control volume.

The total vapour volume $V_{vap,2D}$, defined in equation (10), is a convenient parameter for understanding the transient evolution of the cavitating flow. The total vapour volume is calculated at each time step. After the start-up phase the growth and shedding of the vapour sheet and the collapse of the shed vapour cloud induce a self-oscillatory behavior, which is approximately periodic in time. The graphics of the analyzed variables (Fig.8) show a periodic signal, even if the identified periods are very different from each others. The use of Vvap,2D is the easiest way to identify the periodicity of the vapour formation and collapse during a life cycle. Furthermore, the time-history of the lift and drag coefficients are compared to those of the total vapour volume to correlate the occurring flow phenomena.The lift and drag coefficient has a cyclic time signal. Even if the periods are pretty clear, the fluctuations in a given period are very different from one to another. This phenomenon is mainly due to the dynamic of the shed cavities which are driven by the main flow-field downstream of the cavity closure. The growth and collapse mechanism of the cavity is driven by a cyclic phenomenon, whereas the dynamic of the cavity, when it is swept away can have very different non reproducible behavior. The vapour can be attached to the wall or far from it. This different ways of cavities shedding have an important influence on the pressure field at the hydrofoil wall, and there by on the lift and drag values.

In this section the cycles illustrated in figures 15 and 16 are considered. The solution for the volume fraction α above the hydrofoil is presented for a number of equidistant time-intervals during the cycle.

The vapour cavity is characterized by a thick main cavity and an important shed cavity volume in a disorganized way, such as the cavity can be divided into many small cavities. For one period of cavity creation and collapse, one can remark two distinct life cycles highlighted by the lift and drag signals.

Firstly, starting from the maximum cavity length, the closure region is in the small adverse pressure gradient and the reentrant jet is too thin, such as the cavity closure region is continuously broken into small vapour volumes. Secondly, as the reentrant jet reaches the region of the high pressure gradient, the reentrant jet is more important and the whole cavity is extracted and shed downstream.

3.3.3. 3D configuration(NACA66mod)

These excellent results obtained for 2-D cavitating flow confirm the correct assumptions of the proposed numerical model in terms of vaporization and condensation processes, and to verify it's performance with 3-D considerations, we present in the next some numerical results compared to experimental measurement [16] for cavitating flow around hydrofoil NACA66 (Fig 10).

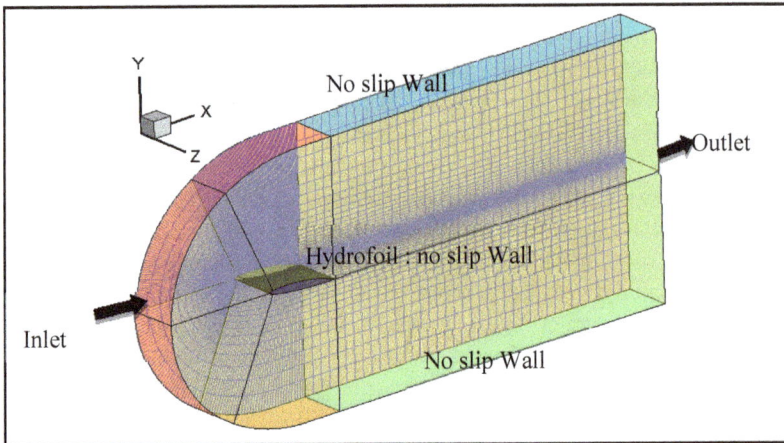

Figure 10. Domain mesh,C-type; y$^+$=50; i=6°

In this section, a numerical simulation was performed to further study the performance of the proposed model and unsteady three-dimensional, the choice is justified because of the availability of experimental measurements. The profile used is NACA66 (mod) -312 a = 0.8 with a string of 0.15m, the width of the profile is 0.19m and is placed at 6 ° incidence relative to the upstream flow, not in an infinite medium -viscous. It has a thickness of approximately 12% and a camber on 2% to 45% and 50% of the leading edge along the rope. Theoretical points of this section have been interpolated by B-spline technique using the software mesh.

The study area is in 3D, it is made larger than the experimental test section to avoid numerical problems mainly due to reflections from the boundary conditions. The estate consists of 156 000 cells, the mesh topology is structured with a C-type conditions with wall on its border, the hydrofoil with the no-slip condition. (see Fig. 8). The steady state solution was used as an initial condition, the turbulence model used is k-ε RNG.

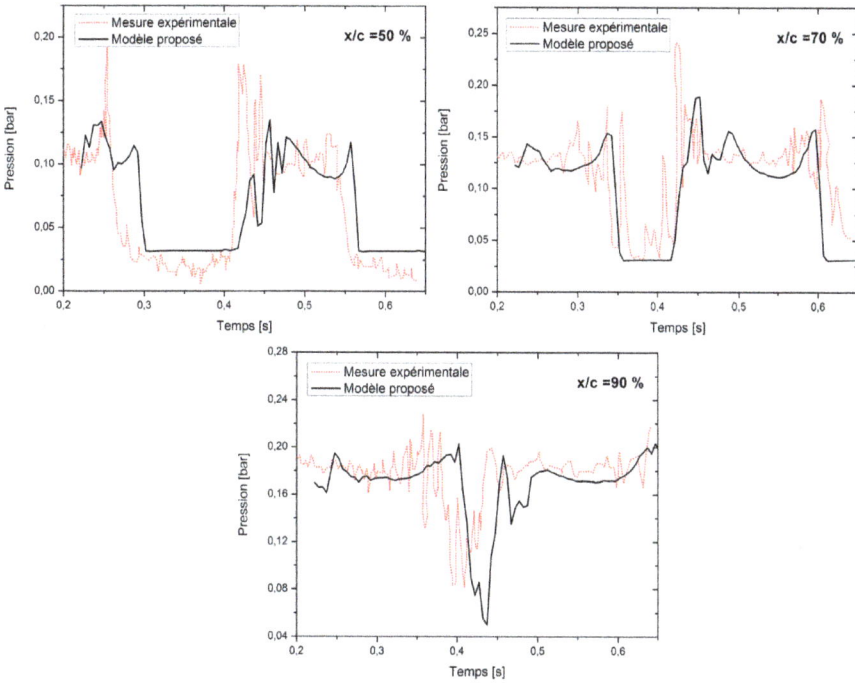

Figure 11. Domain grid and boundary condition at the left, at the right, the validation of the pressure measurement [16] respectively at 50%,70% and 90% of the chord length with the proposed model, C_{ref}= 5.3 m/s.

The proposed model was validated quantitatively and qualitatively compared in two-dimensional distribution of pressure coefficient and velocity profiles calculated with experimental measurements. We propose in the following one-dimensional confrontation, the application as we have described above is the NACA66 (mod), the advantage of selecting this application is the availability of measuring pressure along the profile for two cycles of detachment from the pocket of steam. We have reported three points of pressure measurement located respectively 50%, 70% and 90% chord.

The obtained numerical result presented by figure9 show that the model reproduce correctly the typical behaviour of partial cavity with development of re-entrant jet and the periodic shedding of cavitation clouds. Firstly, starting from the maximum cavity length, the closure region is in the small adverse pressure gradient and the re-entrant jet is too thin, such as the cavity closure region is continuously broken in to small vapour volumes. Secondly, as the re-entrant jet reaches the region of the high pressure gradient, the re-entrant jet is more important and the whole cavity is extracted and shed downstream.

Time = 0.2998 [s]

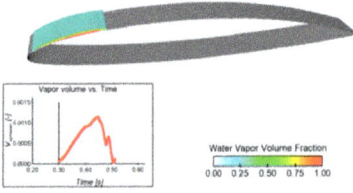

Time = 0.3211 [s]

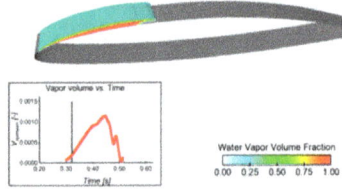

Time = 0.3424 [s]

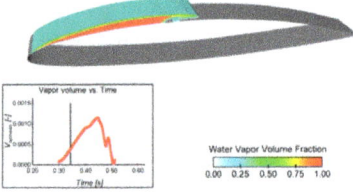

Time = 0.3637 [s]

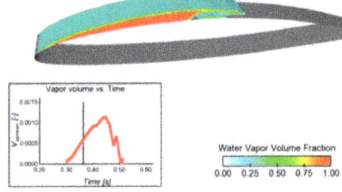

Time = 0.3850 [s]

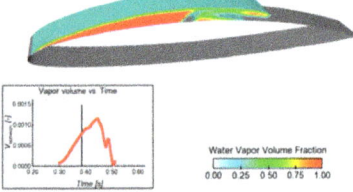

Time = 0.4063 [s]

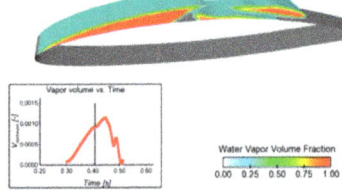

Time = 0.4276 [s]

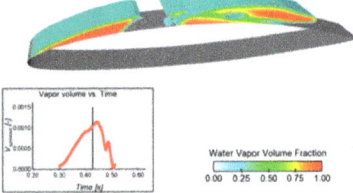

Time = 0.4489 [s]

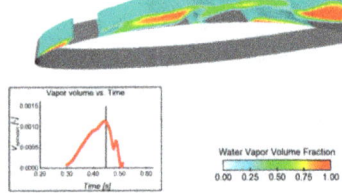

Time = 0.4702 [s]

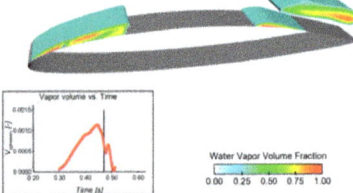

Time = 0.4915 [s]

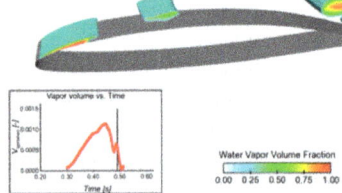

Figure 12. Volume fraction of vapor for $\sigma=1.$, for life cycle (period T=0.2s).

This different ways of cavities shedding have an important influence on the pressure field at the hydrofoil wall, and there are characterized by the temporal evolution of the lift coefficient Cl compared to experimental result [16]

4. Conclusion

A comprehensive theoretical approach is done, and detailed formulations of the proposed model are presented. Influence of the numerical parameter has been widely studied. A comparative study is made for the steady flow. Good agreement with measurements was obtained for the proposed model. We have shown the importance of liquid compressibility on the pocket of vapour and how it is modeling in the proposed model. Computations with the proposed model is compared with experimental data and other numerical models, it shows the ability of the model to reproduce the steady-state developed cavitation flow fields. Finally, the unsteady behaviour of the cavitating flow depends strongly on the turbulence model, a k-ε RNG model is adopted in this steady.

Author details

Hatem Kanfoudi, Hedi Lamloumi and Ridha Zgolli
Laboratory of Hydraulic and Environmental Modelling, National Engineering School of Tunis, Tunis, Tunisia

5. References

[1] A. Alajbegovic, H.A. Grogger, and H. Philipp, "Calculation of transient cavitation in nozzle using the two-fluid model," in *Proc. ILASS-Americas'99 Annual Conf.*, 1999, pp. 373 – 377.

[2] Yuan, W. Sauer, J. Schnerr, and H. G.,: Mec. Ind., 2001, vol. 2, pp. 383 - 394.

[3] Y Chen and S.D Heiste, "Modelling Hydrodynamic Nonequilibrium in Cavitating Flows," in *Tr. ASME Journal Fluids Eng*, vol. 118, 1996, pp. 172 – 178.

[4] Y Chen and S.D Heister, "Two-phase modelling of cavitated flows," in *In ASME Cavitation and Multiphase Forum*, vol. 24, Reno, Nevada, 1995, pp. 799 – 806.

[5] R.F. Kunz, D.A Boger, and D.R. Stinebring, "A preconditioned Navier-Stokes method for two-phase flows with application to cavitation prediction," *Computer & Fluids*, vol. 29, pp. 849 – 875, 2000.

[6] V. Ahuja, A. Hosangadi, and S. Arunajatesan, "Simulations of caviting flows using hybrid unstructured meshes," *J. Fluids Engng*, vol. 123, pp. 331 – 339, 2001.

[7] A.K. Singhal, M.M. Athavale, H. Li, and Yu Jiang, "Mathematical basis and validation of the full cavitation model," , vol. 124, 2002, pp. 617 – 624.

[8] A Kubota, H. Kato, and H. Yamaguchi, "A new modelling of cavitating flows: a numerical study of unsteady cavitation on a hydrofoil section," *Journal of Fluid Mechanics*, pp. 240: 59-96, 1992.

[9] J. H. Ferziger and M. Peric, *Computational Methods for Fluid Dynamics*. Berlin, Germany: Springer, 1996.

[10] ANSYS CFX,., 2010, p. 69.

[11] JE Bardina, P.G Huang, and T.J. Coakley, "Turbulence Modeling Validation, Testing and Development," *NASA Technical Memorandum 110446, AIAA*, 1997.

[12] PH. Dupont, "Etude de la dynamique d'une poche de cavitation partielle en vue de la prediction de l'érosion dans les turbomachines hydrauliques," 1993.

[13] Youcef Ait Bouziad, *physical modelling of leading edge cavitation: computational methodologies and application to hydraulic machinery*.: Lausanne, EPFL, 2006.

[14] O. Coutier-Delgosha, R. Fortes-Patella, J.L. Reboud, M. Hofmann, and B. Stoffel, "Experimental and Numerical Studies in a Centrifugal Pump with Two-Dimensional curved Blades in Cavitating condition," vol. 125, no. Transactions of the ASME, pp. 970–978, 2003.

[15] J.L. Reboud, O. Coutier-Delgosha, B. Pouffary, and R. Fortes-Patella, "Numerical Simulation of Unsteady Cavitating Flows: some Applications and Open Problems," in *In Fifth International Symposium on Cavitation*, 2003.

[16] J.B. Leroux, J.A. Astolfi, and J.Y. Billard, "EXPERIMENTAL STUDY OF UNSTEADY AND UNSTABLE PARTIAL CAVITATION," in *9éme journée de l'hydrodynamique*, 2003.

Chapter 4

Bubble Rise Phenomena
in Non-Newtonian Crystal Suspensions

N.M.S. Hassan, M.M.K. Khan and M.G. Rasul

Additional information is available at the end of the chapter

1. Introduction

1.1. Bubble trajectory

The characteristics of bubble motion in non-Newtonian liquids are still not well understood because many parameters influence the terminal rise velocity, trajectory and shape of bubbles (Frank, 2003). As the bubble motion is a complex problem, the degree of the complexity increases with bubble size (Kulkarni and Joshi, 2005). When bubble rises through liquid, the most resistance will be imposed directly on top and the bubble first moves along a straight vertical path and then develops a zigzag motion which consequently can change into a spiralling motion, at the same incidence as the preceding zigzag (Ellingsen and Risso, 2001; Mougin and Magnaudet, 2002; Shew and Pinton, 2006; Zenit and Magnaudet, 2009).

In most reported studies, very small bubbles (less than 1 mm) rise through water maintaining their spherical shape due to surface tension. The trajectory of these bubbles follows a straight line until it completes its journey (Clift *et al.*, 1978; Duineveld, 1995). On the other hand, considerable deformations are observed for bubbles with diameters larger than 1 mm (Ellingsen and Risso, 2001). This deformation occurs due to the increase in the variations of hydrostatic and dynamic pressure over the bubbles' surface (Magnaudet and Eames, 2000). Therefore, large bubbles cannot remain spherical and deform into oblate spheroids first and then become ellipsoidal, and with further increase in size they switch into a spherical or ellipsoidal cap. Bubble motion such as velocity and trajectory also change with the increase in bubble size (Hassan *et al.*, 2010a).

The bubble is not always rising in straight path. When the bubble size increases, a straight path turns into zigzag or spiral in fluids of small Morton number. Then the path becomes nearly straight again for a spherical cap bubble. Only a straight path is observed (Yang,

2006) in liquids of large Morton number. Aybers and Tapucu (1969) reported different types of trajectories such as zigzag, helical or spiral and rocking motions. Haberman and Morton (1954) also observed rectilinear (Re < 300), spiral and rocking motions. They indicated that the spiral path could be either clockwise or counter-clockwise, depending on the conditions of bubble release. The major axis of the bubble is always directed perpendicular to the direction of motion. Saffman (1956) observed only zigzag bubble rise motions as the bubble rises in water when the radius of the bubble was less than 1 mm, but bubbles of larger radius showed either zigzag or spiral motions. Feng and Leal (1997) verified various possible trajectories for different shape regimes. A single bubble can follow a zigzag path at Re ≈ 600, accompanied with vortex shedding behind the bubble. Under the same experimental conditions, Yoshida and Manasseh (1997) reported that the bubbles can also follow a spiral trajectory without vortex shedding. Tsuge and Hibino (1997) reported that the trajectories of rising spherical and ellipsoidal gas bubbles at higher Reynolds numbers are identical.

The trajectories of bubbles are strongly influenced by the bubble deformations and the surrounding fluid flow (Ellingsen and Risso, 2001). Bubble deformations and fluid flow could be explained with dimensionless groups such as the Reynolds Number (Re), the Weber number (We), the Morton number (Mo) and bubble aspect ratio (E) (Hassan et al., 2008c; Hassan et al., 2010a). In fluid mechanics, Re gives a measure of the ratio of inertial forces to viscous forces and consequently quantifies the relative importance of these two types of forces for given flow conditions. On the other hand, We is often useful in analyzing fluid flows where there is an interface between two different fluids, especially for multiphase flows such as bubble rise in liquids. It can be thought of as a measure of the relative importance of the fluid's inertia compared to its surface tension. The quantity is useful in analyzing the formation of droplets and bubbles. The dimensionless number such as Mo , is also used together with the Eötvös number (Eo) to characterize the shape of bubbles or drops moving in a surrounding fluid or continuous phase. Eötvös number is considered as proportional to buoyancy force divided by surface tension force. Theses dimensionless groups are defined as:

$$\text{Re} = \frac{\rho_{liq} d_{eq} U_b}{\mu_1} \tag{1}$$

$$We = \frac{\rho_{liq} U_b^2 d_{eq}}{\sigma} \tag{2}$$

$$Mo = \frac{g \mu_{liq}^4 \Delta\rho}{\rho_{liq}^2 \sigma^3} \tag{3}$$

$$E = \frac{d_w \text{ or semi major axis}}{d_h \text{ or semi minor axis}} \tag{4}$$

Where U_b is the bubble rise velocity, m/sec; d_{eq} is the bubble equivalent diameter, m; ρ_{liq} is the density of the liquid, kg/m^3; μ_1 is the liquid viscosity, Pa.s; σ is the surface tension of the liquid, N/m; g is the aacceleration due to gravity, m/s^2 and $\Delta\rho$ is the density difference between liquid and air bubble, kg/m^3.

Usually, Re controls the liquid flow regime around the bubble and We, Mo and E characterise the bubble deformations and bubble shapes. Therefore, the influences of Re, We and Mo are seen as important for elucidation of the bubble trajectories.

Several studies on the behaviour of bubble rise trajectories in water are available in the literature (Haberman and Morton 1954; Saffman 1956; Ellingsen and Risso, 2001; Duineveld, 1995). However, no study exists on bubble trajectories in non-Newtonian crystal suspensions. This study explores the characteristics of the bubble trajectory in non-Newtonian crystal suspensions and compares these results with water and different concentration polymeric solutions.

1.2. Bubble shapes

The shape of the bubbles greatly influences the bubble rise velocity and it has a significant role in determining the rates of heat and mass transfer and coalescence. Normally, a motionless bubble has a spherical shape because surface tension minimises surface area for a given volume. When a bubble has motion, different forces exist such as drag caused by the liquid, viscosity of the liquid, pressure difference between the top and bottom of the bubble as well as the wall effects.

Mainly, three types of shape such as spherical, ellipsoidal and spherical-cap or ellipsoidal cap in free motion under the influence of gravity are observed in Newtonian liquids.

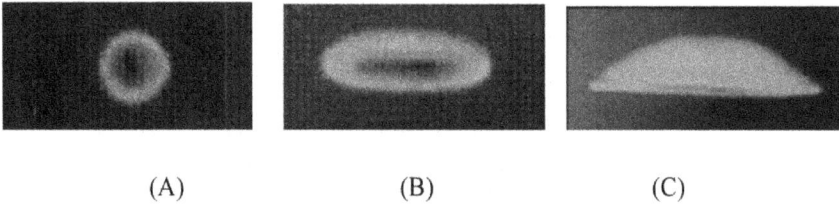

(A)   (B)   (C)

Figure 1. Different types of bubble shape in Newtonian fluid

The shapes of bubble are related to the Re. At low Re, the bubble retains its shape as a sphere because interfacial forces and viscous forces are much more important than inertia forces. Most bubbles of small size fall into this category. The spherical shape of the bubble is shown in the Figure 1(A).

The next category of bubbles is termed "ellipsoidal"; these are oblate with a convex interface around the surface when viewed from the inside. The liquid viscosity may affect the bubble shape, stretching the bubble out laterally, so that actual shapes may differ considerably from true ellipsoids. However, the general shape is comparable to an ellipsoid which is shown in Figure 1 (B).

Large bubbles have a flat base or a spherical wedge, which may look very similar to segments cut from a sphere. They are heavily distorted from the equilibrium shape of a sphere. In this case, the Reynolds number is high and these bubbles are termed as spherical-cap or ellipsoidal-cap. This type of bubble is shown in Figure 1 (C).

Beside these shapes, some other shapes are also noted in the literature (Clift *et al.*, 1978). A bubble that has an indentation at the rear is known as dimpled. Large spherical or ellipsoidal-caps may also trail thin envelopes of dispersed fluid referred to as skirts. Bubbles in different shape regimes are shown in Figure 2.

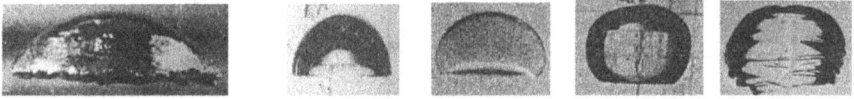

Figure 2. Photograph of bubbles in different shape regime (Clift *et al.*, 1978)

In non-Newtonian liquids, many more shapes of bubble have been reported (Calderbank, 1967; De Kee *et al.*, 1996; Chhabra, 2006). Among those, a distinctive feature of bubble shapes observed in non-Newtonian liquids is the appearance of a "pointed" tail prior to the transition to hemispherical caps. Experimental observations were reported by De Kee and Chhabra (1988) and De Kee *et al.* (1990) in aqueous carboxyl-methylcellulose and polyacrylamide solutions that had bubble shapes changing from spherical to prolate tear drop, then to oblate cusped, to oblate and finally to Taylor-Davies (Davies and Taylor, 1950) type spherical caps with the increase of bubble size. Shape behaviours also depend on the rheological and physical properties of the liquid. A large number of different shapes are observed in rheologically complex liquids depending upon the physical properties of the dispersed and continuous phases and the size of bubbles.

1.2.1. Bubble shapes in Newtonian fluids

It is usually established that the shapes of the bubbles rising in Newtonian fluid can be generalized in terms of the magnitudes of the following dimensionless parameters (Grace, 1973, Grace *et al.*, 1976; Clift *et al.*, 1978; Chhabra, 2006) such as Re, Eo, Mo, viscosity ratio,

$X = \dfrac{\mu_a}{\mu_{liq}}$, and density ratio, $\gamma = \dfrac{\rho_a}{\rho_{liq}}$.

For the case of a bubble with a negligible inner density, the Mo as defined in equation (3) can be simplified to:

$$Mo = \frac{g\mu_{liq}^4}{\rho_{liq}^2 \sigma^3} \tag{5}$$

Based on these dimensionless groups, researchers (Grace *et al.*, 1976; Bhaga, 1976; Bhaga and Weber, 1981; Grace and Wairegi, 1986) have constructed the so called "shape maps" as shown in Figure 3.

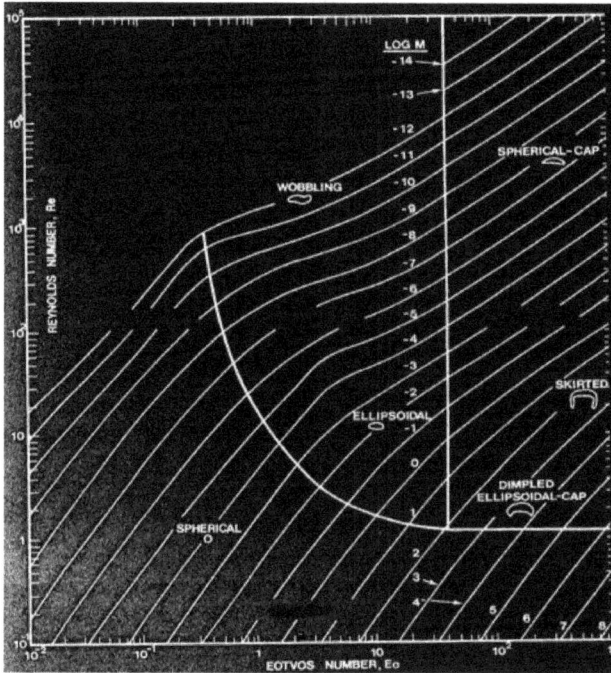

Figure 3. Shape regimes for bubbles (Clift *et al.*, 1978)

This map is a quite useful qualitative method in measuring the shapes of bubbles for Newtonian liquids, based on visual observations. The shape changes from one form to another in which usually occurs over a range of conditions in comparison to the sharp boundaries shown in all such maps (Chhabra, 2006). Spherical shape is observed at very small values of Mo, Eo and Re but on the other hand, ellipsoidal bubbles are encountered at relatively high Re and moderate Eo. Finally, the spherical cap shape occurs only at moderately high Eo and Re. Furthermore, Tadaki and Maeda (1961) developed a simple quantitative shape map which has also proved to be very successful for computing the bubbles shapes in Newtonian liquids. This approach incorporates eccentricity *(E)* which is defined as the maximum width of the bubble divided by the maximum height of the bubble.

1.2.2. Bubble shapes in non-Newtonian fluids

Many investigators have presented the quantitative information on bubble shapes in non-Newtonian fluids (Carreau *et al.*, 1974; De Kee and Chhabra, 1988; Miyahara and Yamanaka,

1993; De Kee *et al.*, 1996; Chhabra, 2006). It is found that bubbles rising in stagnant non-Newtonian inelastic and viscoelastic fluids remain spherical in shape up to larger volumes than in Newtonian media (De Kee and Chhabra, 1988; De Kee *et al.*, 1986, 1996, 2002). For predicting the bubble shapes in non-Newtonian fluids, the quantitative approach involved with eccentricity is found more useful compared to qualitative (shape maps) approach. General observations of the shape of the bubbles shows that the bubbles moving in non-Newtonian fluids are tear drop shaped at small Re in viscoelastic liquids. It is reported that the bubble shapes tend to be spherical, oblate spheroidal and finally spherical capped with increasing Re. The quantitative description of the bubble shapes in terms of the process and rheological parameters can be seen in terms of an eccentricity (E) or aspect ratio defined as the ratio of the maximum width to the maximum height. Literature suggests that prolate shaped (E < 1) and oblate shaped (E > 1) bubbles should be treated separately (Acharya *et al.*, 1977). A quantitative relationship between the bubble eccentricity and the relevant dimensionless groups has been attempted by Miyahara and Yamanaka (1993).

Again, a new dimensionless group G_1 which is a measure of the ratio of the elastic stresses to surface tension stresses can be simplified (Acharya *et al.*, 1977) as:

$$G_1 = \frac{K\left(\dfrac{U_b}{R}\right)^n}{\sigma R} \tag{6}$$

The aspect ratio (E < 1) for prolate shaped bubbles correlated empirically with this new dimensionless group G_1 as:

$$E = 0.616(G_1)^{-0.168} \tag{7}$$

The above correlation (7) could be used at least in the range of $0.68 < E < 1$.

At large bubble volumes or high Reynolds numbers, when the bubble moves in Newtonian fluid tending to make the bubble shape an oblate spheroid (E > 1), the fluid inertia causes major distortions from sphericity. The tendency towards distortion is opposed by surface tension stresses. In this region, We can be defined as the ratio of the internal stresses to the surface tension stresses; this controls the bubble deformation. In this context for high Reynolds number flow, another new dimensionless group G_2 has been simplified (Acharya *et al.*, 1977).

$$G_2 = \frac{Wi}{Re\,We} \tag{8}$$

The Weissenberg number (Wi) is defined by:

$$Wi = \dot{\gamma}\lambda \quad \left(\text{Shear rate * relaxation time}\right) \tag{9}$$

The aspect ratio correlated empirically with the G_2 for oblate shaped bubble is given by:

$$E = 1 + 0.00083G_2^{-0.87} \tag{10}$$

Equation (10) applies over the range of variables $1 \leq E \leq 1.5$ and $0.001 \leq G_2 \leq 0.015$.

For Newtonian and non-Newtonian fluids, bubble diameter (width and height) and eccentricity are determined experimentally and this experimental data are used for the calculation of different dimensional groups. Bubble shapes in Newtonian fluids can be predicted in comparison with the so called "shape maps". At the end, different theoretical bubble shapes are computed in comparison with above correlation and data from the relevant literature.

1.3. Path and shape instability of a rising bubble

The path instability of a bubble is a fascinating phenomenon of bubble trajectory when bubble rises in fluids (Hassan et al., 2010a). It is one of the most challenging and least understood aspects of bubble dynamics (Magnaudet and Eames, 2000). Many researchers have investigated experimentally and numerically the path instability of bubbles (Haberman and Morton, 1954; Saffman, 1956; Aybers and Tapucu, 1969; Wegener and Parlange, 1973; Mougin and Magnaudet, 2002; Mougin and Magnaudet, 2006; Brenn et al., 2006; Yang and Prosperetti, 2007; Zenit and Magnaudet, 2008; Fernandes et al., 2008; Adoua et al., 2009, Hassan et al., 2010a). Many investigators (Haberman and Morton, 1954; Saffman, 1956; Aybers and Tapucu, 1969; Duineveld, 1994, 1995; Maxworthy, 1996; de Vries, 2001; de Vries et al., 2002) indicated that the onset of path instability of a bubble is in general occurring at different stages, and the condition for bubble deformation is that We must exceed a critical value ($We > We_{cr}$). Duineveld (1994) determined that the critical We at which a bubble starts to change its path is in the range $We_{cr} \approx 3.3$ in clean water. Saffman (1956) observed the onset of path instability in water occurs once the equivalent bubble diameter d_{eq} is 1.4 mm for a straight path, $1.4\ mm \leq d_{eq} \leq 2.0\ mm$ for a zigzag trajectory, and $d_{eq} = 2.0\ mm$ for both zigzag and spiral trajectories, and also estimated the corresponding critical Re to be $Re_{cr} \approx 400$ for path instability. Aybers and Tapucu (1969) indicated a rectilinear path at the equivalent diameter of $d_{eq} = 1.34\ mm$, a spiral trajectory for $1.34\ mm \leq d_{eq} \leq 2.0\ mm$, zigzag trajectory changing to helical trajectory for $2.0\ mm \leq d_{eq} \leq 3.60\ mm$, and only zigzag trajectory for $3.60\ mm \leq d_{eq} \leq 4.2\ mm$. The onset of path instability is generally assumed to occur at a critical Weber number (Duineveld, 1994; Maxworthy, 1996 and Tsuge and Hibino, 1997). Tsuge and Hibino (1997) proposed an empirical relation for determining We_{cr} which is presented in equation (11).

The shape of a bubble is also an important variable for predicting the bubble rise characteristics and it also depends on dimensionless groups We (Churchill, 1989) as the bubble rise velocity, U_b . The bubble shape is assumed to be stable for low We and becomes unstable for larger We . Therefore, the We is an important parameter to determine the bubble deformation. Many investigators (Haberman, 1954; Saffman,1956; Hartunian and

Sears, 1957; Aybers and Tapucu,1969; Duineveld, 1995) indicated that the onset of both shape and path instability of a bubble are in general occurring at different stages, and the condition for bubble deformation is that We exceeds a critical value ($We > We_{cr}$). The literature suggests that the critical We at which a bubble starts to deform is in the range, $We_{cr} \approx 3\ to\ 4$ (Saffman, 1956). Deane and Stokes (2002) indicated that $We_{cr} = 4.7$ must be exceeded for the deformation of the bubble to occur, and they found the critical Reynolds number $Re_{cr}(450)$ that must be exceeded for bubble shape oscillations and elongation and subsequent deformation to occur in water. Tsuge and Hibino (1997) proposed an empirical relation for determining We_{cr} as:

$$We_{cr} = 21.5\,Re_{cr}^{-0.32},$$

(11)

$$Re_{cr} = 9.0\,Mo^{-0.173}$$

Churchill (1989) and Chhabra (2006) indicated that the shape of the bubble is also represented quantitatively by the aspect ratio or eccentricity ratio and is given by:

$$E = \frac{d_w\ or\ major\ axis\ diameter}{d_h\ or\ minor\ axis\ diameter}$$

(12)

Moreover, bubble deformation is accurately measured by the deformation parameter, D regardless of any liquid medium (Whyte et al., 2000) which is given by:

$$D = \frac{d_w - d_h}{d_w + d_h}$$

(13)

The literature indicates that the path and shape instability is well understood in water as well as some non-Newtonian fluids. As there is no massecuite data available for path and shape instability, there is a need to describe the path and shape transition in high viscous massecuite equivalent fluids for this current research.

2. Rheological characterisation

Rheology is the study of the science of deformation and flow of matter. One of the key tasks of rheology is to establish the relationships between characteristics such as shear stress and strain and its derivatives (Bhattacharya, 1997). Polymer solutions often demonstrate non-Newtonian flow behaviour. Such fluids commonly have a viscosity that depends on the shear rate and display elastic effects like normal stresses that greatly affect their response to deformation. Usually, when polymers are under low shear, the viscosity is high. As the shear rate increases, the viscosity decreases. Generally the viscosity increases again as the shear force reduced. This behaviour is called shear thinning.

The natures of Polyacrylamide (PAM), Xanthan gum (XG) and crystal suspensions were usually known to demonstrate shear thinning behaviour, as the viscosity of these materials decreases with higher applied shear rates (Hassan *et al.*, 2010b). The "Power Law" model was generally used to describe the non-Newtonian behaviour of these materials.

The rheological properties for three (0.025%, 0.05% and 0.1%) concentrations of the various polymer solutions and crystal suspensions were measured using an ARES Rheometer with bob and cup geometry. For rheological test, the diameters used for bob and cup were 25 mm and 27 mm respectively producing a gap of 1 mm between them. The samples were poured into the cup, and then lowered the bob until the upper surface bob was 1 to 2 mm below the surface of the sample. It is noted that the upper surface of the bob must be between zero and five mm below the upper surface of the cup. The rheological properties of these solutions are summarised in Table 1. The range of shear rates used to determine fluid rheology was 1-100 s^{-1} (Hassan *et al.*, 2007a, 2007b; Hassan *et al.*, 2008a, 2008b; Hassan *et al.*, 2010b; Hassan, 2011).

Fluid Type	Viscosity, $Pa.s$ at 1 s^{-1} (shear rate)	Viscosity, $Pa.s$ at 100 s^{-1} (shear rate)	Fluid consistency, K, $Pa.s^n$	Power law index, n
Polyacrylamide, 0.025% (by weight)	0.004	0.003	0.005	0.85
Polyacrylamide, 0.05% (by weight)	0.035	0.007	0.035	0.62
Polyacrylamide, 0.1% (by weight)	0.10	0.013	0.09	0.60
Xanthan Gum, 0.025 % (by weight)	0.008	0.003	0.007	0.77
Xanthan Gum, 0.05% (by weight)	0.039	0.006	0.039	0.71
Xanthan Gum, 0.1% (by weight)	0.11	0.012	0.10	0.57
Xanthan gum crystal suspension: 0.05% xanthan gum + 1% polystyrene (by weight)	0.11	0.006	0.11	0.63

Table 1. Rheological properties of non-Newtonian solution

The fluid consistency index (K) and power law index (n) for all solutions were determined from the response curve of shear rate against shear stress. The storage modulus (G'), the loss modulus (G'') of polymer solutions and crystal suspension solutions were tested (Hassan, 2011). The storage modulus (G') and the loss modulus (G'') of polymeric solutions and crystal suspensions were also investigated. G' measures the energy stored and recovered per cycle, when various fluids are compared at the same strain amplitude. It is a measure of the elasticity of the system. G'' measures the energy dissipated or lost for each cycle of sinusoidal deformation, when fluids are compared at the same strain amplitude. It is the measured viscous response of the system.

A comprehensive explanation of fluid flow curves for different polymeric solutions and selection of crystal suspended non-Newtonian fluid are presented in later sections. The values of n, K, G', G'' and the response curves (shear stress against shear rate and viscosity against shear rate) of non-Newtonian fluids were compared with the literature data of the massecuite fluid for determining the massecuite equivalent non-Newtonian crystal suspension fluid (Hassan, 2011)

2.1. Viscosity of polymer solutions

The viscosities as a function of shear rate for 0.025%, 0.05% and 0.1% concentrations of PAM solutions tested are presented in Figure 4. Examination of Figure 4 indicates that these three concentrations of PAM solutions exhibit non-Newtonian shear-thinning flow behaviour over the shear rates tested. It shows that the viscosity of these polymer solutions increases with the increase in PAM concentration with the 0.1% concentration has higher viscosity than the other two concentrations.

Viscosities as a function of shear rate for 0.025%, 0.05% and 0.1% concentrations of XG solutions are plotted in Figure 5. The same phenomenon as seen for PAM was observed for the three concentrations of XG solutions. The viscosity in XG solutions for all concentrations was found to be relatively higher in comparison with the PAM solutions, and its shear thinning effect (slope) was more pronounced for the entire range presented.

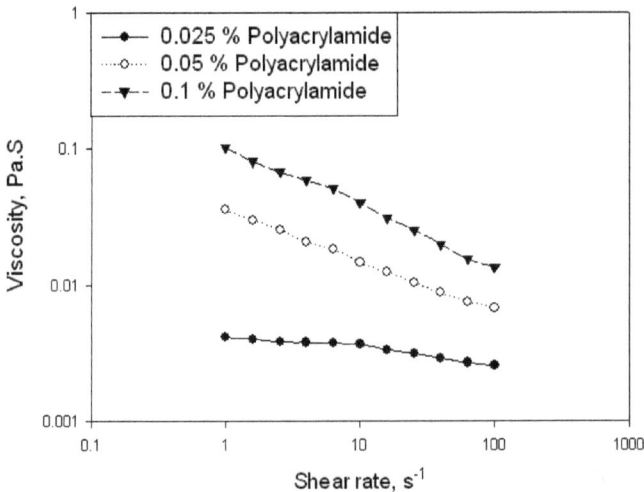

Figure 4. Viscosity versus shear rate of different concentrations of polyacrylamide solutions

2.2. Power law and fluid consistency index

Figure 6 presents n and K values in terms of fluid concentrations for different liquids. The figure shows that the K value for PAM solution concentrations was observed to be less than that in the corresponding XG solution concentrations. XG showed a higher viscous property (K) in comparison with PAM solution.

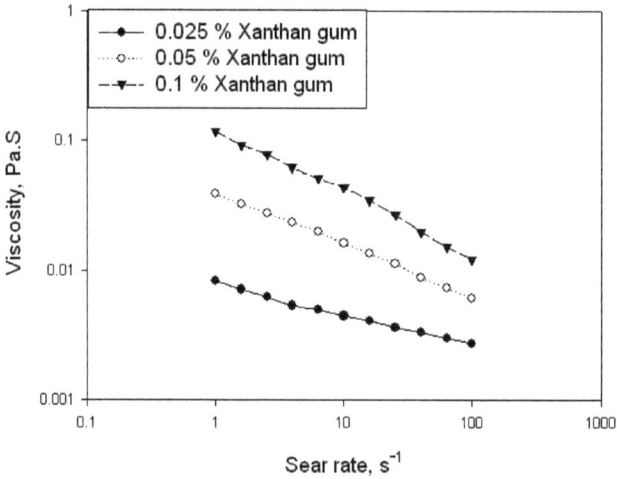

Figure 5. Viscosity versus shear rate of different concentrations of xanthan gum solutions

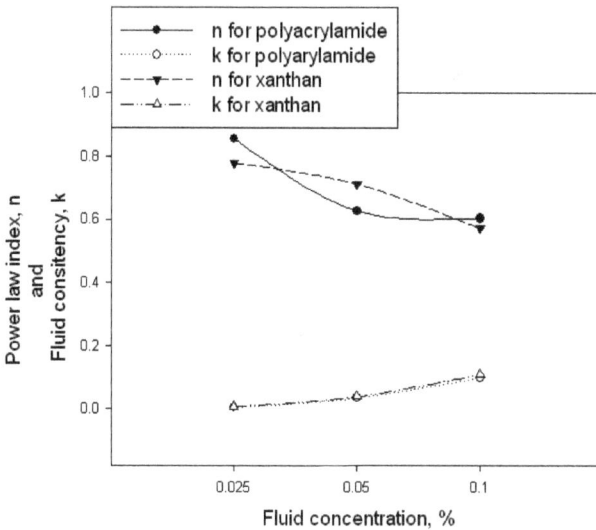

Figure 6. Power law index (n) and fluid consistency index (K) as a function of fluid concentration

It is also seen that the n generally decreases with the increase in liquid concentration for the three fluids. The value of n was observed lowest for 0.1% concentration of PAM and was highest for 0.025% PAM solution. On the other hand, 0.05% XG has the highest n value in comparison with the other liquid corresponding to the same concentration.

After analysing all the fluids used in this work, it is clear that the value of K for XG and PAM solutions increases with increase in concentration. This is expected since K is a direct function of the viscosity whilst the value of n which is less than unity, signifies the degree of shear thinning effect.

Two concentrations of 0.025%, and 0.05% of PAM and XG were used for bubble rise experiments as the higher concentrations of these fluids are not visually clear.

2.3. Viscoelastic property of polymer solutions

Storage modulus (G') or elastic response and loss modulus (G") or viscous response for 0.1% concentration of PAM solutions as a function of frequency are shown in Figure 7. The relevance and the definition of G' and G" in relation to the current study are discussed in Section 2.

PAM solution shows both viscous and elastic responses, with the elastic effects being more dominant than the viscous effects. This phenomenon can be observed in Figure 7 where PAM solutions with 0.1% concentrations showed more elastic response at higher shear rates. G' and G" of 0.025% and 0.05% concentrations of PAM solutions are not shown since they showed a similar phenomena to that of 0.1% concentration.

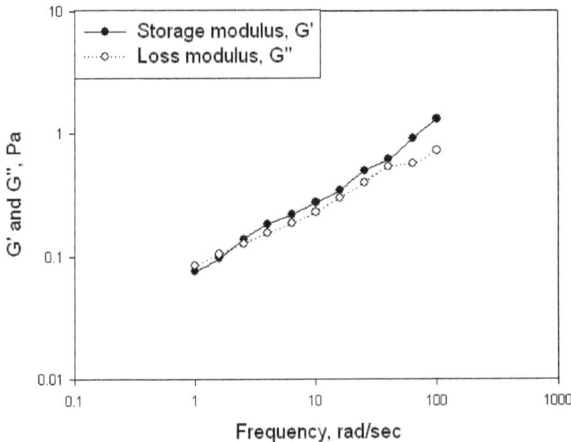

Figure 7. Storage modulus (G') and loss modulus (G") versus frequency for 0.1% polyacrylamide

G' and G" for three concentrations of xanthan gum solutions as a function of frequency are illustrated in Figures 8, 9 and10 respectively.

XG solutions have shown both viscous and elastic responses, with viscous effects more dominant than elastic effects at the range of lower frequency. It is seen from Figure 8 that XG solution with 0.025% concentration exhibited more viscous response in comparison with the elastic response. However, it exhibits similar elastic and viscous responses at the higher frequency range.

In Figure 9, the viscous response was more pronounced than the elastic response at the lower and higher frequency range. On the other hand, examination of Figure 10 indicates that the elastic response was more pronounced than the viscous response at the higher frequency range.

Therefore it can be concluded that the viscous effects of the XG solutions were more pronounced than elastic effects, and this behaviour was similar to that observed with high grade massecuites.

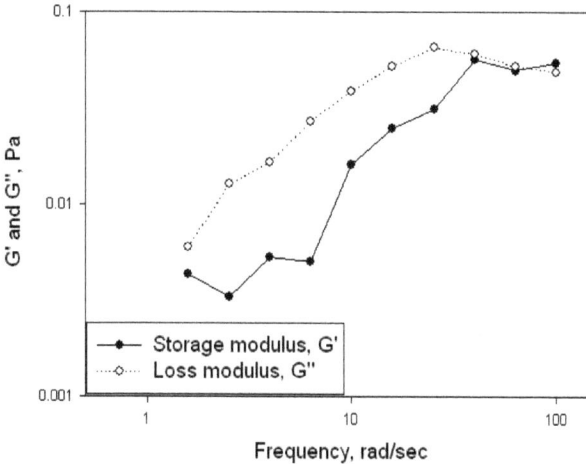

Figure 8. Storage modulus (G′) and loss modulus (G″) versus frequency for 0.025% xanthan gum

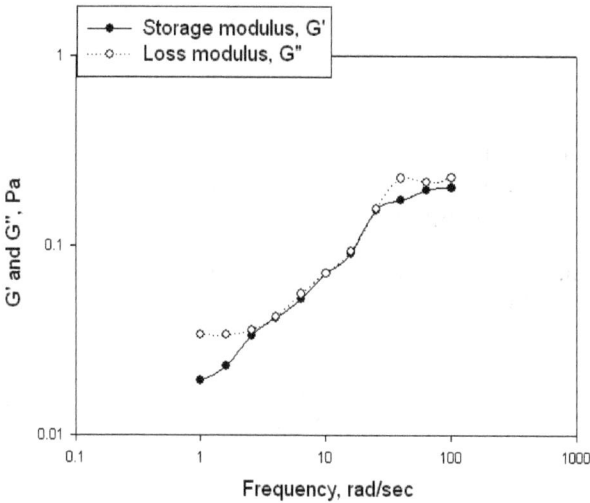

Figure 9. Storage modulus (G′) and loss modulus (G″) versus frequency for 0.05% xanthan gum

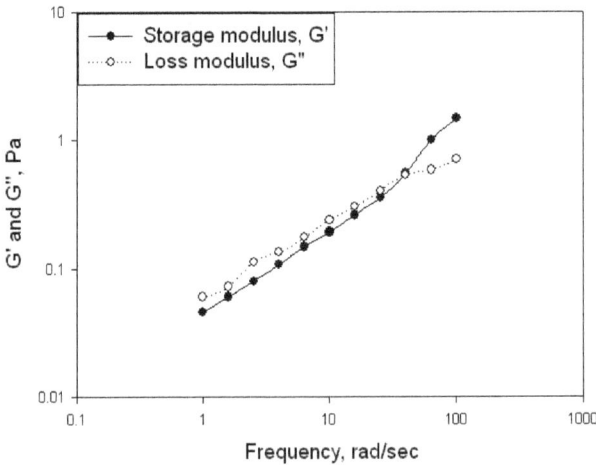

Figure 10. Storage modulus (G') and loss modulus (G") versus frequency for 0.025% xanthan gum

2.4. Selection of crystal suspension

The uses of bubbles are very significant in vacuum pan operation in the sugar industry. Vacuum pans (large cylindrical vessels with vertical heating surfaces) are used to process the massecuite - a fluid made from sugar crystals and mother sugar syrup (molasses) (Hassan et al., 2010a, 2010b; Hassan, 2011). Usually, massecuites show non-Newtonian shear thinning behaviour at lower shear rates; that is the viscosity is highest at low shear rates and decreases with increasing shear rates (Ness, 1983; Broadfoot and Miller, 1990). But massecuites are not optically clear and, in non-industrial environments, massecuites endure problems due to degradation during storage and changing rheological properties under different temperature conditions.

The properties of massecuites depend on concentration of sugar crystals and their purity. Generally, massecuites are described as exhibiting shear thinning behaviour (Broadfoot *et al.*, 1998; Broadfoot and Miller, 1990; Adkins, 1951). It has been shown that the power law model does give a good fit to experimental data (Ness, 1983; Broadfoot and Miller, 1990; Bojcic *et al.*, 1997; Broadfoot, 1998). Temperature has also been shown to affect viscosity through changing the values of K and n (Broadfoot *et al.*, 1998; Ness, 1983; Broadfoot and Miller, 1990; Awang and White, 1976). When determining the effect of viscosity on bubble rise in three phase materials such as massecuites (solid-liquid-gas; sugar crystal – molasses – vapour), the characteristics of the bubble rise are dependent on the liquid viscosity (molasses) rather than the mixture viscosity (molasses plus crystals). Literature also suggests that the viscosity of molasses is less than the mixture viscosity of massecuites (Rackemann, 2005; Ness, 1983; Broadfoot and Miller, 1990; Adkins, 1951; Nezhad, 2008; Hassan, 2011).

As mentioned earlier, massecuites are not optically clear. Therefore a fluid needs to be prepared to replicate the massecuite with an equivalent non-Newtonian fluid for studying the bubble rise characteristics which is optically clear and has similar rheological properties to that of the massecuite.

Figure 11 shows the viscosity – shear rate flow curve of a high grade massecuite (Type A). Also shown in Figure 11 are the shear rate flow curves for 0.05% of XG with different percentages of crystal.

Observation from Figure 11 indicates that different crystal suspended XG solutions exhibit non-Newtonian shear-thinning flow behaviour. The viscosity of these crystal suspensions increases with the increase in crystal concentration, with the shear thinning effect (slope) more pronounced over the range of shear rates presented. The XG solution with 25% crystal concentration has the highest viscosity and the lowest power law index in comparison with the other concentrations used in this study. However, the 5%, 10% and 25% crystal suspensions were not optically clear enough to visually observe the bubble rise. After analysing all the solutions in Figure 11, it is seen that the flow curve of XG solution (0.05% XG by weight) with 1% crystal content exhibits similar rheological characteristics to that of high grade massecuite for the range of shear rate presented; also it was optically clear. In other words, the viscosity of the crystal suspension is not exactly the same, but it shows close similarity with power law index, fluid consistency and shear thinning flow behaviour to that of high grade massecuite and molasses.

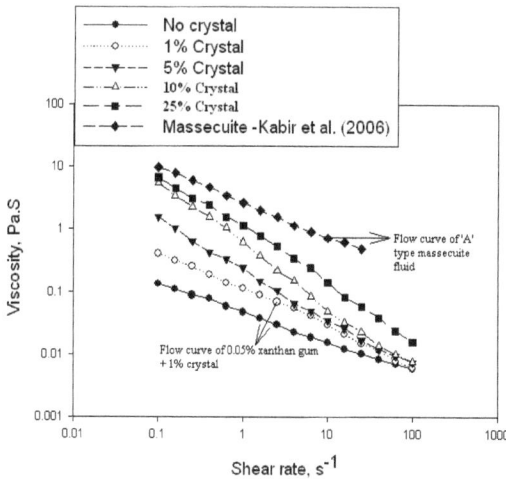

Figure 11. Viscosity versus shear rate of 0.05% xanthan gum solutions with different percentages of crystal content and massecuite of 40% crystal content at 65 °C (Hassan *et al.,* 2010b, Kabir *et al.,* 2006)

However, XG crystal suspension showed both viscous and elastic responses with the viscous effects more dominant than the elastic effects. This phenomenon is shown in Figure 12 where XG crystal suspension displays more viscous response at very low and very high shear rates and this is also similar to that high grade of massecuite.

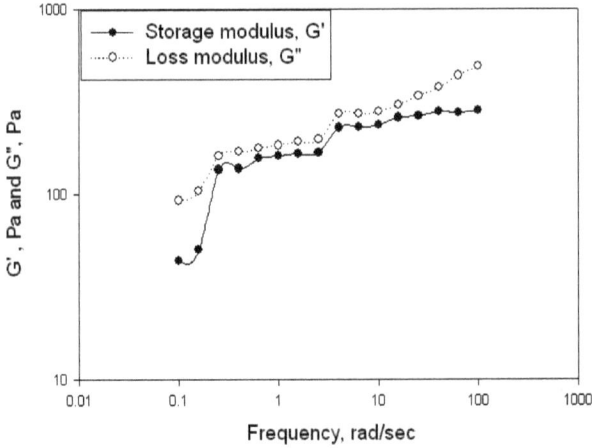

Figure 12. Storage modulus (G′) and loss modulus (G″) versus frequency for 0.05% xanthan gum + 1% crystal (Hassan *et al.*, 2010b)

3. Experimental setup

The experimental set-up consisted of transparent cylinders to hold fluids and bubble insertion mechanism and video camera lifting apparatus. A schematic of the experimental set-up is shown in Figure 13.

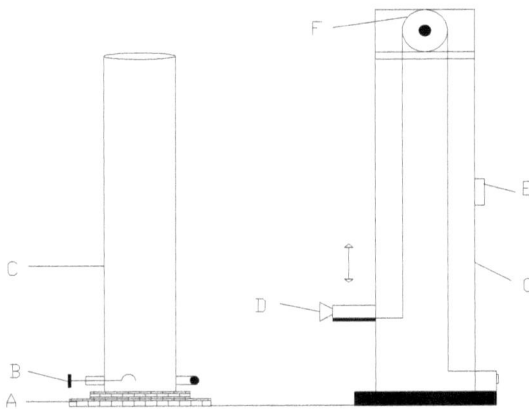

Figure 13. Schematic diagram of experimental apparatus

where A = Sturdy base; B = Rotating spoon; C = Cylindrical test rig (125mm or 400 mm diameter), D = Video camera; E = Variable speed motor; F = Pulley; and G = Camera lifting apparatus.

Two-test rigs were designed and fabricated for investigating the bubble rise characteristics in Newtonian and non-Newtonian fluids. Rig 'A' consists of a polycarbonate tube approximately 1.8 m in height and 125 mm in diameter as shown in Figure 14.

The polycarbonate material was chosen for optical clarity to enable precise visual observation. Rig A has a sturdy base and is capped with suitable connections to allow the rig to operate under partial vacuum conditions. It also contains two holes near the base. One is to facilitate the removal of the fluid contained in it and the other is to insert air bubbles into the test rig. The insertion mechanism, shown in Figure 15, consists of a ladle or spoon with a small pipe running down the centre that has a capability to control the injection of air. The air was injected through this pipe into the upside-down ladle using a syringe. The cup was then twisted to allow the bubble to rise.

Rig B (shown in Figure 14) consists of an acrylic tube of 400 mm in diameter and 2.0 m in height with the same connections for bubble insertion and fluid draining in test Rig A. Rig B was employed to examine the larger sizes of bubble to eliminate the wall effect.

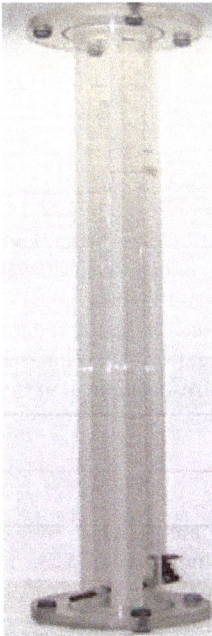

Test Rig 'A' (polycarbonate)   Test Rig 'B' (acrylic)

Figure 14. Photographs of test rig A and B

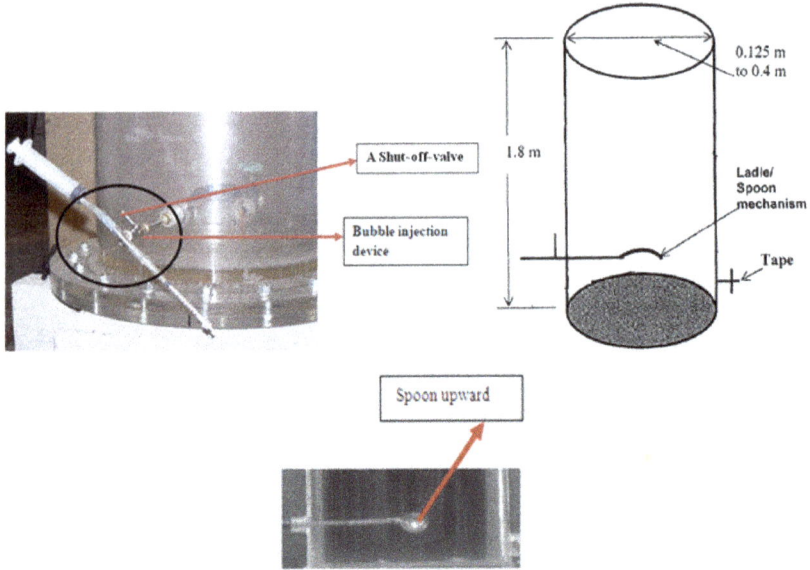

Figure 15. Bubble insertion mechanisms

4. Experimental procedure

4.1. Determination of bubble shape

The still images were taken in order to determine the bubble shape. A charge-coupled device (CCD) digital camera was used to capture the 2D bubble shapes (height and width) using the commercial software SigmaScan Pro 5.0 and Adobe Photoshop: DV Studio version 3.1E-SE. These images were analysed in order to determine the size and the aspect ratio of each bubble. Only a complete good quality recorded images was taken into account. Blurred images were ignored as the accuracy of the bubble size determination depends on the quality of the picture.

4.2. Determination of bubble diameter

Bubble equivalent diameter was measured from the still images which were obtained from the video clips. The still images were analysed using commercial software "SigmaScan Pro 5.0" and the bubble height (d_h) and the bubble width (d_w) were measured in pixels. The pixel measurements were converted to millimetres based on calibration data for the camera. The bubble equivalent diameter d_{eq} was calculated (Lima-Ochoterena and Zenit, 2003) as:

$$d_{eq} = \left(d_h \times d_w^2\right)^{\frac{1}{3}} \tag{14}$$

For this measurement it was assumed that the bubble is axi-symmetric with respect to its minor axis direction as shown in Figure 16.

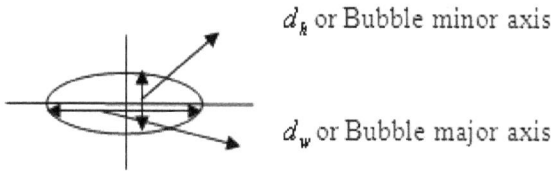

d_g or Bubble minor axis

d_w or Bubble major axis

Figure 16. Bubble major and minor axes

This equivalent diameter was used for the calculation for bubble eccentricity, bubble deformation parameter, Reynolds number and bubble drag co-efficient.

4.3. Determination of bubble trajectory

Bubble trajectory was computed from the still frames obtained from the video image. The still frame was opened in "SigmaScan Pro" commercial software, which was capable of showing pixel location on an image.

The pixel coordinates (X and Y) of the bubble's centre were noted and recorded into the spreadsheet. The X and Y coordinates correspond to the distance from the left and top edges respectively. The pixel line running through the centre of the bubble release point was known. The deviation of the bubble centre from the release point was computed by subtracting the X-location of the bubble centre from the X-location of the bubble release point.

A linear equation was developed for the determination of the rigs X pixel coordinate for a given Y coordinate. This was used in association with the Y of the bubble's centre for the correction of any changes which might occur based on the bubble's location (e.g. if the bubble is at the top of the image, the tube's centre is slightly to the left than if the bubbles were at the lower portion of the image).

Still frames from the digital video camera captures images where the camera can see two of the markers. The number of pixels that separates these two markers was used in association with their real separation in millimetres, which was 100 mm. Consequently, a conversion factor was made to change distances in pixels to distance in millimetres. Using this conversion factor, the deviation of the bubble from the tube's centre was converted into millimetres.

5. Results and discussions

The results of the behaviour of bubbles' motions and their transition, and shapes encountered and their oscillations are reported in the next section. Different bubble volumes (0.1 mL, 2.0 mL, 5.0 mL and 10.0 mL) were used at 1.0 m height of the liquid column for the calculation of bubbles' trajectories and shapes.

5.1. Bubble trajectory in Newtonian and non-Newtonian Fluids

The trajectory experimental results for 5.76 mm diameter bubbles are shown in Figure 17 for different liquids when measured over a height of 1 m from the point of air injection. Figure 17 shows the deviation of the bubble from its release point as it rises through the liquids. The general trend is for the bubble to remain close to the line of the release centre when the bubble is released and, as it rises through the liquids, it spreads out progressively as the height increases.

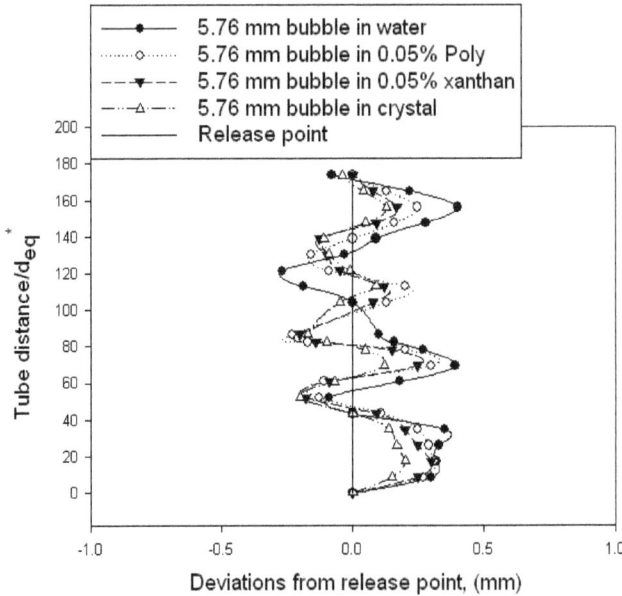

Figure 17. Experimental rise trajectories for 5.76 mm diameter (0.01 mL) bubbles in different liquids

It is seen from the experiments that the smaller bubble of 5.76 mm diameter deviated horizontally more in water than in the other three liquids, and the bubble shows a zigzag motion in all liquids. But the zigzag motion was produced more in water compared to other two liquids. The horizontal movement in 0.05% PAM and XG polymer solutions was observed to be less than that in water, and was the least in crystal suspension. This phenomenon agrees well with the experimental findings of Saffman (1956) and was reported by Hassan *et al.* (2008a; 2008b; 2008c; 2010a). The lesser horizontal deviations observed in the case of polymers and crystal suspensions are due to the higher effective viscosity of these liquids than that of water and a smaller Re of a given bubble. Hence, the vorticity produced at the bubble surface by the shear-free condition is also smaller, since this surface vorticity is an increasing function of Re until it becomes independent of Re for a large enough value. This smaller surface vorticity produces a less unstable wake resulting in smaller horizontal motions.

Smaller bubbles less than 2 mm in diameter rise in water in a straight or rectilinear path (Clift *et al.*, 1978; Duineveld, 1995), but the linear trajectory in all liquids was not observed as the bubble equivalent diameter of this study was equal to or more than 5.76 mm. However, the path of a rising bubble is not always straight. The shape of the bubble was found to be stable for very low *We* , and it turned into an oscillatory or unstable shape for larger *We* . At low Re and *We* for smaller bubbles, the rising bubble showed a zigzag trajectory. The zigzag motion is due to an interaction between the instability of the straight trajectory and that of the wake. The periodic oscillation of the wake is somewhat less in the crystal suspension due to the increase in viscosity. When a bubble rises in liquids, the bubble experiences both a lateral force and a torque along its path and two counter-rotating trailing vortices appear behind the bubble; hence its path, changes from a straight trajectory to a zigzag trajectory (Mougin and Magnaudet, 2006).

The trajectory experimental results for 15.63 mm diameter bubbles are illustrated in Figure 18.

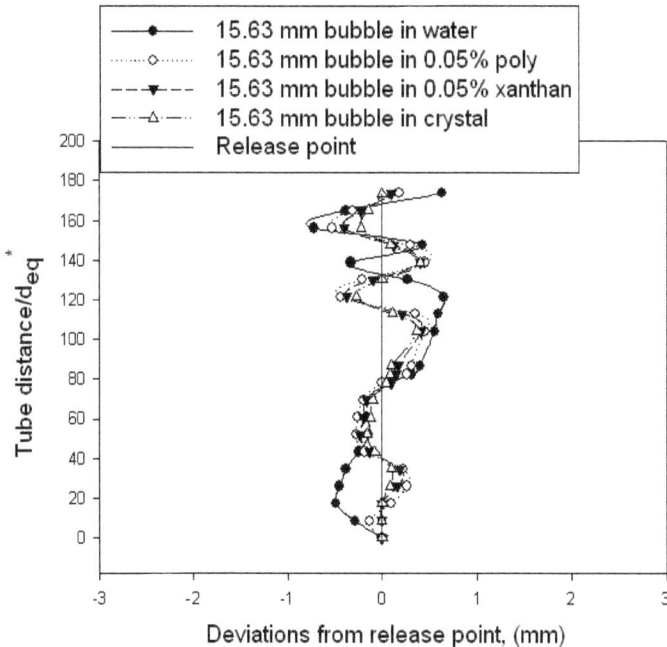

Figure 18. Experimental rise trajectories for 15.63 mm diameter (2 mL) bubbles in different liquids

With increasing bubble size, the bubble shape changes from spherical to ellipsoidal, the bubble surface oscillations transform from a simple oscillation to higher order forms, and the trajectory changes from a simple zigzag to more spiral trajectories. The bubble of 15.63 mm in water initially deviated horizontally, then followed a spiral motion and finally attained a zigzag path. However, the 15.63 mm bubble initially followed a straight path, then it followed only a zigzag motion in polymer solutions and crystal suspension until it finished its journey.

The trajectory experimental results for 21.21 mm bubbles are plotted in Figure 19. As seen from the experiments, the 21.21 mm bubbles initially choose the zigzag motion, and finally switch to a spiral path for all liquids. The transition from zigzag to spiral motion was also observed by Aybers and Tapucu (1969) as the bubble size increases. The transition from zigzag to spiral is also consistent with observations found in the literature (Saffman, 1956). For bubbles of size 15.63 mm and 21.21 mm, path instability occurs as the bubble size increases.

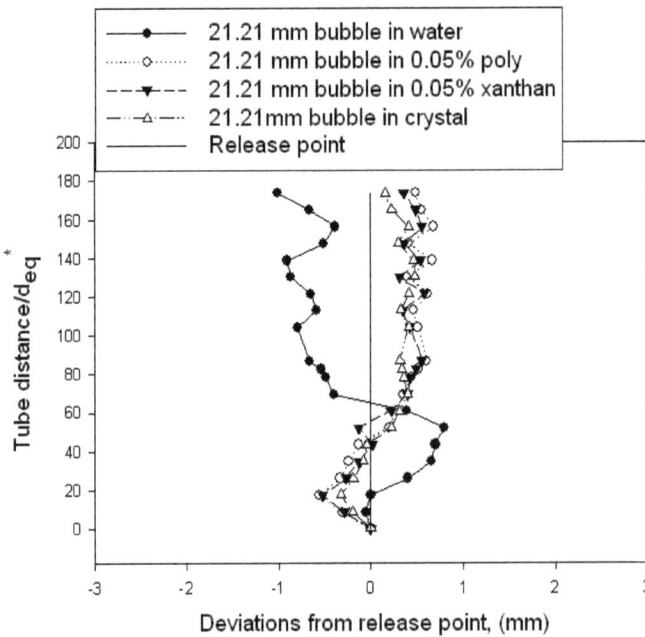

Figure 19. Experimental rise trajectories for 21.21 mm diameter (5 mL) bubbles in different liquids

No zigzag motion was observed for the larger bubble of 26.84 mm as seen in Figure 20 for water. The trajectory of these bubbles totally changed into a spiral motion. These bubbles initially choose a straight path and finally, they switch to a spiral path. It is seen from Figure 20 that the 26.84 mm bubble shows more spiral motion than the 21.21 mm bubble.

The crystal suspended xanthan gum produced more spiral motion in comparison with water and polymer solutions. The larger bubbles however experience more resistance on top and deform as their size increases. Furthermore, it can be confirmed that path instability was seen for ellipsoidal bubbles of sizes 2 mL and 5 mL only, and not for the 10 mL bubble.

This is consistent with the findings by other researchers (Haberman and Morton, 1954; Aybers and Tapucu, 1969; Wegener and Parlange, 1973; Shew *et al.*, 2006; Hassan *et. al.*, 2010a).

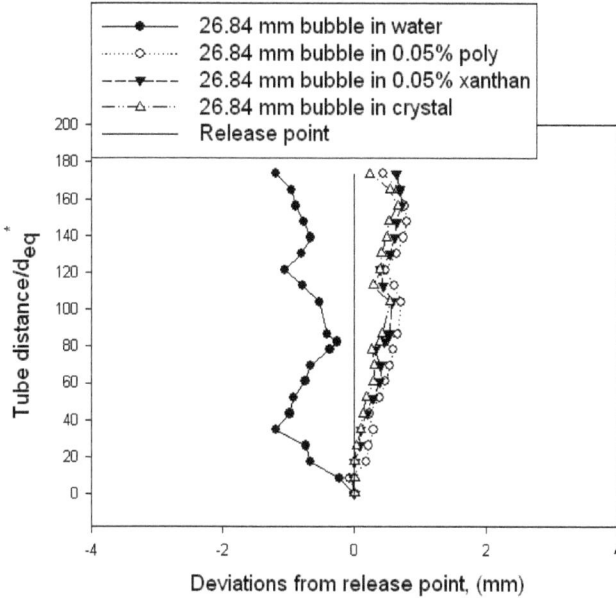

Figure 20. Experimental rise trajectories for 26.84 mm diameter (10 mL) bubbles in different liquids

5.2. Influences of Re , We and Mo on bubble trajectory

The observed Weber number as a function of bubble equivalent diameter in water, polymer solutions and crystal suspension is plotted in Figure 21. It is observed that We increases with increase in bubble equivalent diameter. As seen in Figure 21, the horizontal lines indicate the critical Weber number (We_{cr}) that must be exceeded for path transition to occur. In this study, the values of We_{cr} were measured for water, $We_{cr} \approx 3$; 0.05% xanthan gum, $We_{cr} = 6.31$, 0.05% polyacrylamide $We_{cr} = 6.13$ and crystal suspension, $We_{cr} = 7.9$ (Hassan, 2011). The equation 11 was used to calculate the We_{cr} .It is seen from Figure 21 that the measured We_{cr} in water shows consistent results with published literature (Hartunian and Shears, 1957; Aybers and Tapucu, 1969; Duineveld, 1994; 1995). Observations from Figure 21 also indicate that the We_{cr} at which the transition to oscillatory path behaviour commences is the least for water and the most for crystal suspension. Therefore, the path transition initiation of bubble rise in crystal suspension is less rapid than in water. The reason for this is that the Mo is different for these four liquids and is much higher for the crystal suspension than for water. As a result the observed change suggests that the *We* is not entirely responsible to characterise the bubble path transition. Hence, for a given *We*, the deformation is much less in crystal suspension than in water, and the vorticity produced is also less in crystal than that in water.

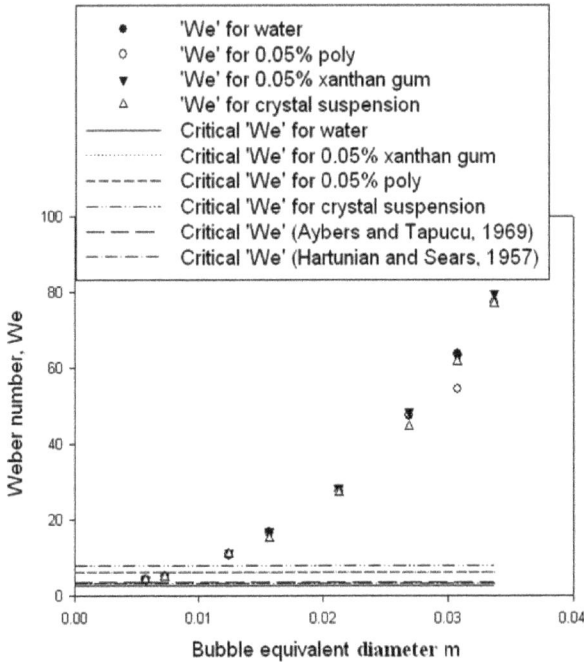

Figure 21. Weber number as a function of the bubble equivalent diameter (Hassan *et al.,* 2010a)

The effects of Re and *We* values on bubble trajectory in crystal suspension are discussed below for three regions. The characteristics of bubble rise corresponding to Re and *We* are also summarised in Table 2. At low Re and *We* for smaller bubbles (5.76 mm), the rising bubble shows a zigzag trajectory. For the intermediate region, the bubble of size 15.63 mm shows both zigzag and spiral trajectories. It is noted that path instability occurred in this region due to increase in bubble size and unstable wake structure. At moderately high *We* and Re, the bubbles deform and change from spherical to ellipsoidal and experience more surface tension and inertia force which induces both zigzag and spiral trajectories. At very high Re and *We*, larger bubbles (> 21.21 mm) produce a spiral motion in all liquids as the effect of wake shedding influences the bubble to induce a spiralling rising motion.

Bubble size, mm	Re	We	We_{cr}	Bubble shape	Region of Re and We values	Zigzag motion	Spiral motion
5.76	45.78	4.33		Nearly Spherical	Low	√	×
15.63	102.06	15.10		Ellipsoidal	Intermediate	√	√
21.21	150.37	27.30	7.9	Ellipsoidal or Ellipsoidal cap	Moderately high	√	√
26.84	207.91	44.60		Ellipsoidal or Spherical cap	Very high	×	√

Table 2. The characteristics of bubbles corresponding to Re *and We* for crystal suspension (Hassan, et al., 2010a; Hassan, 2011)

5.3. Bubble shapes encountered in different liquids

5.3.1. Bubble shapes in water

Grace *et al.*, (1976) constructed the so called "shape maps" as shown in Figure 3 in earlier section. It has been proven to be a quite useful qualitative method for measuring the shapes of bubbles for Newtonian liquids, based on dimensionless quantities such as Re and *Eo*, with *Mo* being a parameter with specific values for specific liquids. The shapes are predicted in comparison with the "shape maps" on the basis of visual observations. In this study, bubble shapes predicted in water at three different heights are presented in Table 3, and *Eo* against Re with the *Mo* parameter for water is shown in Figure 22 As seen from Figure 22, *Eo* increases with increase in Re with low *Mo* liquids. Furthermore, Re with the variation of bubble equivalent diameter is shown in Figure 23. As seen from Figure 23, the bubble equivalent diameter also increases with increase in Re. It is also observed that *Eo* increases with increase in bubble equivalent diameter. Therefore it can be concluded that Re and *Eo* have a greater influence on bubble shapes.

Observations from Table 3 and Figure 23 indicate that nearly spherical shapes were observed at small values of *Eo* and *Re* at three different heights. Ellipsoidal bubbles were encountered at moderate Re and *Eo*. Usually, ellipsoidal bubbles are termed as oblateness with a convex interface around the surface (Clift *et al.*, 1978). The transition from spherical to ellipsoidal is due to the surface tension and inertia effects on bubble surface. Finally, the ellipsoidal cap or spherical cap shape occurred only at very high Re and high *Eo*. In addition, a wobbling shape was also found at 0.5 m height for 5.0 mL bubble in high Re region. Wobbling shapes were mainly found at sufficiently high Re and *Eo*. This shape is mainly occurred due to the un-symmetric vorticity created in the bubble wake and the instability of bubble interface. In this region the bubble shows oscillating behaviour as the bubble velocity increases with increase in bubble size. The wobbling bubbles were also noticed by Clift *et al.* (1978) in high *Re* region of the ellipsoidal regime of Grace's (1976) shape maps.

a. Different bubble shapes at 0.1 m height			
0.1 mL (5.76 mm)	2 mL (15.63 mm)	5 mL (21.21 mm)	10 mL (26.84 mm)
•	◠	◢	◣
b. Different bubble shapes at 0.5 m height			
0.1 mL	2 mL	5 mL	10 mL
•	◄	◢	◣
c. Different bubble shapes at 1 m height			
0.1 mL	2 mL	5 mL	10 mL
▬	◤	◄	◣

Table 3. Bubble shapes observed in water at three different heights

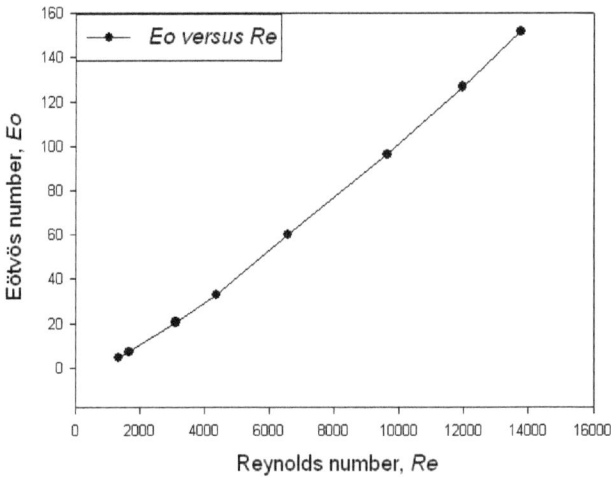

Figure 22. *Eo* versus *Re* with Mo as parameter ($Mo = 2.502 \times 10^{-11}$) for water

At higher *Eo* and *Re*, the bubbles usually exhibit noticeable deformation, and therefore they become ellipsoidal and consequently they have a spherical cap shape. These shapes encountered in water are consistent with the published literature of Grace (1976), Clift *et al.*, (1978) and Bhaga and Weber (1981).

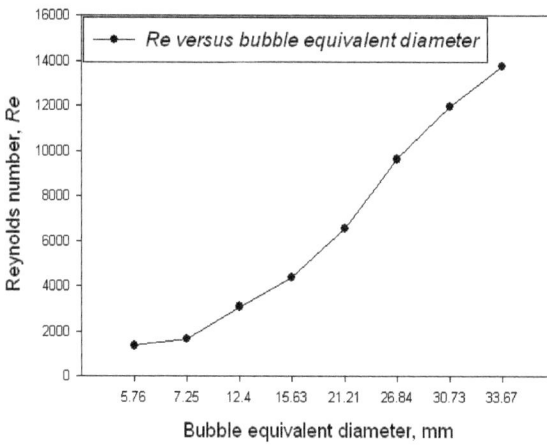

Figure 23. *Re* versus bubble equivalent diameter

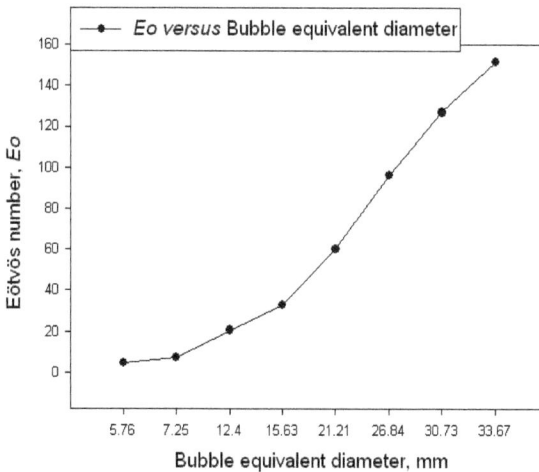

Figure 24. *Eo* versus bubble equivalent diameter

5.3.2. Bubble shapes in polymeric solutions

Tables 4 and 5 summarise the different bubbles shapes observed in 0.05% XG and 0.05% PAM. As seen from Tables 4 and 5, different bubble shapes were encountered apart from the spherical, ellipsoidal and spherical-cap. Usually, the bubble shapes are greatly dependent on *Re*, *Eo* and the aspect ratio or eccentricity of bubble, *E* . The bubble shapes in terms of bubble eccentricity (which is defined as the ratio of maximum bubble width to the maximum bubble height) are also dependent on rheological properties of the liquids. The literature suggests that prolate shaped (E < 1) and oblate shaped bubbles (E > 1) are usually observed in non-Newtonian viscoelastic liquids (Acharya, 1977). The bubble shapes observed in this study are mainly oblate shaped as the eccentricity was found to be greater than 1.

The bubbles usually deformed as their size increased and the rate of deformations were more pronounced at higher values of Re, Eo and E. As evidenced from Figures 25 and 26, Re increases with the increase of Eo and E, meaning that for higher Re and Eo bubbles become more flat due to the higher bubble velocity.

At lower values of Re, Eo and E, bubbles shapes were observed to be nearly spherical in 0.05% XG and 0.05% PAM. With a further increase in Re, Eo and E, the bubbles become flat and appear to be an oblate shape for both liquids. As seen from Table 5, the bubble appears to have a skirted shape at 1.0 m height for the 2.0 mL bubble in PAM solution. The skirt region near the rear of the bubble becomes thin, and with further increase of Re, Eo and E, the skirt would turn out to be thinner and unstable, as is evidenced in Table 5 for the 5.0 mL bubble size corresponding to the same liquid. For larger bubble size (10.0 mL), it turns into an ellipsoidal cap shape at very high Re, Eo and E. The ellipsoidal cap shaped bubble was also noticed in 0.05% XG solutions at 1.0 m height for the 10.0 ml bubble. Besides their usual shapes, wobbling bubbles were also encountered in 0.05% XG solution for 2.0 mL and 5.0 mL of bubbles. This is due to the bottom part of the bubble experiences deformations resulting from the unstable and non symmetrical bubble wake. In particular, the irregular pairs of secondary vortices are also created in the wake as a result of the separation of the boundary layer from different sides of the bubble surface. These observed shapes are consistent with the published literature (Clift et al., 1978; Bhaga and Weber, 1981).

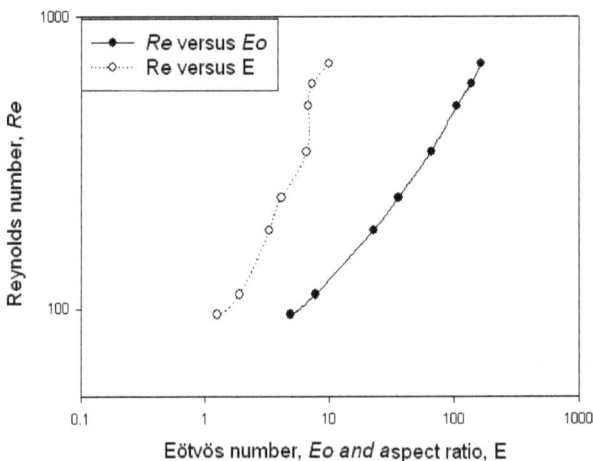

Figure 25. Re versus Eo and E for 0.05% xanthan gum

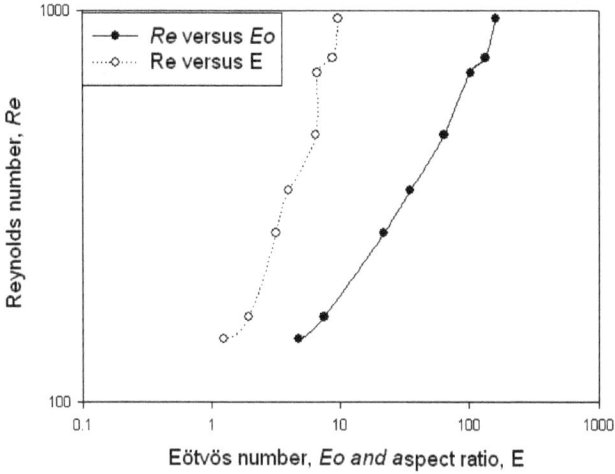

Figure 26. *Re* versus *Eo* and E for 0.05% polyacrylamide

a. Different bubble shapes at 0.1 m height			
0.1 mL	2 mL	5 mL	10 mL
b. Different bubble shapes at 0.5 m height			
0.1 mL	2 mL	5 mL	10 mL
c. Different bubble shapes at 1 m height			
0.1 mL	2 mL	5 mL	10 mL

Table 4. Bubble shapes observed in 0.05% xanthan gum at three different heights

Different bubble shapes at 0.1 m height			
0.1 mL	2 mL	5 mL	10 mL
Different bubble shapes at 0.5 m height			
0.1 mL	2 mL	5 mL	10 mL
Different bubble shapes at 1 m height			
0.1 mL	2 mL	5 mL	10 mL

Table 5. Bubble shapes observed in 0.05% polyacrylamide at three different heights

5.3.3. Bubble shapes in crystal suspension

Re as a function of Eo and E, is illustrated in Figure 27. As expected, Re increases with the increase in Eo and E. The bubbles become more oblate as their size increases. As seen from Table 6, the shapes encountered for larger bubbles (5.0 mL and 10.0 mL) in crystal suspension were mostly ellipsoidal and ellipsoidal cap.

The shape instability was visualised at different heights of the liquid from the still images as are illustrated in Table 6. Usually, the shapes of the bubbles in non-Newtonian liquids are dominated by the magnitudes of inertial, viscous, surface tension, gravity and buoyancy forces (De Kee and Chhabra, 1988). Table 5.5 shows that the bubble volume of 0.1 mL was close to spherical in shape. The 0.1mL bubble was slightly elongated sideways at 0.5 m height and it looked nearly elliptical in shape at 1 m height of the liquid column. On the other hand, the 2.0 mL bubble was continuously distorted, changing into different shapes at different heights, and finally formed into an ellipsoidal shape. This transition from one shape to another is mainly dependent on the rheological properties of the crystal suspension. The shapes reported for larger bubbles (5.0 mL and 10.0 mL) were mainly oblate (flattened from top to bottom end), spherical-capped or ellipsoidal-capped. The larger bubbles (10.0 mL) also differed slightly from the base oblate shape at different heights, and finally looked to be at spherical capped at 1.0 m height.

5.3.4 Shape deformation in crystal suspensions

The bubble deformation parameter D is plotted as a function of the dimensionless bubble diameter in Figure 28. It is experimentally observed that the bubble deformation increases with increase in bubble diameter. As expected, D increases with increase in bubble diameter for other liquids as well. Therefore, Figures for D versus d_{eq} in other liquids are not shown for brevity of the study.

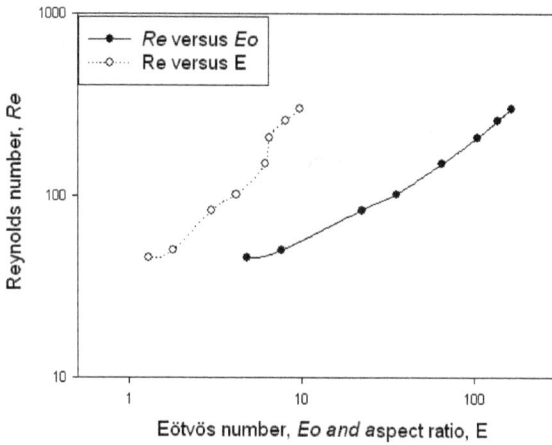

Figure 27. *Re* versus *Eo* and E for crystal suspension

a. Different bubble shapes at 0.1 m height			
0.1 mL	2 mL	5 mL	10 mL

b. Different bubble shapes at 0.5 m height			
0.1 mL	2 mL	5 mL	10 mL

c. Different bubble shapes at 1 m height			
0.1 mL	2 mL	5 mL	10 mL

Table 6. Bubble shapes observed in crystal suspension at three different heights

As seen from Figure 28, this deformation is shown to be minimal in smaller sizes of bubble diameter. However, these bubbles were not fully spherical. As the bubble diameter increases, D increases which is caused by the deformation in bubble shape from close to spherical to the spherical-cap shape bubble observed for the largest bubble of 20 mL (dimensionless equivalent diameter of 31.26).

The bubble shape is assumed to be stable for low We and becomes unstable for larger We. Therefore, the We value is an important parameter to determine the bubble deformation. As shown in Figure 21, the bubble shape deformation normally occurs at the critical Weber number, We_{cr}. The values of We_{cr} for all liquids were calculated and are presented in Figure 21. It is observed from Figure 21 that the We_{cr} must exceeds 7.9 for crystal suspension for bubble shape deformation and elongation and subsequent deformation to occur. As the bubbles' size increases further, they attain a spherical-cap shape at higher We numbers ($We > 27.3$). The initiation of shape deformation during bubble rise is less rapid in crystal suspension than in water.

All bubble shapes encountered in this research were compared with the available published data and the results are found to be in good agreement with the experimental predictions of Clift *et al.*, (1978); Bhaga and Weber, (1981).

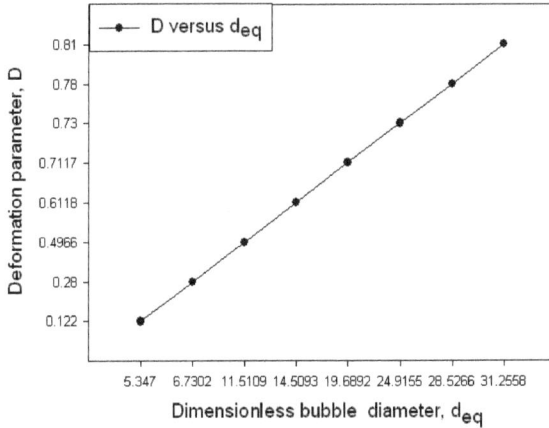

Figure 28. Deformation parameter as a function of dimensionless bubble equivalent diameter

6. Conclusion

A comprehensive comparison of experimental results of the bubble trajectory and shapes are made for water, polymeric solutions and a crystal suspension.

It was seen from this study that the trajectories of bubbles were significantly influenced by the bubble deformations and the surrounding liquid flow, and Re usually controlled the fluid flow regime around the bubble. The influences of Re, We and Mo were seen as important for explaining the bubble trajectories. It was visually observed that the smallest bubbles (5.76 mm) exhibited the most horizontal movement in water and the least in crystal suspension. These smaller bubbles showed a zig-zag trajectory in all fluids. This zig-zag motion was more pronounced in water than in other two fluids.

At intermediate and moderately high Re and We, for bubble sizes of 15.63 mm (2 mL) and 21.21 mm (5.0 mL), path transition occurred and they produced two distinct unstable zigzag and spiral trajectories for all fluids. It was also found that the path transition occurred more rapidly in water than in the other two fluids.

It was observed that larger bubbles (> 21.21 mm) produced a spiral motion in all fluids at very high Re and We. It was also established from experimental observations that larger bubbles ($\geq$ 10.0 mL or 26.84 mm) produced more spiral motion in crystal suspension than in the other fluids. It was observed that the bubble shapes are greatly dependent on Re, Eo and the aspect ratio or eccentricity of bubble, E. The dimensionless numbers such as Mo together with the Eötvös number (Eo) are also used to characterise the shape of bubbles moving in a surrounding fluid.

Generally, spherical, ellipsoidal and spherical cap shapes were observed in all three fluids. Apart from these, other bubble shapes were also encountered in those fluids. A skirted bubble shape was also found in PAM solution. Wobbling bubbles were also reported in water and crystal suspension.

The experimental results substantiated that the path oscillation and the shape deformation of bubble rise was less rapid in crystal suspension than in water.

Author details

N.M.S. Hassan*, M.M.K. Khan and M.G. Rasul
Power and Energy Research Group,
Institute for Resource Industries and Sustainability (IRIS), Faculty of Sciences,
Engineering and Health Central Queensland University, Rockhampton, Queensland, Australia

7. References

Acharya, A., Mashelkar, R. A. and Ulbrecht. J., 1977, Mechanics of bubble motion and deformation in non Newtonian media, Chem. Eng. Sci. 32, 863-872.

Adoua R, Legendre D and Magnaudet J 2009 Reversal of the lift force on an oblate bubble in a weakly viscous linear shear flow *J. Fluid Mech.* 628 23–41.

Adkins, B. G., 1951, Notes on the viscosity of molasses and massecuites, Proc. Qld. Soc. Sugar Cane Technol 18, 43-52.

Awang M., White, E. T., 1976, Effect of crystal on the viscosity of massecuites, Proc. Qld Soc. Sugar Cane Technol., 43, 263-270.

Aybers, N. M., and Tapucu, A., 1969, The motion of gas bubble rising through stagnant liquid, Warme und Stoffubertraguang 2, 118-128.

Bhaga, D. and Weber, M. E., 1981, Bubbles in viscous liquids: shapes, wakes and velocities., Journal of Fluid Mechanics., 105, 61-85.

Bhaga, D., 1976, PhD. Thesis, McGill Univ., Montreal, Canada.

Bhattacharya, S. N., 1997, Rheology fundamentals and measurements, printed at the Royal Melbourne Institute of Technology, Melbourne Australia.

Brenn G, Kolobaric V and Dusrst F 2006 Shape oscillations and path transition of bubbles rising in a model bubble column *Chem. Eng. Sci.* 61 3795–805.

Broadfoot, R. and Miller, K. F., 1990, Rheological studies of massecuites and molasses, International Sugar Journal 92, 107-112.

Broadfoot, R., Miller, K. F. and McLaughlin, R. L., 1998, Rheology of high grade massecuites, Proc. Aust. Soc. Sugar Cane Technol 20, 388 -397.

---

* Corresponding Author

Bojcic, P., Khan. M. M. K. and Broadfoot, R., 1997, Fitting loss for a power-law viscous fluid at low Reynolds number, Proceedings of the 7th Asian Congress of Fluid mechanics, Chennai, 725-728.

Calderbank, P. H., 1967, Gas absorption from bubbles, The Chem. Engr.,45, CE209, 109-115.

Carreau, P. J., Device, M. and Kapellas, M., 1974, Dynamique des bulles en milieu viscoelastique, Rheol. Acta 13, 477-489.

Chhabra, R. P., 2006, Bubbles, drops, and particles in non-Newtonian fluids, Taylor & Francis Group, CRC Press.

Churchill, S. W., 1989, A theoretical structure and correlating equation for the motion of single bubbles, Chemical Engineering and Processing 26, 269-279.

Clift, R., Grace, J. R. and Weber, M. E., 1978, Bubbles, Drops and Particles, Academic Press, 1978, republished by Dover, 2005.

Davies, R. M. and Taylor, G. I., 1950, The mechanics of large bubbles rising through liquids in tubes, Proc. of Roy. Soc. London, 200 Ser. A, 375-390.

De Kee, D. and Chhabra, R. P., 1988, A photographic study of shapes of bubbles and coalescence in non-Newtonian polymer solutions, Rheol. Acta 27, 656-660.

De Kee, D., Carreau, P. J. and Mordarski, J., 1986, Bubble velocity and coalescence in viscoelastic liquids, Chem. Eng. Sci. 41, 2273-2283.

De Kee, D., Chhabra, R. P. and Dajan, A., 1990, Motion and coalescence of gas bubbles in non-Newtonian polymer solutions, J. Non-Newtonian Fluid Mech. 37, 1-18.

De Kee, D., Chhabra, R. P. and Rodrigue, D., 1996, Hydrodynamic of free-rise bubbles in Non-Newtonian polymer solutions, Handbook of Applied Polymer Processing Technology, 87-123.

De Vries A W G 2001 Path and wake of a rising bubble *PhD Thesis* The University of Twente, the Netherlands.

De Vries, A. W. G., Biesheuvel, A. and Van Wijngaarden, L., 2002, Notes on the path and wake of a gas bubble rising in pure water, Int. Jnl. Multiphase Flow 28, 1823-1835.

Deane, G. B. and Stokes, M. D., 2002, Scale dependence of bubble creation mechanisms in breaking waves, Nature 418, 839-844.

Duineveld, P. C., 1994, Bouncing and coalescence of two bubbles in water, PhD Thesis, University of Twente.

Duineveld, P. C., 1995, The rise velocity and shape of bubbles in pure water at high Reynolds number, Journal of Fluid Mechanics, 292, 325-332.

Ellingsen K and Risso F 2001 On the rise of an ellipsoidal bubble in water: oscillatory paths and liquid induced velocity *J. Fluid Mech.* 440 235–68.

Feng, Z. C. and Leal, L. G., 1997, Nonlinear bubble dynamics, Ann. Rev. Fluid Mech. 29, 201-243.

Fernandes P C, Ern P, Risso F and Magnaudet J 2008 Dynamics of axisymmetric bodies rising along a zigzag path *J. Fluid Mech.* 606 209–23.

Frank, X., Li, H. Z. and Funfschilling, D., 2005, An analytical approach to the rise velocity of periodic bubble trains in non-Newtonian fluids, Eur. Phys. J. E 16, 29-35.

Frank, X., Li, H. Z., Funfschilling, D., Burdin, F. and Ma, Y., 2003, Bubble motion in non-Newtonian fluids and suspensions", Can J Chem Eng., 81:483–490.

Grace, J. R. and Wairegi, T., 1986, Properties and characteristics of drops and bubbles, Encyclopedia of. Fluid Mech., 3, 43-57.

Grace, J. R., 1973, Shapes and velocities of bubbles rising in infinite liquids, Trans. Inst. Chem. Engg. 51, 116-120.

Grace, J. R., Wairegi, T. and Nguyen, T. H., 1976, Shapes and velocities of single drops and. through immiscible liquids, Trans. Inst. Chem. Eng. 54, 167-173.

Haberman, W. L. and Morton, R. K., 1954, An experimental study of bubbles moving in liquids, Trans ASCE 2799, 227-252.

Hartunian, R. A. and Sears, W. R., 1957, On the instability of small gas bubbles moving uniformly in various liquids, Journal of Fluid Mech. 378, 19-70.

Hassan, N. M. S., Khan, M. M. K. and Rasul, M. G., 2007a, Characteristics of air bubble rising in low concentration polymer solutions, WSEAS TRANSACTIONS on FLUID Mechanics, ISSN: 1790-5087, 2 (3), 53-60.

Hassan, N. M. S., Khan, M. M. K., Rasul, M. G. and Rackemann, D. W., 2007b, An experimental study of bubble rise characteristics in non–Newtonian (power-law) fluids, Proceedings of the 16th Australasian Fluid Mechanics Conference, Gold Coast, Australia, 1315-1320.

Hassan, N. M. S., Khan, M. M. K. and Rasul, M. G., 2008a, An investigation of bubble trajectory and drag co-efficient in water and non-Newtonian fluids, WSEAS Transactions on Fluid Mechanics, Special issue: Sustainable Energy and Environmental Fluid Mechanics, ISSN: 1790-5087, 3(3), 261 -270.

Hassan, N. M. S., Khan, M. M. K, Rasul, M. G. and Rackemann, D. W. 2008b, An Experimental investigation of bubble rise characteristics in a crystal suspended non – Newtonian Fluid, Proceedings of the XVth International Congress on Rheology. 80th Annual Meeting, American Institute of Physics (AIP), Monterey, California, USA, ISBN 978-07354-0550-9, 743-745.

Hassan, N. M. S., Khan, M. M. K and Rasul, M. G., 2008c, Air bubble trajectories in polymeric Solutions and crystal Suspensions, Proceedings of the 4th BSME-ASME International Conference on Thermal Engineering, Dhaka, Bangladesh, ISBN 984-300-002844-0, 970- 975.

Hassan, N. M. S., Khan, M. M. K and Rasul, M. G., 2010a, Modelling and experimental study of bubble trajectory in non-Newtonian crystal suspension, Fluid Dynamics Research, IOP Publishing, 42, 065502.

Hassan, N. M. S., Khan, M. M. K., Rasul, M. G. and Rackemann, D. W., 2010b, Bubble rise velocity and trajectory in xanthan gum crystal suspension, Applied Rheology, International Journal, 20, 65102.

Hassan, N. M. S., 2011, Bubble rise phenomena in various non-Newtonian Fluids, PhD Thesis, Central Queensland University, Australia.

Kabir, M. A., Slater, A. R., Khan, M. M. K. and Rackemann, D., 2006, Characteristics of bubble rise in water and polyacrylamide solutions, International Journal of Mechanical and Materials Engineering, 64-69.

Kulkarni, A. A. and Joshi, J. B., 2005, Bubble formation and bubble rise velocity in gas-liquid systems: A. review, Ind. Eng. Chem. Res., 44, 5873-5931.

Lima Ochoterena, R. and Zenit, R., 2003, Visualization of the flow around a bubble moving in a low viscosity liquid, Revista Mexicana De Fisica 49, 348-352.

Maxworthy, T., Gnann, C., Kurten, M. and Durst, F., 1996, Experiments on the rise of air bubbles in clean viscous liquids. J. Fluid Mech. 321, 421-441.

Miyahara, T. and Yamanaka, S., 1993, Mechanics of motion and deformation of a single bubble rising through quiescent highly viscous Newtonian and non-Newtonian Media, Journal of Chemical Engineering of Japan, 26 (3), 297-302.

Magnaudet J and Eames I 2000 The motion of high-reynolds-number bubbles in inhomogeneous flows *Annu. Rev.Fluid Mech.* 32 659–708.

Mougin, G. and Magnaudet, J., 2002, Path instability of a rising bubble. Phys. Rev. Lett. 88 (2002), paper number 014502.

Mougin G and Magnaudet J 2006 Wake-induced forces and torques on a zigzagging/spiralling bubble J. Fluid Mech. 56 7185–94.

Munson, B. R., Young, D. F. and Okiishi, T. H., 2009, Fundamentals of Fluid Mechanics, 6th Edition, John Wiley & Sons, Inc.

Ness, J. N., 1983, On the measurement of massecuite flow properties, Proc. Int. Soc. Sugar Cane Technol 18, 1295-1303.

Nezhad, Hazi, A., 2008, Materials science experimentations of raw materials, intermediates and final products in the sugar beet manufacturing process, Dr.-Eng. Thesis, The Technical University of Berlin.

Rackemann, D. W., 2005, Evaluation of circulation and heat transfer in calandria tubes of crystallisation vacuum pans, M.Sc. Eng. Thesis, James Cook University.

Saffman, P. G., 1956, On the rise of small air bubbles in water, Journal of Fluid Mechanics, Digital Archive, 1: 249-275.

Shew, W. L., and Pinton, J. F., 2006, Viscoelastic effects on the dynamics of a rising bubble, J. Stat. Mech., Online at stacks.iop.org/JSTAT/2006/P01009, doi: 10.1088/1742-5468/2006/01/P01009.

Tadaki, T. and Maeda, S., 1961, On the shape and velocity of single air bubbles rising in various liquids, Chem. Eng. (Tokyo), 25, 254.

Tsuge, H. and Hibino, S. I., 1997, The onset conditions of oscillatory motion of single gas bubbles rising in various liquids, Journal of Chemical Engineering of Japan 10, 66-68.

Wegener P P and Parlange J Y 1973 Spherical-cap bubbles *Annu. Rev. Fluid Mech.* 5 680–9

Whyte, D. S., Davidson, M. R., Carnie, S. and Rudman, M. J., 2000, Calculation of droplet deformation at intermediate Reynolds number using a volume of fluid technique, ANZIAM Journal 42, C1520-C1535.

Yang B 2006 Numerical studies of single gas and vapor bubble flows *PhD Thesis* The John Hopkins University, Baltimore, MD.

Yang B and Prosperetti A 2007 Linear stability of the flow past a spheroidal bubble *J. Fluid Mech.* 582 53–78.

Yoshida, S. and Manasseh, R., 1997, Trajectories of rising bubbles, 16th Japanese Multiphase Flow Symposium, Touha, Hokkaido.
www.wseas.us/e library/transactions/fluid/2008/MGR-08.pdf

Zheng, Li and Yapa, P. D., 2000, Buoyant velocity of spherical and non-spherical bubbles/droplets, Journal of Hydraulic Engineering, 126(1), 852-854.

Zenit R and Magnaudet J 2008 Path instability of rising spheroidal air bubbles: a shape controlled process *Phys. Fluids* 20 1.

Zenit R and Magnaudet J 2009 Measurements of the stream wise vorticity in the wake of an oscillating bubble *Int. J. Multiph. Flow* 35 195–203.

Chapter 5

Incorporation of Computational Fluid Dynamics into a Fluid Mechanics Curriculum

Desmond Adair

Additional information is available at the end of the chapter

1. Introduction

In this chapter, the development, implementation and evaluation of a suitable curriculum for students to use computational fluid dynamics (CFD) as part of a fluid mechanics course at intermediate undergraduate level are described. CFD is the simulation of fluids, heat transfer, etc., using modelling, that is, mathematical physical problem formulation, and numerical methods which includes, discretization methods, solvers, numerical parameters, and grid generation. Historically only Analytical Fluid Dynamics (AFD) and Experimental Fluid Dynamics (EFD) have been taught at the undergraduate level but inclusion of CFD is now possible and desirable, with the advancing improvements of computer resources.

The use of simulation can now be found in many areas of engineering education, for example for electronics laboratories (Campbell et al. 2004; Feisel & Rosa, 2002), for chemical reactions (Qian and Tinker, 2006) and for diesel engine simulation (Assanis & Heywood, 1986). Simulation has also been effective in fluid mechanics and heat transfer (Devenport & Schetz, 1998; Zheng & Keith, 2003; Rozza et al., 2009). Some work in developing an educational user-friendly CFD interface and package has already been carried out (Pieritz et al., 2004) where the general aspects and simplification of the three main processes of CFD, the pre-processor, the solver and the post-processor were carefully considered. An electronic learning system while using an existing CFD package (Hung et al., 2005), and, integration of CFD with experimentation in the area of flow control research (Ardag et al., 2009) have been reported.

1.1. CFD in engineering practice

Why should CFD be included in an undergraduate fluid mechanics curriculum? The simple answer is that CFD has now a major component of professional life in engineering practice and to prepare students properly they must get exposure to all aspects of their chosen profession. In the areas of analysis and design, simulation based design is commonly used instead of the traditional "build and test", as it is much more cost effective than EFD and a substantial database is provided for diagnosing the adjacent flow field. Simulations can readily be done of physical flow phenomena that are difficult to measure, for example, full scale situations, environmental effects and hazards. With the introduction of CFD into a curriculum, it is possible to educate the young engineer as to the pros and cons of using the three areas, AFD, EFD and CFD and to be in a position to know which area to use according to the problem or project confronting them. Importantly, the engineer should learn not be prejudiced against using any of the three areas. So there is an increasing need to integrate computer-assisted learning and simulation, in the guise of CFD, into undergraduate engineering courses, both as a learning tool and as initial professional training.

1.2. General concerns about introducing simulation into a curriculum

Issues of concern arise when simulation is being introduced into a curriculum. These include learning vs. research objectives, usability vs. predetermined objectives and student demographics (Stern et al., 2006). A proper balance should be sought between these competing objectives, for example, it is just as important that a student be taught the practical and systematic ways of using a CFD package in a general sense, as well as achieving a specific result. There is much evidence from previous studies that: the use of simulation enhances the curriculum (Feisel & Rosa, 2002; Rozza et al., 2009); there is increased learning efficiency and understanding (Keller et al., 2007; Kelsey, 2001; LaRoche et al., 2002); there is effectiveness of new and hands-on learning methods (Patil et al., 2009); and, it is effective to use a combination of physical and simulation laboratories (Stern et al., 2006). Importantly, user-friendly interfaces must be designed so that students do not struggle with the mechanics of performing simulations to the detriment of understanding concepts. Also, when developing a curriculum which incorporates CFD, care must be taken to include flows of current interest while including diverse learning objectives. It must be remembered that CFD has become a widely used tool in fluids engineering covering many industries including Aerospace, Chemical Processing, Automotive, Hydraulics, Marine and Oil & Gas, and, hence choices have to be made when developing a curriculum.

In engineering practice, a current pacing element is the lack of personnel capable of using CFD. Until quite recently, most engineers using CFD software in industry and research centres had completed post-graduate degrees where CFD courses were taught either formally or informally. Now, as CFD becomes more pervasive in engineering practice and engineers are expected to use CFD without post-graduate education, teaching CFD at the undergraduate has become more usual and necessary. An obvious advantage of integration of CFD software into an undergraduate lecture and laboratory course is that analytical and experimental results can be compared with CFD results. The teaching approach would be to use interactive exercises to compliment traditional fluid mechanics course, and some success has already been noted in the previously mentioned studies above.

1.3. Specific concerns about introducing CFD into a curriculum

There are many issues, which if not carefully considered and implemented, can lead to teaching and learning difficulties. For example, which is best? - demonstration using CFD software, or allowing the students hands-on experience and the ability to investigate for themselves. Also, CFD could very well detract from a deeper knowledge of fluid mechanics, as, for example, boundary conditions, inlet conditions etc. are often built into the CFD package, and can be set without any real understanding. Students soon realize that they can get results, and reasonably plausible results, by mechanically following instructions, and not having much understanding of methodology and/or procedures. There is no doubt that when a student first uses CFD, a lot of new knowledge and required skills descends on them from many directions hence rendering to a steep learning curve. Without careful planning this curve can become overwhelming. Lastly, because CFD software is relatively less expensive than laboratory equipment there is a danger that it may replace laboratory experiments when this is not appropriate.

The questions in the above paragraph need to be assessed and evaluated when developing a curriculum. Here, the emphasis is on the development, implementation and evaluation of an effective curriculum for students to learn CFD, keeping in mind the issues of the previous paragraph, as part of a fluid mechanics course at intermediate undergraduate level. The objectives are to supplement and enhance the traditional course and to prepare students to use CFD effectively in engineering practice. The software chosen here is a commercial industrial software, and exposes students to the same or similar software they may use as professionals in industry. The software package provides students with a "Virtual Reality" interface, which allows for relative ease in setting up flows and the ability to visually reinforce concepts in fluid flow and heat transfer during the post-processor stage.

1.4. Outline of the chapter

In Section 2 of this chapter, basic computational fluid dynamics elements introduced to students in two lectures at the start of their course to introduce CFD elementary theory, methodology and procedures are outlined. In Section 3 the concept of the CFD interface is explained and in Section 4 the course/laboratories, learning objectives, applications, coursework and the integration of the CFD laboratories into the existing fluid mechanics course are described. Section 5 presents evaluation design, results and discussion, in the form of three investigations, one as a controlled experiment comparing the CFD group with a controlled group, one measuring the student learned knowledge and skills regarding the CFD interface and one eliciting student views on using CFD by questionnaire. Section 6 gives conclusions drawn and possible work for the future.

2. Basic computational fluid dynamics elements

This section outlines essential elementary CFD theory, which must be introduced students before they encounter hands-on experience in the laboratory. The following are extracts from the initial lectures given to the CFD student group.

2.1. Overall idea of CFD

CFD is used to replace the continuous problem domain with a discrete domain using a grid. In the continuous domain, each flow variable is defined at every point in the domain, whereas in the discrete domain, each flow variable is defined only at the grid nodes or sub-grid nodes. In a CFD solution, the relevant flow variables would only be directly solved at the grid nodes with values between obtained by interpolation. The governing partial differential equations and boundary conditions are defined in terms of the continuous variables, velocity, pressure, etc. These can be approximated in the discrete domain leading to a large set of coupled, algebraic equations in the discrete variables. Setting up this discrete system, and solving it involves a large number of repetitive calculations, hence the use of computers.

2.2. Discretization using the finite-volume method

To keep the explanation simple, consider the following 1D equation.

$$\frac{du}{dx} + u = 0; \qquad 0 \le x \le 1; \quad u(0) = 1 \tag{1}$$

A typical discrete representation of the above equation is shown on the following grid,

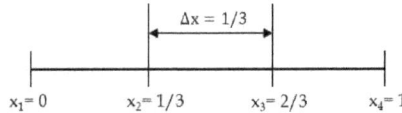

Figure 1. Discrete representation

This grid has four equally spaced grid nodes with Δx being the spacing between successive nodes, and since the governing equation is valid at any grid node then,

$$\left(\frac{du}{dx}\right)_i + u_i = 0 \tag{2}$$

where the subscript i represents the value at grid node x_i. In order to get an expression for $(du/dx)_i$ in terms of u at the grid nodes, u_{i-1} is expanded in a Taylor's series,

$$u_{i-1} = u_i - \Delta x \left(\frac{du}{dx}\right)_i + O(\Delta x^2) \tag{3}$$

The error in $(du/dx)_i$ due to neglecting terms in the Taylor series is called the truncation error, and, since the truncation error is $O(\Delta x)$ this discrete representation is termed first-order accurate.

The following discrete equation then ensues,

$$\frac{u_i - u_{i-1}}{\Delta x} + u_i = 0 \tag{4}$$

which is an algebraic equation.

When dealing with two-dimensional geometry the grid used in the CFD laboratories here will consist of relatively simple rectangles, or a Cartesian grid. In the finite-volume method, such a rectangle is called a "cell". For 2D flows, triangular cells are often used. For the 3D flows used in the laboratories here the grid will have cuboid cells. It should be noted that it is also common to use hexahedrals, tetrahedrals or prisms. In the finite-volume approach, the integral form of the conservation equations is applied to the control volume defined by a cell to get the discrete equations for the cell. The integral form of the continuity equation for steady, incompressible flow is shown below, where the integration is over the surface S of the control volume and $\vec{n}$ is the outward normal at the surface. This really means from this equation that the net volume flow into the control volume is zero.

$$\int_S \vec{V} \cdot \vec{n} dS = 0 \tag{5}$$

Consider the rectangular cell shown,

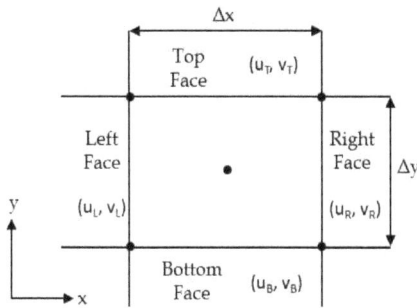

Figure 2. Cartesian control volume

The velocity at face i is taken to be $\vec{V}_i = u_i \hat{i} + v_i \hat{j}$. Applying the mass conservation Equation (5) to the control volume defined by the cell gives,

$$-u_L \Delta y - v_B \Delta x + u_R \Delta y + v_T \Delta x = 0 \tag{6}$$

This is the discrete form of the continuity equation for the cell. It is equivalent to summing up the net mass flow into the control volume and setting it to zero. Discrete equations for the conservation of momentum and energy for the cell can also be obtained.

Rearranging Equation (4) gives,

$$-u_{i-1} + (1 + \Delta x)u_i = 0 \tag{7}$$

Applying this equation to the 1D grid (Fig. 1) shown earlier at grid points i = 2, 3, 4 gives,

$$\begin{bmatrix} -1 & 1+\Delta x & 0 & 0 \\ 0 & -1 & 1+\Delta x & 0 \\ 0 & 0 & -1 & 1+\Delta x \end{bmatrix} \begin{bmatrix} u_2 \\ u_3 \\ u_4 \end{bmatrix} = \begin{bmatrix} 0 \\ 0 \\ 0 \end{bmatrix} \tag{8}$$

The discrete equation cannot be applied at the left boundary ($i = 1$) since u_{i-1} is not defined. Instead, a boundary condition must be applied here.

In a general situation, the discrete equations are applied for the cells in the interior of the domain. For grid cells at or near the boundary, a combination of discrete equations and boundary equations are applied. This leads to a system of simultaneous algebraic equations. Boundary conditions are very important to have a well-defined problem and it should be remembered that a wrong boundary condition will give a totally wrong result.

In a practical CFD application, depending on the size of the grid and the number of variables to be solved for, the number of unknowns in a discrete system may run into thousands or even millions so the matrix inversion needed to solve the system of equations needs to optimized. Also due to the truncation error of Equation (3), it is clear that as the number of grid points is increased, and Δx is reduced, the error in the numerical solution would decrease and the agreement between numerical and exact solutions would be better. When the numerical solutions obtained on different grids agree to within a level of tolerance specified by the user, they are referred to as "grid converged" solutions. The concept of grid convergence applies to the finite-volume approach also when the numerical solution, if correct, becomes independent of the grid as the cell size is reduced. It is very important that the effect of grid resolution on the solution is always investigated.

2.3. Basic equations of CFD

The Navier-Stokes and continuity equations provide the foundations for modelling fluid motion. The laws of motion that apply to solids are valid for all matter including liquids and gases. A principal difference, however, between fluids and solids is that fluids distort without limits. Analysis of a fluid needs to take account of such distortions. The Navier-Stokes equations can be derived by considering the dynamic equilibrium of a fluid element. They state that the inertial forces acting on a fluid element are balanced by the surface and body forces. For incompressible flow, that is when the fluid density is constant, and ignoring body forces, the Navier-Stokes equations can be written as,

$$\frac{\partial u}{\partial x} + \frac{\partial v}{\partial y} + \frac{\partial w}{\partial z} = 0 \tag{9a}$$

$$\rho\left(\frac{\partial u}{\partial t} + u\frac{\partial u}{\partial x} + v\frac{\partial u}{\partial y} + w\frac{\partial u}{\partial z}\right) = -\frac{\partial p}{\partial x} + \mu\left(\frac{\partial^2 u}{\partial x^2} + \frac{\partial^2 u}{\partial y^2} + \frac{\partial^2 u}{\partial z^2}\right) \tag{9b}$$

$$\rho\left(\frac{\partial v}{\partial t} + u\frac{\partial v}{\partial x} + v\frac{\partial v}{\partial y} + w\frac{\partial v}{\partial z}\right) = -\frac{\partial p}{\partial y} + \mu\left(\frac{\partial^2 v}{\partial x^2} + \frac{\partial^2 v}{\partial y^2} + \frac{\partial^2 v}{\partial z^2}\right) \tag{9c}$$

$$\rho\left(\frac{\partial w}{\partial t} + u\frac{\partial w}{\partial x} + v\frac{\partial w}{\partial y} + w\frac{\partial w}{\partial z}\right) = -\frac{\partial p}{\partial z} + \mu\left(\frac{\partial^2 w}{\partial x^2} + \frac{\partial^2 w}{\partial y^2} + \frac{\partial^2 w}{\partial z^2}\right) \tag{9d}$$

In the above equations, u, v, w are the velocity components in the x, y, z directions, ρ is the density, p is the pressure, and μ is the viscosity.

Turbulence is of fundamental interest to engineers because most flows encountered in engineering are turbulent. This happens because it is difficult to keep the flow laminar, or by intention as turbulence is essential for the engineering application. However, for turbulent flows, the variation of quantities with time is so random that its detailed variation can be of little, if any, engineering relevance so averaged quantities with time are therefore calculated using the Reynolds-averaged Navier-Stokes equations shown below.

$$\frac{\partial \bar{u}}{\partial x} + \frac{\partial \bar{v}}{\partial y} + \frac{\partial \bar{w}}{\partial z} = 0 \tag{10a}$$

$$\rho \left(\frac{\partial \bar{u}}{\partial t} + \bar{u}\frac{\partial \bar{u}}{\partial x} + \bar{v}\frac{\partial \bar{u}}{\partial y} + \bar{w}\frac{\partial \bar{u}}{\partial z} \right) = -\frac{\partial \bar{p}}{\partial x} + \frac{\partial}{\partial x}\left(\mu\frac{\partial \bar{u}}{\partial x} - \rho\overline{u'u'} \right)$$

$$+ \frac{\partial}{\partial y}\left(\mu\frac{\partial \bar{u}}{\partial y} - \rho\overline{u'v'} \right) + \frac{\partial}{\partial z}\left(\mu\frac{\partial \bar{u}}{\partial z} - \rho\overline{u'w'} \right) \tag{10b}$$

$$\rho \left(\frac{\partial \bar{v}}{\partial t} + \bar{u}\frac{\partial \bar{v}}{\partial x} + \bar{v}\frac{\partial \bar{v}}{\partial y} + \bar{w}\frac{\partial \bar{v}}{\partial z} \right) = -\frac{\partial \bar{p}}{\partial y} + \frac{\partial}{\partial x}\left(\mu\frac{\partial \bar{v}}{\partial x} - \rho\overline{u'v'} \right)$$

$$+ \frac{\partial}{\partial y}\left(\mu\frac{\partial \bar{v}}{\partial y} - \rho\overline{v'v'} \right) + \frac{\partial}{\partial z}\left(\mu\frac{\partial \bar{v}}{\partial z} - \rho\overline{v'w'} \right) \tag{10c}$$

$$\rho \left(\frac{\partial \bar{w}}{\partial t} + \bar{u}\frac{\partial \bar{w}}{\partial x} + \bar{v}\frac{\partial \bar{w}}{\partial y} + \bar{w}\frac{\partial \bar{w}}{\partial z} \right) = -\frac{\partial \bar{p}}{\partial z} + \frac{\partial}{\partial x}\left(\mu\frac{\partial \bar{w}}{\partial x} - \rho\overline{u'w'} \right)$$

$$+ \frac{\partial}{\partial y}\left(\mu\frac{\partial \bar{w}}{\partial y} - \rho\overline{v'w'} \right) + \frac{\partial}{\partial z}\left(\mu\frac{\partial \bar{w}}{\partial z} - \rho\overline{w'w'} \right) \tag{10d}$$

All the instantaneous quantities were replaced by their corresponding time-averaged quantities. Also, due to the averaging process extra terms appear, for example, $-\rho\overline{u'u'}$, $\rho\overline{u'w'}$ appear. These terms behave like stress terms and require further equations if the system is to be solved. This will be further discussed in Section 2.6 (Turbulence Modelling) below.

2.4. Boundary conditions

When solving the Reynolds-averaged Navier-Stokes equations and continuity equation, appropriate initial conditions and boundary conditions need to be applied. Boundary conditions are a required component of the mathematical model and direct the motion of the flow. They can be used to specify fluxes into the computational domain, with boundaries and internal surfaces represented by face zones and boundary data assigned to these zones.

Different types of boundary conditions can be applied at surfaces. When using a Dirichlet boundary condition, one prescribes the value of a variable at the boundary, e.g. $u(x)$ = constant and when using a Neumann boundary condition, one prescribes the gradient normal to the boundary of a variable at the boundary, e.g. $\frac{\partial u(x)}{\partial n}$ = constant. It should be noted that at a given boundary, different types of boundary conditions can be used for different variables.

A wide range of boundary conditions types permit the flow to enter and exit the solution domain, for example, general (pressure inlet, pressure outlet), incompressible flow (velocity inlet, outflow), compressible flow (mass flow inlet, pressure far-field), and special (inlet vent, outlet vent, intake fan, exhaust fan). The boundary location and shape should be selected such that flow either goes in or out. This is not mandatory, but will typically result in better convergence. There should not be large gradients of variables in the normal direction to the boundary near inlets and outlets as this indicates an incorrect problem specification. Also grid skewness near a boundary should be minimized.

2.5. Basic numerical solvers

In a practical problem, as mentioned above, a matrix would be extremely large, so needing a prohibitively large amount of memory to invert it directly. Therefore the matrix would be inverted using an iterative scheme instead. Iteration serves two purposes, namely, it allows for efficient matrix inversion with greatly reduced memory requirements and it is necessary to solve nonlinear equations. In steady problems, a common and effective strategy used in CFD codes is to solve the unsteady form of the governing equations and march the solution in time until the solution converges to a steady value. In this case, each time step is effectively an iteration, with the guess value at any time level being given by the solution at the previous time level. The finite-difference equation at a grid point is arranged so that the quantity to be calculated is expressed in terms of values at the neighbouring grid points, including guessed values. Then as we sweep from say left to right on the grid, successive values of the variable are updated, including any guessed values. However, since guessed values at some of the neighbouring points were used, only an approximate solution for the matrix inversion during each iteration is obtained. However as each iteration ensues across the grid, the values of the variable at each grid point converges towards the exact solution, making the error introduced due to guessing tend to zero. This iterative type of matrix inversion allows for efficient matrix inversion with greatly reduced memory requirements and it is necessary when solving nonlinear equations.

How do we judge when the solution is converged. Basically it is when the difference between the value of the variable being solved at the present iteration step and it's value solved for in the previous iteration step, referred to as the residual, is small enough. It is very common to use the summation of the residuals at each grid point normalised by the average of the variable.

2.6. Turbulence modelling

There are two different states of flow, laminar and turbulent. Laminar flows are characterized by smoothly varying velocity fields in space and time and these flows arise when the fluid viscosity is sufficiently large to damp out any perturbations to the flow that may occur due to boundary imperfections or other irregularities. These flows occur at low-to-moderate values of the Reynolds number. Turbulent flows, on the other hand, are characterized by large, nearly random instabilities that grow until nonlinear interactions cause them to break down into finer and finer eddies that eventually are dissipated by the action of viscosity.

For turbulent flow there is a deviation of the velocity from the mean value defined as,

$$u' \equiv u - \bar{u} \tag{11}$$

Due to this deviation, or more commonly called the fluctuation in velocity, and when the Navier-Stokes equations are averaged to become the Reynolds-Averaged Navier-Stokes equations, which are the equations which govern the mean velocity and pressure in the CFD package used here, extra terms are introduced called Reynolds stresses. This presents a problem in that there are more unknowns than there are equations leading to the necessity to model these extra terms to "close" the equations. There have over this last 50-60 years been many ways suggested as the best solution for closure, none of which are completely satisfactory. For the CFD laboratories here, the k-ε and the LVEL (Agonafer et al., 2008; Launder & Spalding, 1972) methods will be used.

3. Computational fluid dynamics teaching laboratories

3.1. Introduction

For the CFD laboratories the interface is a "three-dimensional" fully interactive environment. This interface uses, "virtual reality", is easy-to-use, and allows a student to simulate a flow from beginning to end without resorting to specialized codes. The code is also used widely in the professional engineering, so giving students useful skills which contribute to their preparation for the workplace. The virtual reality environment is designed as a general purpose CFD interface consisting of the VR-Editor (pre-processor), the VR-Viewer (post-processor) and the solver module which performs the flow simulation calculations. The VR-Editor allows the student to set the size of the computational domain, define the position, size and properties of objects to be introduced into the domain, specify the material(s) which occupies the domain, specify the inlet and outlet boundary conditions, specify the initial conditions, select a turbulence model when necessary, specify the position and fineness of the computational grid, and specify other parameters influencing the speed of convergence of the solution procedure. On setting up a particular flow, the characteristics of the flow are calculated using the solver module. The progress of the calculations is clearly monitored until convergence is reached or the iteration limit is reached. In the VR-Viewer, the results of a flow simulation are displayed graphically. The post-processing capabilities of the VR-Viewer used here are, vector plots, contour plots, iso-surfaces, streamlines and x-y plots.

3.2. Interface design specifications

The CFD laboratory is designed so that practical procedures are user-friendly and easy to implement, and also to show students that CFD methodology needs to be systematic and rigorous. The complete process, at this level of CFD can, if so desired, be completely automated with the students going through a step-by-step process seamlessly from the set-up of the problem, through the solving to the display of the results. However it is very important that the laboratory also mirrors what is found in engineering practice, where a systematic approach is found. Careful consideration must be given to the areas listed in Table 1 below.

Geometry	Solid and other fluid boundaries
Physics	Incompressible/compressible fluid, which quantities to be solved for, closure of the equations, initial and boundary conditions
Grid	The choice here is Cartesian meshing or orthogonal meshing. The Cartesian mesh can be automatically generated or built manually.
Numerics	Convergence monitoring, selection of numerical scheme.
Post-processing	Flow visualization, analysis, verification, validation using published experimental or empirical data.

Table 1. Areas for systematic consideration

To contribute to the student's self-learning, a hierarchical system of predefined active options within the virtual reality environment are introduced for later simulations.

3.3. Interface design features

The design of the CFD interface (shown on Fig. 3) was chosen and designed to have features which systematically informs, is vocationally sound and is easy to use.

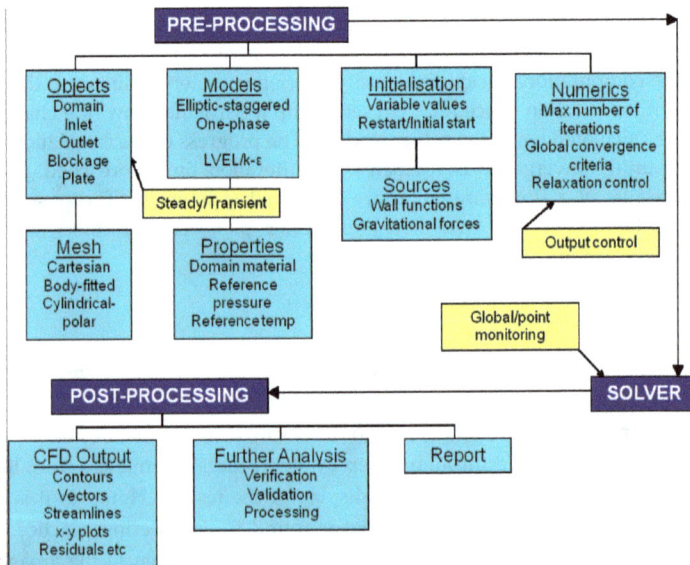

Figure 3. Summary of CFD interface

Each simulation process follows exactly how modelling is carried out in engineering practice, with the students setting up, solving and analysing the particular problem step-by-step.

The students interact with the software using mouse and keyboard input. There was no requirement for advanced computer language skills enabling the students to concentrate on the methodology and procedures of CFD. An important feature of the CFD interface is that it is stand-alone. By this is meant that grid generation, solving and post-processing are all combined in the virtual reality environment.

Also important is that the results obtained by students could be easily moved to Microsoft Office during the reporting stage. As the software package is built on the Windows OS using PCs with relatively low computer power, it was important that the CFD interface was built on fast and accurate solvers, as is found for this software. Because students are given a limited time in the laboratory, and also in order to keep their level of interest heightened, results should come back reasonably quickly. The post-processor was designed to plot contours, vectors, streamlines and, when needed, to make animations. Students had access to colour printers to produce figures for their reports.

4. Integration of CFD laboratory into fluid mechanics curriculum

4.1. Existing fluid mechanics undergraduate course

The CFD laboratory was integrated into a fourth semester course, for students of mechanical engineering. Traditionally the course used two lectures, each 3 hours in length, per week for theoretical fluid mechanics with four experimental laboratories, from Week 3 to Week 10, giving hands-on experience and demonstration of fundamental principles over the twelve week semester. The laboratories were given 3 hours per week to complete. With the introduction of CFD, the course was restructured to consist of two lectures, each three hours per week for theoretical mechanics, and, two CFD laboratories and two experimental laboratories. Again these laboratories were from Week 3 to Week 10, and the CFD laboratories alternated with the experimental laboratories. Two textbooks were also added to the required reading list (Ferziger & Peric, 1996; Tu et al., 2007).

The main learning outcomes are to understand the equations that govern fluid flow (conservation of mass, momentum and energy) and be able to apply them to a range of practical problems, including:

• predicting drag forces on bluff, streamline bodies and flat plates;
• analysing the flow in pipe systems;
• analysing performance of radial flow pumps and turbines; and,
• matching pumps and turbines for particular applications.

The unit also aims to develop skills in working effectively with others through the laboratory component of the unit. One or two seminars were help early in the semester, to discuss expectations regarding laboratory practice and reporting. The CFD student group was initiated to CFD as discussed next.

4.2. Necessary initiation of students to CFD

With the introduction of CFD into the course two extra lectures, each 2 hours in length, entitled "An Introduction to Computational Fluid Dynamics (CFD)" were presented to prepare students to learn CFD methodology and procedures.

4.2.1. Complementary nature of CFD and experimental fluid mechanics

During the first lecture, the students are introduced to the idea that theoretical fluid mechanics, experimental fluid mechanics and computational fluid dynamics are complementary in modern engineering practice. As with the experimental laboratories, students are then introduced to CFD general methodology and procedures. The students learn as to when and why CFD is used, and the breakdown of CFD into three processes namely the pre-processor, the solver and the post-processor. This is followed by the idea of a finite domain, subdividing the domain into control volumes (Cartesian grid), selecting the quantities to be calculated (and hence which conservation equations to be used), geometrical aspects, fluid and solid properties, sources within the domain (e.g. gravitational force), some of the numerical details (including initialization, the iterative process and how to achieve convergence), the use of different boundary conditions (solid wall, inlet, outlet), and finally how to close the conservation equations using turbulence modelling (k-ε, LVEL) (Agonafer et al., 2008; Launder & Spalding, 1972). Lastly, and importantly, the students are taught to be critical of their results and how to examine if what they are getting is what they might have expected.

4.2.2. CFD methodology and procedures

In the second lecture, the important part is a demonstration with full facilities for students to have 'hands-on' experience as the demonstration proceeds. Several simple three-dimensional flows are used as exemplars to give an overall view of the CFD process. The virtual reality environment, (Fig. 4), which facilitates the VR-Editor (pre-processor), the VR-Viewer (post-processor) and the solver module, which performs the simulation calculations, is introduced.

It is demonstrated how the VR-Editor is used to set the size of the computational domain, defining the position, size and properties of objects introduced into the domain, specifying the material which occupies the domain, specifying inlet and outlet boundary conditions, specifying initial conditions, selecting a turbulence model (if appropriate), specifying the fineness of the computational grid and setting parameters which influence the speed of convergence of the solution procedure. Viewing movement controls are also introduced for zooming, rotation, and, vertical and horizontal translations. It is then shown how to create a new simulation and to set data for geometry, automatic meshing, the quantities to be solved and the turbulence model. Objects are introduced as required together with any inlets and outlets. Finally the solver parameters are specified for the total number of iterations, the position of a 'probe', partially needed for monitoring convergence as the solver proceeds, is set, and the students are encouraged to have a final look at the automatic mesh and positions of objects before starting to solve.

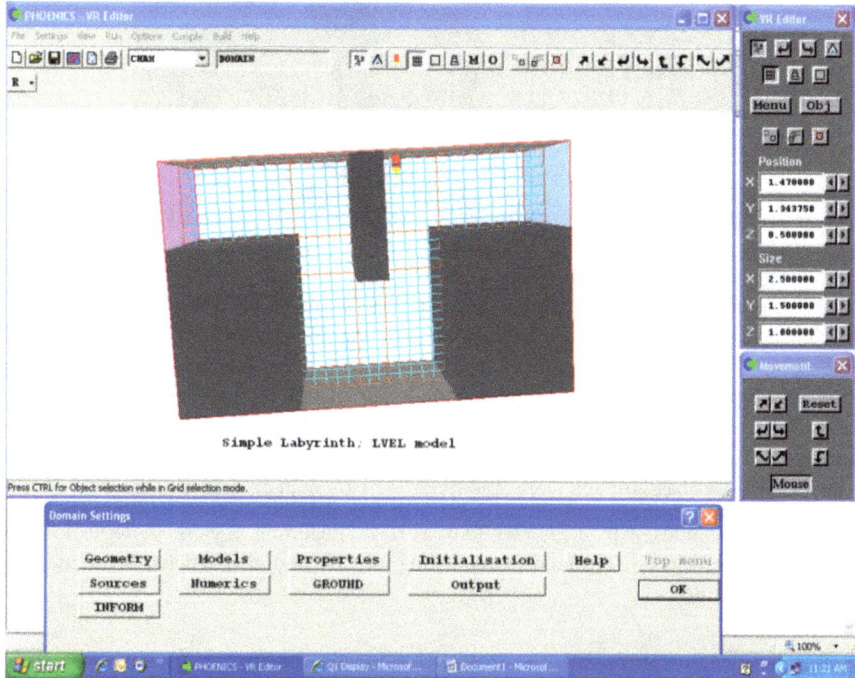

Figure 4. General view of VR-Editor (pre-processor)

As a converged solution is approached, the variables at the monitoring point become constant while with each successive sweep through the domain, the values of the errors are shown to decrease steadily.

On completion of the solver, the results of the flow-simulation can be viewed using the post-processor called the VR-Editor. This can display vector plots, contour plots, iso-surfaces, streamlines and x-y plots as demonstrated on Fig. 5. For ease of use the VR-Viewer is close to the design of the VR-Editor but with clearly marked toggle-buttons for vector display, etc.

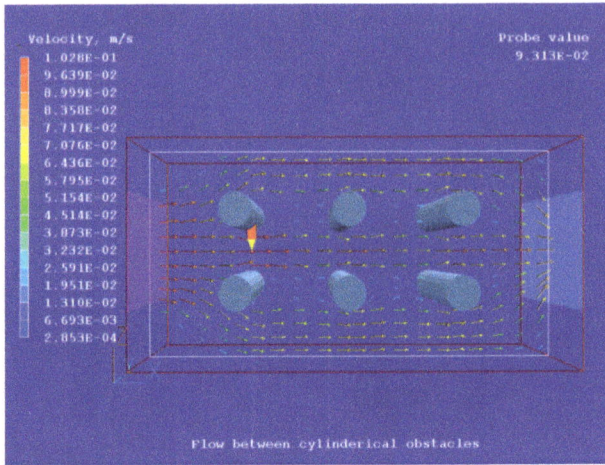

Figure 5. Typical vector-plot output of the post-processor

The students are encouraged to display the quantities calculated in various ways and also to experiment with each of the control buttons for zooming, rotating and obtaining meaningful views of selected planes. The students are shown how to print their results and save both input and output files. A second example is then demonstrated where instead of using the automated mesh option, students are taught to think why and where fine and coarse meshing is appropriate and how to implement it. Also included is how to produce x-y plots, an important part of assignment work, and reporting system in professional life.

This lecture concludes with a broader look at CFD, including a more in-depth look at numerical methods in CFD, turbulence modelling for CFD and grid-generation alternatives to the Cartesian grid.

4.3. Replacing laboratory experiments with CFD simulation

From the original four experimental laboratories, "Predicting drag forces on bluff, stream-line bodies and flat plates" and "Analysing the flow in pipe systems" were chosen to become CFD laboratories while "Analysing performance of radial flow pumps and turbines" and "Matching pumps and turbines for particular applications" remained as experimental. The four laboratories were conducted sequential from the beginning to the end of the semester. The students were expected to be self-guided to encourage self-learning for both the CFD and experimental laboratories, although a tutor and technician were in attendance. Detailed step-by-step notes were provided for all four laboratories.

To overcome some of the issues mentioned in the introduction concerning integration of CFD into the fluid mechanics course, it was decided to allow students, as much as possible, to have hands-on experience and investigations and assignments by themselves as opposed to demonstration. When setting boundary conditions, especially near-wall conditions, students had to explain such topics as the need for grid refinement, so getting them to think rather than just mechanically do. Also the students were encouraged to gradually develop a 'feeling' for when pressure or mean velocity gradients were high within the flow and hence generate higher mesh densities, so again getting away from a mechanical approach. The steep learning curve met during the early days of the course was partly alleviated by getting the students to stick strictly to the procedures of Fig. 3.

5. Teaching and learning evaluation

The evaluation process was subdivided into three investigations, one in the form of a controlled experiment comparing the knowledge of the group with CFD in their course with those of a controlled group using only the conventional experimental laboratories, the second measuring student knowledge and skill outcomes for the CFD interface, and, the third in the form of an online questionnaire eliciting the views of students on using CFD.

5.1. Controlled experiment

To investigate the effectiveness of introducing CFD laboratories into the fluid mechanics course, a controlled experiment applying a pre-test-post-test control group design was conducted (Pfahl et al., 2004). The students had to undertake two tests, one before the respective course (pre-test) and one after the respective course (post-test) with the introduction of CFD laboratories then being evaluated by comparing within-student post-test to pre-test scores, and by comparing the scores between students in the CFD group (A), i.e. those who were taught using the course containing CFD laboratories, to those students in the control group (B), i.e. taught using the conventional method of experimental laboratories only. The various possibilities of the methods of teaching the students are summarized using Fig. 6.

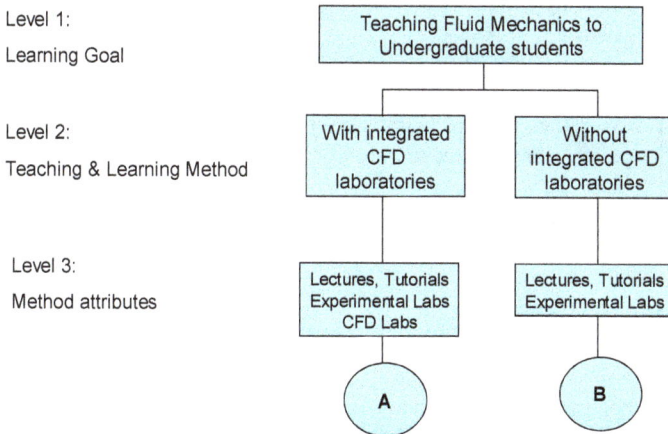

Figure 6. Course arrangements

To measure the performance of the two groups, four constructs were used with each construct represented by one dependent variable. Each dependent variable has the hypothesis,

1. There is a positive learning effect in both groups (A: CFD group, B: control group). This means post-test scores are significantly higher than pre-test scores for each dependent variable.
2. The learning is more effective for group A than for group B, either with regard to the performance improvement between pre-test and post-test (the relative learning effect), or with regard to post-test performance (absolute learning effect). The absolute learning effect is of interest because it may indicate an upper bound of the possible correct answers depending on the method of teaching.

The design starts with random assignment of students to the CFD group (A) and control group (B) with the members of both groups completing a pre-test and post-test. The pre-test measured the performance of the two groups before the courses and the post-test measured the performance of the two groups after the courses. The students did not know that the post-test and pre-test questions were identical and neither were they allowed to retain the pre-test questions with the correct answers only given to the students after the experiment.

The students were in the fourth semester of an engineering course with the number of students in group A, $N_A = 46$, and in group B, $N_B = 35$. The personal characteristics of the students are summarized in Table 2.

Characteristics	
Average age	22.3 years
Percentage female	21%
Preferred learning style(s)	
Reading with exercise	16%
Lecture	11%
Tutorial	30%
Laboratory	18%
Working in groups (with peers)	25%
Opinion of most effective learning style(s)	
Reading with exercise	9%
Lecture	10%
Tutorial	32%
Laboratory	18%
Working in groups	31%

Table 2. Personal characteristics

The initial testing was conducted after a short introduction as to the purpose of the experiment and general organizational issues. The pre-test was then carried out with the data for the dependent variables collected. Following the pre-test, the students were placed in either the control group or the experimental group and all students participated in both the pre-test and post-test. After completing their courses, both groups of students performed the post-test using the same questions as during the pre-test, thus providing data on the dependent variables for the second time. In addition the students were asked to answer questions about subjective perceptions.

Dependent variables J.1 Interest in Fluid Mechanics ('Interest') J.2 General knowledge of Fluid Mechanics('Understand general') J.3 Understanding of 'simple' Fluid Mechanics ('Understand simple') J.4 Understanding of 'difficult' Fluid Mechanics ('Understanding difficult')   *Subjective perceptions* S.1 Available time budget versus time need ('Time pressure') S.2 Course evaluation (useful, engaging, easy, clear)

Table 3. Experimental variables

Data for two types of variables were collected, the dependent variables (J.1, …, J.4) and the subjective perception variables (S.1, S.2). These variables are listed in Table 3. The dependent variables are constructs used to capture aspects of learning provided by the courses and each was measured using 5 questions. Selected examples of questions used as shown in Table 4.

The results for the dependent variable J.1 were found by applying a five-point Likert-type scale (Likert, 1932) with each answer mapped to the value range R= [0, 1].

The values for variables J.2 – J.4 are average scores derived from five questions for each. Missing answers were marked as incorrect. The data for the subjective perception variables was collected after the post-test. The values for variable S.1 are normalized averages reflecting the time needed for understanding and doing the tasks associated with Weeks 2 – 12.

The descriptive statistics for the experiment are summarized in Table 5. The columns 'Pretest scores' and 'Post-test scores' show the calculated values for mean ($\bar{x}$), median (m) and standard deviation (σ) of the raw data collected, and the column 'Difference scores' shows the difference between the post-test and pre-test scores.

J.1 example question I consider it very important for mechanical engineering students to know as much as possible about fluid mechanics. (1 = fully agree / 5 = fully disagree) Circle number below. Agree 1   2   3   4   5 Disagree
J.2 example question What are the four main forces acting on an aircraft when flying straight and level?
J.3 example question What is flow separation? What causes it? What is the effect of flow separation on the drag coefficient?
J.4 example question Fluid flows out of a large tank into a straight section of pipe with a diameter d. A boundary layer of thickness δ grows along the pipe wall. Transition occurs at $x = 0$ due to a sharp edge at the inlet. The turbulent boundary layer development may be approximated by the flat-plate relation, $\delta/x = 0.391$ $Re^{-0.2}$. Estimate the distance required for the boundary layer to completely fill the pipe for a Reynolds number of 2×10^5, neglecting changes in core velocity U with x.

S.1 example question
 	
I did not have enough time to:
 	
    - complete the tutorials
 	
    - complete the laboratory sessions
 	
    - write-up the laboratory reports
 	
    - complete the post-test

S.2 example question
 	
I consider the explanations/information provided for Laboratory Sessions
 	
                1   2   3   4   5
 	
        Useful   Useless
 	
        Boring   Engaging
 	
        Difficult   Easy
 	
        Clear   Confusing

Table 4. Example questions (pre-test, post-test, subjective perceptions)

	Pre-test scores				Post-test scores				Difference scores			
	J.1	J.2	J.3	J.4	J.1	J.2	J.3	J.4	J.1	J.2	J.3	J.4
Group A												
($\bar{x}$)	0.67	0.61	0.27	0.23	0.69	0.83	0.58	0.35	0.02	0.22	0.31	0.12
(m)	0.71	0.60	0.24	0.24	0.72	0.78	0.49	0.31	0.01	0.18	0.25	0.07
(σ)	0.11	0.34	0.29	0.25	0.09	0.22	0.21	0.27	0.01	0.29	0.25	0.26
Group B												
($\bar{x}$)	0.76	0.52	0.34	0.21	0.77	0.57	0.55	0.37	0.01	0.05	0.21	0.16
(m)	0.75	0.54	0.31	0.18	0.78	0.56	0.53	0.39	0.03	0.02	0.22	0.21
(σ)	0.09	0.18	0.32	0.31	0.13	0.21	0.13	0.22	0.11	0.19	0.24	0.27

Table 5. Scores of dependent variables

Table 6 shows the calculated values for mean, median and standard deviation of the raw data collected on subjective perceptions.

	S.1	S.2
Group A		
($\bar{x}$)	0.54	0.49
(m)	0.52	0.49
(σ)	0.19	0.13
Group B		
($\bar{x}$)	0.46	0.48
(m)	0.41	0.44
(σ)	0.21	0.14

Table 6. Scores of subjective perceptions

The students in control group (B) expressed less need of additional time than those of the CFD group (A), while students of both groups were fairly equal in their perception of their respective course usefulness, engagement, difficulty and clarity.

Standard significance testing was used to investigate the effect of the treatments on the dependent variables J.1 to J.4. The null hypotheses were,

- $H_{0,1}$: There is no difference between pre-test scores and post-test scores within experimental group (A) and control group (B).
- $H_{0,2a}$: There is no difference in relative learning effectiveness between CFD group (A) and control group (B).
- $H_{0,2b}$: There is no difference in absolute learning effectiveness between CFD group (A) and control group (B).

For $H_{0,1}$ and focusing on the CFD group (A), Table 7 shows the results using a one-tailed t-test for dependent samples. Column one specifies the variable, column two represents the Cohen effect size, d, (Cohen, 1988; Ray & Shadish, 1996), column three the degrees of freedom, column four the t-value of the study, column five the critical value for the significance value $\alpha = 0.10$ and column six lists the associated p-value.

Variable	d	df	t-Value	Crit.t$_{0.90}$	p-Value
J.1	0.200	45	1.360	1.301	0.090
J.2	0.770	45	5.220	1.301	0.000
J.3	1.220	45	8.270	1.301	0.000
J.4	0.460	45	3.110	1.301	0.001

Table 7. Results for 'post-test' versus 'pre-test' for group A

It can be seen from Table 7 that all dependent variables achieve a statistically and practically significant result.

Table 8 shows the results of testing hypothesis $H_{0,1}$ for the control group (B) using a one-tailed t-test for dependent samples. The structure of the table is the same as that of Table 7.

Variable	d	df	t-Value	Crit.t$_{0.90}$	p-Value
J.1	0.090	34	0.530	1.307	0.299
J.2	0.260	34	1.540	1.307	0.066
J.3	0.860	34	5.080	1.307	0.000
J.4	0.600	34	3.550	1.307	0.000

Table 8. Results for 'post-test' versus 'pre-test' for group B

It can be seen from Table 8 that the control group (B) achieved statistically and practically significant results for dependent variables J.2 – J.4. For J.1 no significant results can be found.

For $H_{0,2a}$ which states that the difference between post-test and pre-test scores of group A is not significantly larger than the one for group B. Table 9 shows for each dependent variable separately the results of testing hypothesis $H_{0,2a}$ using a one-tailed t-test for independent samples.

Variable	d	df	t-Value	Crit.t$_{0.90}$	p-Value
J.1	0.130	79	1.160	1.292	0.125
J.2	0.690	79	6.130	1.292	0.000
J.3	0.410	79	3.640	1.292	0.000
J.4	-0.150	79	-1.330	1.292	0.906

Table 9. Results for 'performance improvement' (Group A versus Group B)

It can be seen that the hypothesis $H_{0,2a}$ can be rejected only for the variables J.2 and J.3. The value for J.4 indicates that the relative learning effect is superior when the students were exposed to experimental laboratories only.

Table 10 shows for each dependent variable separately the results of testing $H_{0,2b}$ using a one-tailed t-test for independent samples.

Variable	d	df	t-Value	Crit.t$_{0.90}$	p-Value
J.1	-0.720	79	-6.430	1.292	1.000
J.2	1.210	79	10.820	1.292	0.000
J.3	0.170	79	1.520	1.292	0.066
J.4	-0.080	79	-0.715	1.292	0.762

Table 10. Results for 'post-test improvement' (Group A versus Group B)

Again the two variables which show statistically significant results are J.2 and J.3 and hence H$_{0,2b}$ can be rejected for these variables. The variables J.1 and J.4 indicated that more interest is found in the totally experimental course and these students also did better in the more difficult aspects of the course.

5.2. Student knowledge and skill outcomes for the CFD interface

An objective measure of student knowledge and skill outcomes for the CFD interface as applied to the fluid mechanics curriculum was devised. Some of the questions used in the test are shown in Table 11, with the questions directed only at students of CFD group (A).

This test was again run on a pre/post CFD studies basis, i.e. during the first week of the course students completed the pre-test and later in the semester, and after completing the CFD studies, the students completed the post-test. The most intuitive test of students' knowledge and skill outcomes is whether the post-test scores were significantly higher than those of the pre-test scores. Table 12 contains the results for the mean and variance, the number of students (N) taking the test is also shown and the test contained 20 questions.

As can be seen from Table 12, students correctly answer about 36% prior to being instructed in CFD and about 80% average correct for the post-test.

This represents a considerable improvement and is statistically highly significant, i.e.

$$\bar{x}_{post} - \bar{x}_{pre} = 8.52; \quad t(1,33) = 15.3; \quad p < 0.0001$$

It can be seen that the effect is substantial between pre- and post-tests and therefore represents significant improvement in outcomes of the students' knowledge and skills of CFD knowledge and skills. The students, after a relatively brief exposure to and with limited practice of CFD have shown considerable growth in their understanding of CFD concepts, principles and applied problems.

Question No.	Question
1	For flow over a cylinder, what is the cause of the different results found for CFD and in the experimental laboratory? a. The difference is caused by the experimental laboratory uncertainties. b. The difference is caused by the errors from numerical and experimental laboratory uncertainties. c. The difference is caused by the errors from numerical methods. d. The difference is caused by the errors from numerical, modelling and experimental laboratory uncertainties.
2	What is a typical CFD process? a. Geometry → Mesh → Properties → Models → Initiation → Verification → Sources → Numerics → Solver → Post-processing b. Mesh → Geometry → Properties → Models → Initiation → Sources → Numerics → Solver → Post-processing → Verification c. Geometry → Mesh → Properties → Models → Initiation → Sources → Numerics → Solver → Post-processing → Verification d. Geometry → Models → Mesh → Properties → Initiation → Sources → Numerics → Solver → Post-processing → Verification
3	What is the criterion for the validation of a CFD simulation by a laboratory experiment? a. If the difference between the CFD and experimental data is less than the convergence limit. b. If the difference between the CFD and experimental data is less than the CFD data uncertainties. c. If the difference between the CFD and experimental data is less than the experimental data uncertainties. d. If the difference between the CFD and experimental data is less than the combination of the experimental and CFD data uncertainties.

Table 11. Examples of test questions

Pre-test		Post-test		N
$\bar{x}_{pre}$	σ^2_{pre}	$\bar{x}_{post}$	σ^2_{post}	35
7.35	5.28	15.87	4.94	

Table 12. Mean number of correct answers (out of 20)

The success of this introduction could not be assessed using a student comparison performance in CFD laboratories across the years as these data were not available or not in a form that would make for meaningful comparison. This is not necessarily a weakness of the study as it has been suggested (Lucas, 1997) that the measurements of differences in student assessment over time has limited value given the changing nature of student cohorts from one year to another. On the basis of these recommendations a questionnaire was developed to elicit perceptions of the introduction of CFD from students involved in the class.

5.3. Online questionnaire

An anonymous online survey was conducted after students obtained their grades for the laboratory reports to aid formative evaluation of the introduction of CFD. Only students who had completed the course with CFD were surveyed. A questionnaire using 11 statements as listed in Table 13 was designed for this survey. Students were requested to respond to each item in the questionnaire using a five-point scale: strongly agree, agree, neutral, disagree and strongly disagree plus a column for no opinion. An opportunity was also provided for students to comment on their experience at the end of the questionnaire to collect qualitative feedback on their experience so far with CFD.

No.	Question/Statement
1	I found the software easy to use.
2	I have used CFD modelling before.
3	This CFD tool enhances my understanding of the theory course.
4	This CFD tool is a useful addition to the fluid mechanics laboratories.
5	The 'hands-on' aspects of the CFD tool has taught me extra skills.
6	The 'hands-on' aspects of the CFD tool has given me deeper knowledge of fluid mechanics.
7	On using CFD I have learned things that could not be taught through the theory or experimental courses.
8	I now have a knowledge of CFD procedures and methodology.
9	I feel I could now continue to model basic flows.
10	On completion of this course I have run at least one flow simulation with the software provided.
11	I would recommend the CFD laboratory to others.

Table 13. A list of questions/statements used in the survey for students' feedback

Generally, student feedback surveys have a very low response rate (Gamliel & Davidovitz, 2005; Nulty, 2008). However the response rate here was high (>80%) with 5 responses received per question and overall, the results from the survey were positive. The responses to the survey are shown on Fig. 7 and indicate that students felt that they benefited from their exposure to CFD. In additional comments most of the students expressed the view that the amount of material introduced was correct, although some felt that the exercises took a long time to complete correctly. Students were particularly appreciative that they could easily visualize flow using contour and vector plots and generally agreed that the combination of theory, experimental and CFD led to better understanding of fluid mechanics. Students also showed enthusiasm for learning more about CFD.

In addition to the questionnaire of Table 13, the students were asked "Would you recommend that CFD remains in the fluid mechanics course in the future?" To this 82% said yes so showing that they thought CFD as having a positive impact on their studies.

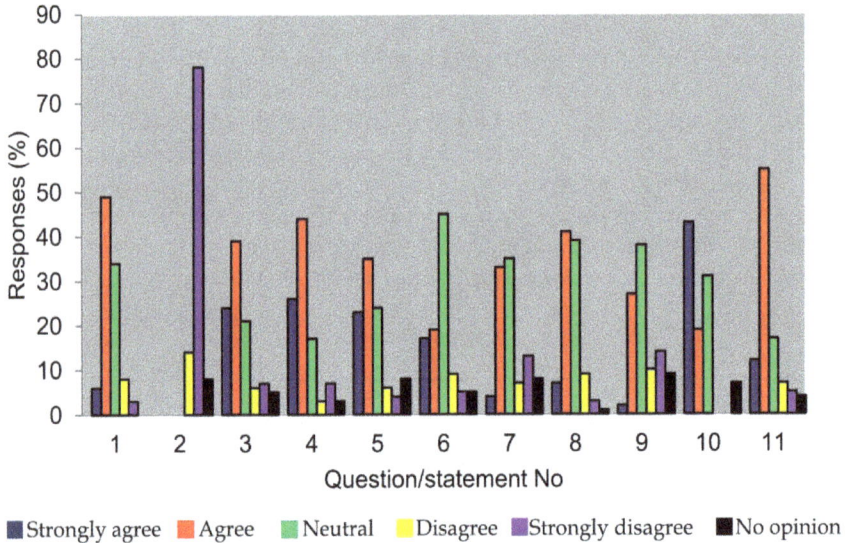

Figure 7. Chart showing survey results (N = 55)

It was noted that the students liked the hands-on and self-discovery approach, although at times some frustration was also noted. Once a demonstration was given there was only an interest to learn by themselves, back up when required by a Teaching Assistant's advice. The traditional view of CFD is that it has a steep learning curve, but with a structures CFD interface and with limited depth imposed it has been demonstrated that the gradient of the curve can be brought to an acceptable level.

Of course, during the skills training at this level, no real mention was made of code development, as the purpose was to develop users of the code only. This can be remedied by a later course which improves the student as a user and starts showing ways of writing new code for special conditions. Actually the software used here has a module which can translate simple instructions into FORTRAN. The concept to represent this software package or any other package as a black box should be remedied as soon as possible in later courses.

6. Concluding remarks

This paper has described the use and efficacy of integrating computational fluid dynamics into a traditional fluid mechanics course. The controlled experiment has shown that the inclusion of CFD laboratories gave students a better appreciation of fluid mechanics in general and the students gained better knowledge of simple concepts. However, the inclusion of CFD laboratories had a detrimental effect on interest when compared to the purely experimental control group and the control group also did better when considering the more difficult aspects of the course. It was found from the study of student knowledge and skill outcomes for the CFD interface that the students could cope with CFD reasonably well, provided the subject is introduced with care. One of the main reasons for the inclusion of CFD was to contribute to the teaching of professional practice skills to intermediate level undergraduate students. It was found that the interface design does provide students with hands-on experience, gained through an interactive and user-friendly environment, and encourages student self-learning. It was noted from the survey that the students liked the hands-on and self-discovery approach, although at times some frustration was also noted.

Author details

Desmond Adair
University of Tasmania, Australia

7. References

Aradag, S., Cohen, K., Seaer, C.A. & McLaughlin, T. (2009). Integration of Computations and Experiments for Flow Control Research with Undergraduate Students, *Computer Applications in Engineering Education*, Vol. 17, No. 1, pp. 1 -11.

Agonafer, D., Liao, D.G. & Spalding, B. (2008). The LVEL turbulence Model for Conjugate Heat Transfer at Low Reynolds Numbers, Concentration, Heat and Momentum Ltd., London, UK.

Assanis, D.N. & Heywood, J.B. (1986). Development and Use of a Computer Simulation of Turbocompounded Diesel System for Engine Performance and Component Heat Transfer Studies, *SAE Transactions*, Vol.2, pp. 451-476.

Campbell, J.O., Bourne, J.R., Mosterman, P.J., Nahni, M., Rassai, R., Brodersen, A.J. & Dawant, M. (2004). Cost-effective Distributed Learning with Electronics Labs, Journal of Asynchronous Learning Networks, Vol. 8, No. 3, pp. 5-10.

Cohen, J. (1988). *Statistical Power Analysis for the Behavioral Sciences* (2nd ed.). Hillsdale, NJ: Lawrence Earlbaum Associates.

Devenport, W.J. & Schetz, J.A. (1998). Boundary Layer Codes for Students in Java, *Proceedings of FEDSM'98, ASME Fluids Engineering Division*, Summer Meeting, Washington DC, June, pp. 21-25.

Feisel, L.D. & Rosa, A.J. (2002). The Role of the Laboratory in Undergraduate Engineering Education, *Journal of Engineering Education*, Vol. 91, No. 1, pp. 121-130.

Ferziger, J. H. & Peric, M. (1996). *Computational Methods for Fluid Dynamics*, Springer, New York.

Gamliel, E. & Davidovitz, L. (2005). Online Versus Traditional Teaching Evaluation: Mode Can Matter, Assessment & Evaluation in Higher Education, Vol. 30, No. 6, pp. 581- 592.

Hung, T.C., Wang, S.K., Tai, S.W. &Hung, C.T. (2005). An Innovative Improvement in Engineering Learning System using Computational Fluid Dynamics Concept, *Computer Applications in Engineering Education*, Vol. 13, No. 4, pp. 306-315.

Keller, C.J., Finkelstein, N.D., Perkins, K.K. & Pollock, S.J. ((2007). Assessing the Effectiveness of a Computer Simulation in Introductory Undergraduate Environments, *AIP Conference Proceedings*, Vol. 883, pp. 121-124.

Kelsey, R. (2001). Brownfield Action: An Integrated Environmental Science Simulation Experience for Undergraduates, *Annual Proceedings of Selected Research and Development*, Atlanta, GA, Nov. 8-12.

LaRoche, R.D., Hutchings, B.J. & Muralikrishnan, R. (2002). FlowLab: Computational Fluid Dynamics (CFD) Framework for Undergraduate Education, *Proceedings of the 2002 ASEE/SEFI/TUB Colloquium*, Berlin, Oct. 1-4.

Launder, B.E. & Spalding, B. (1972). *Mathematical Models of Turbulence*, Academic Press. London.

Likert, R. (1932). A Technique for the Measurement of Attitude, *Archives of Psychology*, Vol. 22, No. 140.

Lucas, U. (1997). Active Learning and Accounting Educators, *Accounting Education*, Vol. 6, pp. 189-190.

Nulty, D.D., The Adequacy of Response Rates to Outline and Paper Surveys: What can be Done? *Assessment & Evaluation in Higher Education*, Vol. 33, No. 3, pp. 301-314.

Patil, A., Mann, L, Howard, P. & Martin, F. (2009). Assessment of Hands-On Activities to Enhance Students' Learning in the First Year Engineering Skills Course, *20th Australasian Association for Engineering Education Conference*, Univ. Of Adelaide, 6-9 Dec., pp. 286-292.

Pfahl, D., Laitenberger, O., Ruhe, G., Dorsch, J. & Krivobokova, T. (2004). Evaluating the Learning Effectiveness of Using Simulations in Software Project Management Education: Results from a Twice Replicated Experiment, Information & Software Technology, Vol. 46, pp. 127-147.

Pieritz, R.A., Mendes, R., Da Sila, R.F.A.F. & Maliska, C.R. (2004). CFD: An educational Software Package for CFD Analysis and Design, *Computer Applications in Engineering Education*, Vol. 12, No. 1, pp. 20-30.

Qian, X. & Tinker, R. (2006). Molecular Dynamics Simulations of Chemical Reactions for Use in Education, *Journal of Chemical Education*, Vol. 81, No. 1, pp. 77-90.

Ray, J.W. & Shadish, W.R. (1996). How Interchangeable are Different Estimators of Effect Size? *Journal of Consulting and Clinical Psychology*, Vol. 64, pp.1316-1325. (see also: (1998) Correction to Ray and Shadish, *Journal of Consulting and Clinical Psychology*, Vol. 66, pp.532.)

Rozza, G., Huynh, D.B.P., Nguyen, N.C. & Patera, A.T. (2009). Real-Time Reliable Simulation of Heat Transfer Phenomena, *ASME-American Society of Mechanical Engineers – Heat Transfer Summer Conference*, Paper HT2009-88212, San Franscisco, USA, July 19-23.

Stern, F., Xing, T., Yarbrough, D.B., Rothmayer, A., Rajagopalan, G., Otta, S.P., Caughey, D., Bhaskaran, R., Smith, S., Hutchings, & B., Moeykens, S. (2006). Hands-On CFD Educational Interface for Engineering Courses and Laboratories, *Journal of Engineering Education*, Vol. 95, No. 1, pp. 63-83.

Tu, J., Yeoh, G. H. & Liu, C. (2007). *Computational Fluid Dynamics – A Practical Approach*, Elsevier, Amsterdam.

Zheng, H. & Keith, J.M. (2003). Web-Based Instructional Tools for Heat and Mass Transfer, *ASEE Conference Proceedings*, June.

Chapter 6

Numerical Simulation on Mechanical Ventilation- Two Case Studies with Different Operation Modes

Yiin-Kuen Fuh, Wei-Chi Huang and Jia-Cheng Ye

Additional information is available at the end of the chapter

1. Introduction

Effectiveness of CFD (Computational Fluid Dynamics) and the idea that the complex dynamics of fluid-related heat transfer and even fire might be studied numerically dates back to the significant improvement of the computer power. In fact, the fundamental conservation equations governing fluid dynamics, heat transfer, and combustion were initially developed in the field of applied physics over a century ago. However, practical mathematical models inherently complex in turbulence are relatively recent due to the explosive development of the Moore's law and computational speed [1].

The first case numerically studies the effect of ventilation rates and bathroom equipment locations on the odor removal efficiency in a bathroom model. One of the most frequently used domestic rooms is bathrooms; therefore, bathrooms are a high priority part of architectural design for many homeowners. The homeowners always want to solve bathroom problems that defecation and urination emit disgusting odors of all kinds, and Dols et al. [2] and Chung and Wang [3] have demonstrated that indoor air quality has a great effect on the productivity and health of human inhabitants. The reports of both of Sandberg et al. [4] and ASHRAE [5] present that indoor air pollutants are normally at higher concentrations than their outdoor counterparts, too. The general solution of bathroom design is the implementation of mechanical ventilation system.

In previous studies, Chung and Chiang [6] observed flow patterns and contaminate particle paths in lavatories by using a numerical simulation, and the diameter of the particle is less than 1 mm. They used a floor exhaust ventilation system for analysis, and the target is to improve the air quality of the lavatories' environment. Their simulation results present only 1.2% and 1.5% of the contaminate particles remain in the men's and women's rooms, respectively. With proper ventilation system, the bathroom may have higher efficiency of purging odors in the bathroom.

Sandberg [7] ,Chung and Hsu [8] defined ventilation efficiency, local air quality index and investigated related impacts of different ventilation patterns arranged by two inlet and outlet diffusers at different locations. In modern society, problem of floor exhaust vent is complex, and tradition ventilation system is mounted a small exhaust fan in the ceiling. Its main function is the suction of moisture air and odors. The air sucked into an exhaust pipe is taken out of house, and then the bathroom becomes negative pressure to let fresh air from outdoors flow into the bathroom.

According to the study of different ventilation patterns, Tung et al. [9] have a novel idea that a mechanical ceiling-supply and wall-exhaust ventilation system remove unpleasant moist air and odors. Although the design is a typical ceiling-exhaust ventilation system, it has a significant difference in the removal of the moisture and odors. The experimental result has showed that doubling the flow rate of supply air from 8.5 to 17 air changes per hour (ACH) leads to a moderate 4% increase in ORE, but a ventilation rate of 8.5 ACH rates is suitable for view of energy-saving. Our study aims to increase ventilation efficiency by numerically investigating different ACH rates and Fire Dynamics Simulator (FDS) [10] is used to analyze the effect of exhaust area. Deployment of local air quality (QI) and odor removal efficiency (ORE) metrics shows great potential to quantitatively evaluate the effectiveness of odor removal in the bathroom [11]. Contaminant source non-uniformity was found to have a great influence on the QI and ORE non-uniformity, with the two tested air distribution methods.

The second case study aims to numerically investigate the operation modes of mechanical ventilation in underground tunnels and compare the effectiveness of CFD with full-scale experiments and existing fire codes in Taiwan. In underground railroad tunnels and subway stations, direct exposure to fire is typically not the most immediate threat to passengers' lives. Instead, smoke inhalation due extreme heat and toxic gases is the main cause of fatality. For fire safety and hazard mitigation reasons, it is of great importance to investigate the mechanisms thermal-induced smoke motion. On the other hand, the current performance-based fire codes in Taiwan for special structures require full-scale tests [12] to experimentally investigate the effectiveness of the smoke control systems in subway stations before opening for public service. One key feature of fire-induced phenomena is called chimney or stack effect [13], which plays significant role in the smoke control of subway stations due to various structures of vertical spaces exist such that the buoyancy force of hot smoke will be greatly enhanced.

Due to land scarcity in greater Taipei area, continuously conversion from existing railroad to underground are performed to increase the land utility at different phases and span many years. In addition, integration issues between several transportation systems such as mass transit and high speed rail are also of great concerns. Fourth phase of Nangkang North Tunnel was completed and in full operation in September, 2009. During the construction period, fire codes in Taiwan for special constructions (e.g. underground stations) are also going through various revisions while internationally, new regulations such as AS 4391 in

Australia [14] and NFPA130 in USA [15] have been validated and updated. In order to verify the compliance of fire codes nationally and internationally, full scale experiments have been performed before actual operation with emphasis on various operation modes of mechanical ventilation. This paper aims to numerically investigate and compare the effectiveness of CFD with experimental results. Typical road tunnel fire with different causes and kinds of fire were numerically analyzed before and the temperature and visibility of tunnel zone were calculated [16]. The simulation results show that the highest central axis section reaches to 1000°C, and the highest temperature of side wall reaches to 989°C under large fire scales. [17] Furthermore, other tunnel fires are also of great concern due to catastrophic fatality and numerically simulated such as Howard Street Tunnel Fire [17], tunnel fire and tunnel lining failure [18], road tunnels and ventilation effects [19]. On the other hand, smoke control of fires in subway stations and stack effects on smoke propagation are simulated and compared with experiments with great accuracy [20, 21]. Ventilation strategies and application to fire in a long tunnel has been considered [22] and ordinary and emergency ventilations are operated by using dedicated supply and extraction ducts. Another key parameter for tunnel fires is the critical speed and simulation results show a good correlation with real fires [23].

According to current fire codes in Taiwan, a recent version which includes performance-based tests is stipulated in 2008 [12] for the purpose of aiding a complete implementation of fire equipments in underground tunnels and subway stations. In chapter 3 of the fire code, one exception is that simulation study is allowed for the difficult-to-test circumstances, but need to be examined by a special committee on a case-by-case basis. In particular, ventilation is a crucial element while managing fire events in tunnels and control strategy for fire-induced smoke is of great concern. Adoption of numerical simulation a validation tool is probably pioneered in NFPA130 2003, version [15] emergency smoke management session of 7.1.2.4. The first full-scale experiment in Taiwan is the Banciao station in 1999 and the actual tests also laid a foundation of performance-based fire codes for future implementation [24]. In what follows various subway stations [25-27] necessitate the full-scale tests and several reports have been reviewed and published before opening to the public service.

2. Theoretical background

In this section, we briefly present the governing equations of FDS and an outline of the general solution procedure which we used for all the simulations. Details of the individual equations are described in original technical guide [1]. The governing equations are essentially a complete set of partial differential equations with appropriate simplifications and approximations noted. The numerical method consists of a finite difference approximation of the governing equations and a procedure for updating these equations in time [1].

2.1. Secondary heading, left justified

The basic conservation equations for mass, momentum and energy for a Newtonian fluid can be found in almost any textbook on fluid dynamics or CFD. FDS adopts a description of the equations, the notation used, and the various approximations employed in Anderson et al. [28]. Note that this is a set of partial differential equations consisting of six equations for six unknowns, all functions of three spatial dimensions and time: the density r, the three components of velocity u = [u;v;w]^T , the temperature T, and the pressure p.

2.1.1. Mass and species transport

Mass conservation can be expressed either in terms of the density, ϱ,

$$\frac{\partial \rho}{\partial t} + \nabla \cdot \rho u = \dot{m}_b^{'''} \tag{1}$$

Here $\dot{m}_b^{'''}$ is the mass production rate by evaporating droplets/particles.

2.1.2. Momentum transport

The momentum equation in conservative form is written:

$$\frac{\partial}{\partial t}(\rho u) + \nabla \cdot \rho u u + \nabla p = \rho g + f_b + \nabla \cdot \tau_{ij} \tag{2}$$

The term uu is a diadic tensor. In matrix notation, with u = [u;v;w]^T , the diadic is given by the tensor product of the vectors u and u^T . The term ∇. ρuu is thus a vector formed by applying the vector operator $\nabla = \left(\dfrac{\partial}{\partial x}, \dfrac{\partial}{\partial y}, \dfrac{\partial}{\partial z} \right)$ to the tensor. The force term f_b in the momentum equation represents external forces such as the drag exerted by liquid droplets. The stress tensor τ_{ij} is defined:

$$\tau_{ij} = \mu \left(2S_{ij} - \frac{2}{3}\delta_{ij}(\nabla \cdot u) \right); \; \delta_{ij} = \begin{cases} 1 \; i=j \\ 0 \; i \neq j \end{cases}; S_{ij}\frac{1}{2}\left(\frac{\partial u_i}{\partial x_j} + \frac{\partial u_j}{\partial x_i} \right) \quad i,j = 1,2,3 \tag{3}$$

The term S_{ij} is the symmetric rate-of-strain tensor, written using conventional tensor notation. The symbol μ is the dynamic viscosity of the fluid.

2.1.3. Energy transport

The energy conservation equation is written in terms of the sensible enthalpy, h_s :

$$\frac{\partial}{\partial t}(\rho h_s) + \nabla \cdot \rho h_s u = \frac{Dp}{Dt} + \dot{q}^{'''} - \dot{q}_b^{'''} - \nabla \cdot \dot{q}^{''} + \varepsilon \tag{4}$$

The sensible enthalpy is a function of the temperature:

$$h_s = \sum_\alpha Y_\alpha h_{s,\alpha} \; ; \; h_{s,\alpha}(T) = \int_{T_0}^{T} c_{p,\alpha}(T')dT' \tag{5}$$

Note the use of the material derivative, $D(\ )/Dt = \partial(\ )/\partial t + u \cdot \nabla(\ )$. The term $\dot{q}'''$ is the heat release rate per unit volume from a chemical reaction. The term $\dot{q}_b'''$ is the energy transferred to the evaporating droplets.

The term $\dot{q}''$ represents the conductive and radiative heat fluxes:¶

$$\dot{q}'' = -k\nabla T - \sum_{\alpha} h_{s,\alpha} \rho D_\alpha \nabla Y_\alpha + \dot{q}_r'' \tag{6}$$

where k is the thermal conductivity.

2.1.4. Equation of state

An approximate form of the Navier-Stokes equations appropriate for low Mach number applications is used in the model.

$$p = \frac{\rho RT}{\overline{W}} \tag{7}$$

Here $\overline{W}$ is the molecular weight of the gas mixture.

2.2. Solution procedure

FDS uses a second-order accurate finite-difference approximation to the governing equations on a series of connected recti-linear meshes. The flow variables are updated in time using an explicit second-order Runge-Kutta scheme. The typical solution algorithm is used as the following to advance in time the density, species mass fractions, velocity components, and background and perturbation pressure. Let $\rho^n, Y_\alpha^n, u^n, \overline{p}_m^n$ and H^n denote these variables at the nth time step. [1]

1. Compute the "average" velocity field $\overline{u}^n$.
2. Estimate ρ, Y_α and $\overline{p}_m$ at the next time step with an explicit Euler step. For example, the density is estimated by

$$\frac{\rho^* - \rho^n}{\delta t} + \nabla \cdot \rho^n \overline{u}^n = 0. \tag{8}$$

3. Exchange values of ρ^* and Y_α^* at mesh boundaries.
4. Apply boundary conditions for ρ^* and Y_α^*
5. Compute the divergence, $\nabla \cdot \overline{u}^*$, using the estimated thermodynamic quantities.
6. Solve the Poisson equation for the pressure fluctuation with a direct solver on each individual mesh:

$$\nabla^2 H^n = -\left[\frac{\nabla \cdot u^* - \nabla \cdot \overline{u}^n}{\delta t}\right] - \nabla \cdot \overline{F}^n \tag{9}$$

7. Estimate the velocity at the next time step

$$\frac{u^* - \bar{u}^n}{\delta t} + \bar{F}^n + \nabla H^n = 0 \tag{10}$$

8. Check the time step at this point to ensure that the stability criteria are satisfied

$$\delta t \max\left(\frac{|u|}{\delta x}, \frac{|v|}{\delta y}, \frac{|w|}{\delta z}\right) < 1 \; ; \; 2 \, \delta t \, v \left(\frac{1}{\delta x^2} + \frac{1}{\delta y^2} + \frac{1}{\delta z^2}\right) < 1 \tag{11}$$

This concludes the "Predictor" stage of the time step. Next the "corrector" stage follows:

1. Compute the "average" velocity field $\bar{u}^*$.
2. Apply the second part of the Runge-Kutta update to the mass variables. For example, the density is corrected

$$\frac{u^* - \bar{u}^n}{\delta t} + \bar{F}^n + \nabla H^n = 0 \tag{12}$$

3. Exchange values of ρ^n and Y_α^n at mesh boundaries.
4. Apply boundary conditions for ρ^n and Y_α^n.
5. Compute the divergence $\nabla \cdot u^{n+1}$ from the corrected thermodynamic quantities.
6. Compute the pressure fluctuation using estimated quantities

$$\nabla^2 H^* = -\left[\frac{\nabla \cdot u^{n+1} - \frac{1}{2}\left(\nabla \cdot \bar{u}^* + \nabla \cdot \bar{u}^n\right)}{\delta t / 2}\right] - \nabla \cdot \bar{F}^*. \tag{13}$$

7. Update the velocity via the second part of the Runge-Kutta scheme

$$\frac{u^{n+1} - \frac{1}{2}\left(\bar{u}^* + \bar{u}^n\right)}{\delta t / 2} + \bar{F}^* + \nabla H^* = 0 \tag{14}$$

8. At the conclusion of the time step, values of H^* and the components of u^{n+1} are exchanged at mesh boundaries via MPI calls.

The details of the predictor-corrector scheme can be found in [1].

3. Mechanical ventilation simulation - Case1

We use FDS to build a virtual-model bathroom dimension which is 2.36 m (length), 2.22 m (width), and 2.36 m (height), shown in Fig. 1(a). Bathroom facilities included a bathtub, a sink, and a toilet in the bathroom model. The red, blue, and purple model represents a bathtub, a sink, and a toilet, respectively. In the ceiling, the red square replaces the supply air vent (a mechanical fan), and dimensions are 0.2 m (length) and 0.4 m (width). On the wall, the red square replaces the exhaust air vent (a free opening), and dimensions are 0.2 m and 0.4 (length & height), respectively. The exhaust airflow rate was 1.15 times the supply airflow rate.

3.1. Toilet position, ACH rate, and exhaust vent area

Malodorous volatile is frequently manufactured from human feces and urine in bathrooms. The odor source is created on the virtual toilet of bathroom model. This paper follows previous study [9] that assume different distances, which is the distance between the odor source and the wall-exhaust vent. As shown in Fig. 1(b), there are cases of three different distances contained 0.8 m (Case 1), 1.1 m (Case 2), and 2.05 m (Case 3) in this study with ventilation condition of 8.5 Air Change per Hour (ACH).

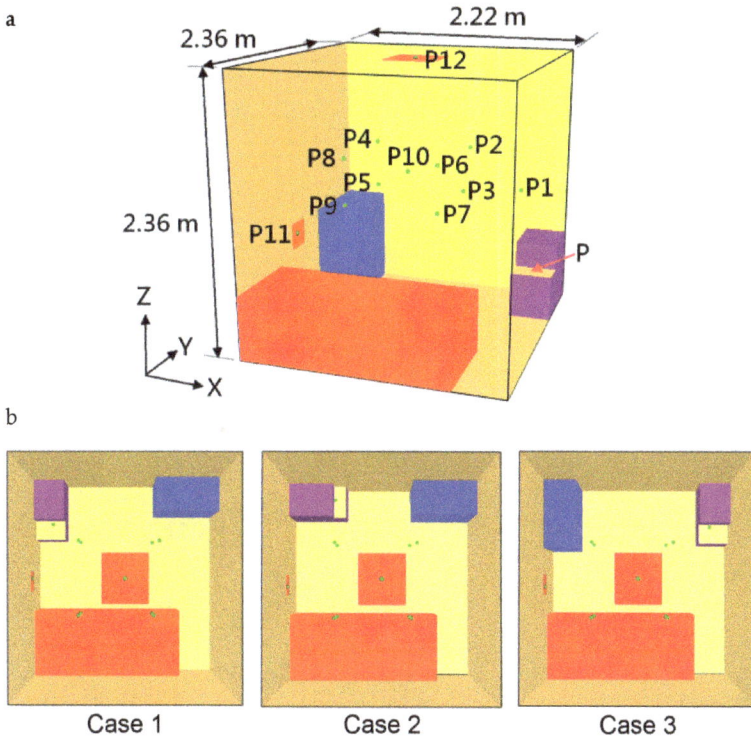

Figure 1. (a) A mock-up bathroom used to analyze the influence of ventilation rates 6.5, 8.5,17, and 24 ACH on the odor removal and shows numerically measured locations of sampling points [9]. (b) In three cases, it reveals related position of bathroom facilities.

According to experiment of the previous study, our study set some issues of four different ACH rates in case 3, and they divide 6.5, 8.5, 17, and 24 ACH rates. Besides, we also change area of wall-exhaust vents contained two cases. One case is 0.24 m (length) and 0.24 m (height), the other is 0.24 m (length) and 0.48 m (height).

3.2. Set concentration sensor position and produce odor source position

By setting a total of 12 concentration sensors, we can gain odor distribution data at specific positions. Odor flow pattern can be obtained and Table 1 shows positions of collected data from P1 to P12. Marked as P and P1–P12 in Fig. 3, ten measuring points (P1–P10) within the bathroom were selected to monitor the concentration levels of odors at these locations, and two extra measuring points were setup outside the room, one (P11) behind the exhaust air vent and another (P12) in front of the supply air vent; moreover, P was the location of a source of odors generated by a person sitting on the toilet in the bathroom.

Sampling location	X-direction (m)	Y-direction (m)	Z-direction (m)
P1			
Case 1	0.30	1.78	1.16
Case 2	0.59	2.08	1.16
Case 3	1.90	1.78	1.16
P2	1.42	1.58	1.56
P3	1.42	1.58	1.16
P4	0.61	1.58	1.56
P5	0.61	1.58	1.16
P6	1.42	0.79	1.56
P7	1.42	0.79	1.16
P8	0.61	0.79	1.56
P9	0.61	0.79	1.16
P10	1.04	1.17	1.40
P11	0.00	1.12	0.76
P12	1.11	1.18	2.36
Oder emission			
P			
Case 1	0.50	2.07	0.40
Case 2	0.50	2.07	0.40
Case 3	1.90	1.86	0.40

Table 1. Locations of sampling points and odor emission from experimental data [9].

3.3. Local air quality index

This study followed the same definition and employed a local air quality index to describe the ventilation system's efficiency in removing odors [9, 29, 30]. A higher score on the local air quality index indicates a better ventilation efficiency due to a lower concentration level. The local air quality index, QI, is defined as follows:

$$QI = \frac{C_e - C_s}{C_p - C_s} \qquad (15)$$

where C_e and C_s are, respectively, concentrations of odors at exhaust and supply air vents. C_p is the concentration of odors at any place P in the bathroom.

3.4. Odor removal efficiency

Again the same terminology of odor removal efficiency was used to express the ventilation efficiency of the whole bathroom [9, 30, 31]. A lower index indicated a greater difficulty in the removal of odors. The odor removal efficiency, ORE, is defined as follows:

$$ORE = \frac{C_e - C_s}{C_b - C_s} \qquad (16)$$

where C_e and C_s are concentrations of odors at exhaust and supply air vents, respectively, and C_b is the average concentration of odors in the bathroom.

3.5. Effects of ventilation rates

As illustrated in Fig. 2(a), the concentration of odors at the 12 observing points (P1–P12) was examined. In these four cases, P1 revealed that odor concentration higher than odor concentration of other pieces points in 6.5 ACH rates, since it's the shortest distance from toilet. We have found the effect of high temperature led to thermally induced buoyancy such that particle concentration of the higher levels observing points are smaller than the lower level counterparts. Points P2, P4, P6, and P8 were positioned under the ceiling-supply vent and at a relative height of 0.66 of the room height [9]; hence, their concentrations were more than the concentration of points P3, P5, P7, and P9. Numerically, the concentration profile of odors in the cases of 8.5, 17 and 24 ACH was similar to the one in the case of 6.5 ACH and this result is in agreement to experiments [9]. Similar agreement between simulation and experiment is found that higher room ventilation rates resulted in lower absolute concentration levels of odors.

As shown in Fig. 2(b), it was numerically observed that changes in the room ventilation rate affected QI in a significant way at 10 points in the bathroom. All numerically monitored points in the case of 6.5 ACH had a smaller QI than those same points in the cases of 8.5, 17, and 24 ACH. It was noted that the experimentally observed QI values were less than 1.0 at many points with low ventilation rate as compared to numerical values of 1.2. Small discrepancy exist while the same conclusion can be drawn that the ventilation rate of 6.5 ACH is not suitable the bathroom ventilation system. Both numerical and experimental results suggest the cases of 8.5, 17, and 24 ACH had better ventilation efficiency due to QI levels were more than 1.2. Furthermore, a higher ventilation rate resulted in less spreading of odors and improved the local air quality in the bathroom.

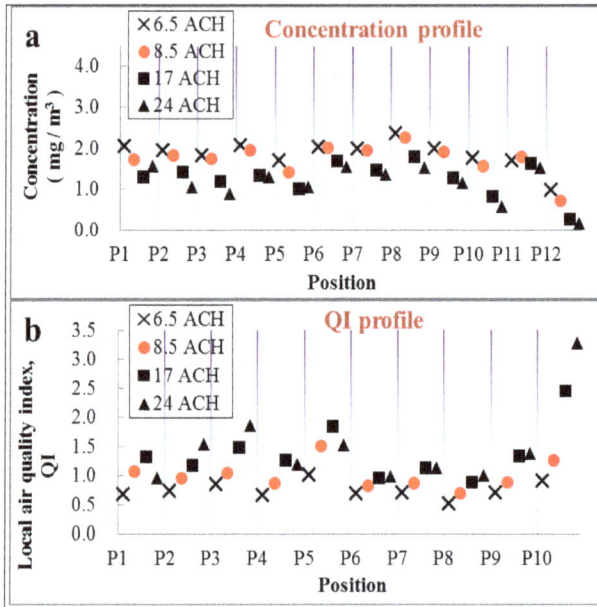

Figure 2. Effect of ventilation rates on (a) Concentration profile and (b) QI profile

3.6. Effects of doubling exhaust-vent area

As illustrated in Fig. 3(a), the concentration of odors at the 12 observing points (P1–P12) was examined. In these four cases of doubling exhaust-vent area, P1 revealed that odor concentration higher than odor concentration of other pieces points in 6.5 ACH rates, since it's the shortest distance from toilet to there. We have found the effect of high temperature led to thermally induced buoyancy such that particle concentration of the higher levels observing points is smaller than the lower level counterparts. Points P2, P4, P6, and P8 were positioned under the ceiling-supply vent and at a relative height of 0.66 of the room height [9] ; hence, their concentration was more than the concentration of points P3, P5, P7, and P9, which were positioned at a relative height of 0.49. As shown in Fig. 3(b), it was numerically observed that the QI values were less than 1.3 at many points with this low ventilation rate, further improvement from 1.2 when doubling the exhaust-vent area.

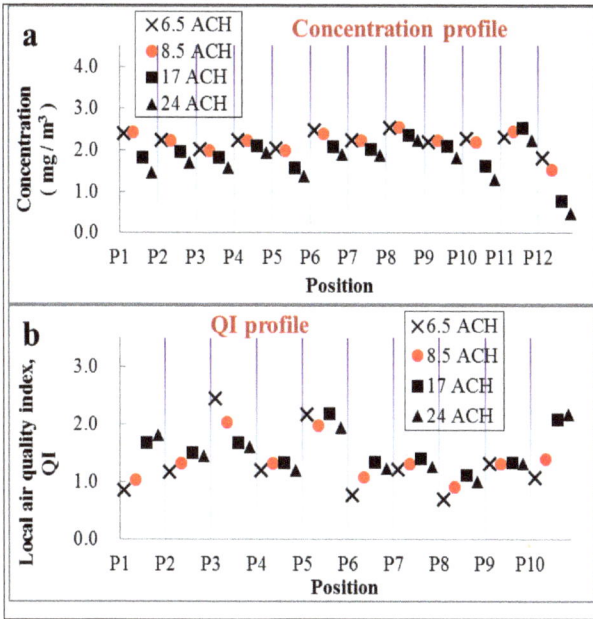

Figure 3. Effect of doubling exhaust-vent area on (a) Concentration profile and (b) QI profile.

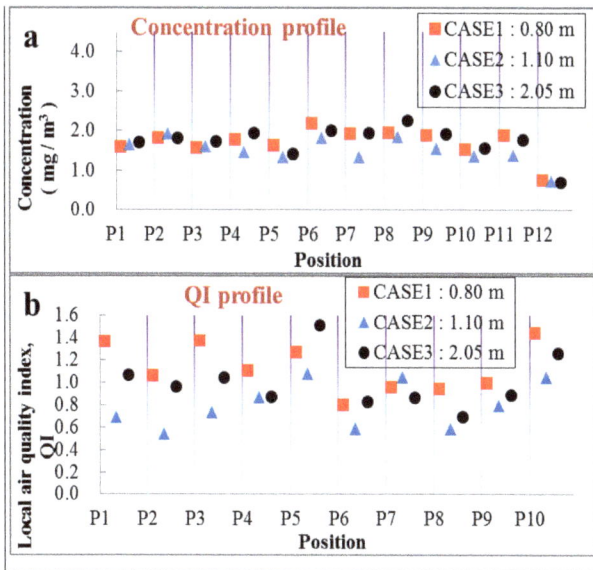

Figure 4. Effect of locations of a toilet on (a) Concentration profile and (b) QI profile.

3.7. Effects of location of the toilet

The present study numerically investigated three different locations of the toilet in the bathroom and compared with experiments [9]. The distance from the exhaust air vent to the toilet (L) was 0.8, 1.1, and 2.05 m for Cases 1, 2, and 3, respectively. In all three cases, as seen in Fig. 4(a), point P1 reported the highest concentration of odors because this was the nearest location to the source of odors, which is in good agreement with reported data [9]. Similarly, the odor concentration at P1 in Cases 1 and 2 was lower than the concentration in Case 3 due to the fact that toilet in Cases 1 and 2 was in close proximity to the exhaust air vent than in Case 3 and thus had lower odor concentration than Case 3.

Fig. 4(b) shows calculation results that changes in L affected QI at 10 points in the bathroom. It was observed numerically that most observed points in Case 1 had the highest QI, and that their QI levels were more than 0.6 [9]. In summary, Case 1 had the best ventilation efficiency among the three cases, which is in consistent with measured data.

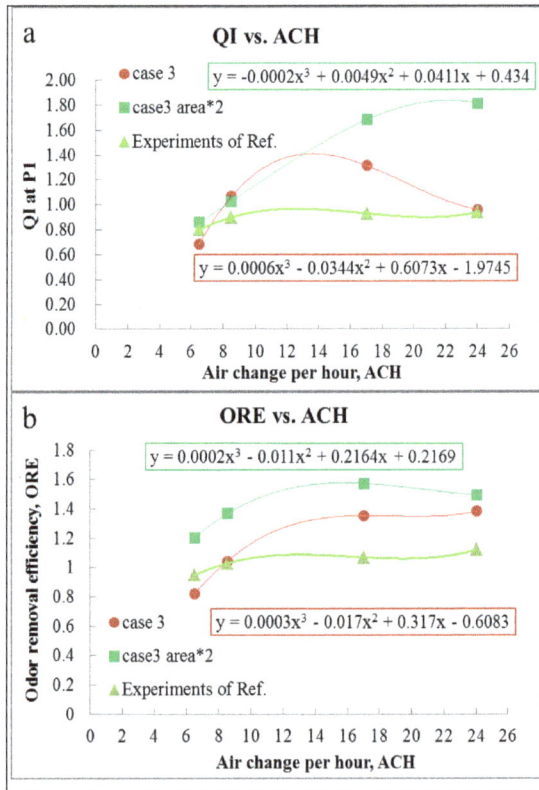

Figure 5. (a) Effectiveness analysis of QI at various ACH and comparison with experiments [9]. (b) Effectiveness analysis of ORE at various ACH and comparison with experiments [9].

3.8. Air change rate and odor removal analysis

A curve fitting performed using 3 order polynomial with numerically simulation's data. Fig. 5(a) show QI values of P1 in the different ACH rates and the curve fitting these QI values. The curve and point plotted in figure present relation between the different ACH rates and QI, and they illustrate trend that what ACH rates have the best QI value. The light green curve, red curve, and green curve are the experimental data from previous study [9], data in case 3, and data with doubling exhaust-vent area in case 3. The red curve is similar to the light green one, and the trends of both are very near. The Green curve has the translation because the mesh of numerical simulations is not fine enough and turbulent flow partly results in this impact.

Fig. 5(b) shows ORE values in the different ACH rates and the curve fitting ORE values. The curve and point plotted in figure present relation between the different ACH rates and ORE, and they illustrate trend that what ACH rates have the best ORE value. The curve and point plotted in figure present relation between the different ACH rates and QI, and they illustrate trend that what ACH rates have the best QI value. The light green curve, red curve, and green curve are the experimental data from previous study [9], data in case 3, and data with doubling exhaust-vent area in case 3. The red curve and green are similar to the light green one, and the trends are very similar. The ORE value with doubling area of exhaust vent is higher other ORE value in any ACH rates. Doubling area of exhaust vents from 0.0576 m2 to 0.1152 m2 leads to a moderate 30% increase in ORE in 8.5 ACH rates. At the same condition, it leads to a moderate 16% increase in ORE in 17 ACH rates.

4. Mechanical ventilation simulation- Case2

Smoke accumulation and migration are two main fire hazards in tunnel fires. The worst case scenario happens when the mechanical ventilation fails. Therefore, how to design a fail-safe system in underground station is of great importance. Typically, effective egress time is calculated as the smoke layer descends till 1.8 meter above the ground. In this study, we simultaneously compare the experiments and CFD simulations regarding two different fire loads of 1 and 5MW for 6 minutes and 20 minutes, respectively.

Fire dynamics simulator, FDS 6.0 [10] is a computational fluid dynamics model of fire-driven fluid flow. The software solves numerically a tform of the Navier-Stokes equations appropriate for low-speed, thermally-driven flow, with an emphasis on smoke and heat transport from fires. Figure 6 shows the schematics of underground railroad station constructed using FDS software. It is noted that two underground levels, B1 and B2, are simulated based on actual dimensions. Moreover, two rectangular openings in 4m*5m*0.3m shown in Fig.1 are used to represent the escalator positions between two underground levels. Total dimensions of 16m*130m*10m space are simulated with a total of more than 3 million computational grids in FDS model. Underground station is shown in blue colour with dimension of 8m*130m*1m while the wall thickness between B1 and B2 is 30cm (as shown in red). Experimentally, two different fire loads of 1 and 5 MW are ignited at the centre of underground station using the standard methods [14, 15] and the smoke descending positions are measured across the station area at 6 and 20 minutes, respectively.

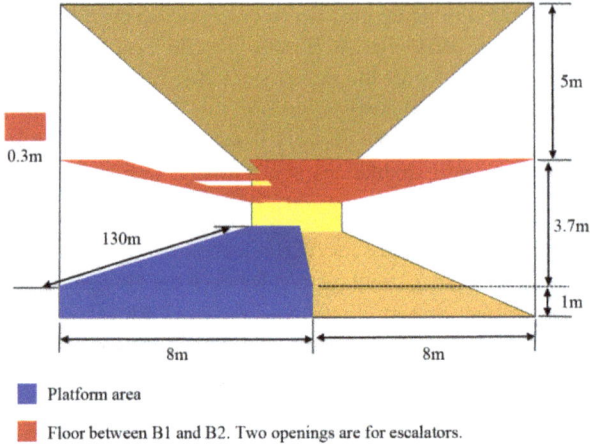

Figure 6. Schematics of underground railroad station constructed using FDS software. Two underground levels, B1 and B2, are simulated and two rectangular openings are used to represent the escalator positions between two underground levels.

4.1. Operation mode 1, all mechanical ventilations are fail or idle

Numerically, smoke patterns under fire load of 1 and 5 MW at time span of 360 and 1200 seconds are shown in Fig 7-8, respectively. It is noted that all mechanical ventilations are set to be in-operation to simulate the worst case scenario of mechanical failures. The particle function in FDS is used as a representation of smoke patterns. According to the fire code of NFPA130, 4 minutes of egression time is required and the key is to avoid smoke descending below 1.8m. For both fire loads at 1 and 5 MW, we can observe that smoke particles occupy less than 5% in area in terms of 1.8m height, especially in the escalator positions. It is considered that the simulation results in compliance with NFPA130 and agrees well with full scale experiments. Accordingly, 20 minutes of smoke accumulation and migration due to forced convection for both 1 and 5 MW fires will result in 50% and 99% smoke layers below 1.8m in escalator areas, which are considered hazardous for egression and unable to escape.

We can summarize the simulation results and compare with previous experiments for 1MW fire load after 6 minutes in table 2. Generally speaking, simulations results are in a good agreement with experimental observations. For the experiment, only visual observations by the inspectors may not be able to detect accumulation of smoke layer, especially when the smoke in little in volume and transparency optically. However, computer simulations enable us to numerically visualize the smoke at any location and scientifically calculate any vital information required. In this case, smoke layer height is of great importance due to egress needs, we can calculate tiny smoke accumulation in the areas located at as front and back station. Numerically, we can find less than 5% in area for smoke height<1.8m at both front and back station and the reason is because the escalator/stair induced chimney effect. Overall, both simulation and experiment are in good agreement and according to the fire code of NFPA130, 4 minutes of egression time is satisfied with 50% allowance time before smoke descending below 1.8m.

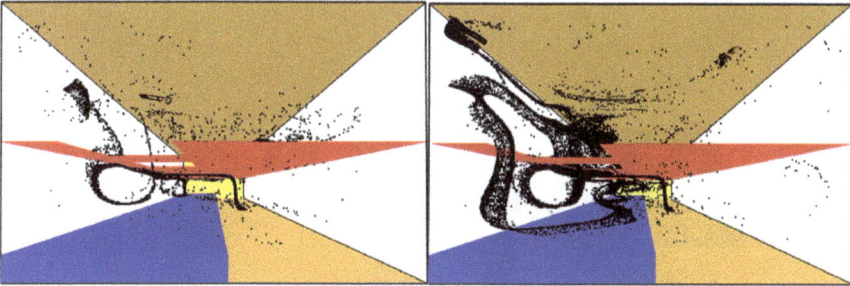

Figure 7. Smoke patterns under fire of 1MW at (a) 360s and (b) 1200s, respectively

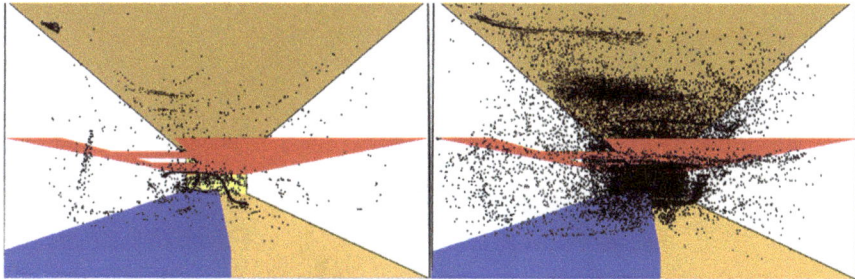

Figure 8. Smoke patterns under fire of 5MW at (a) 360s and (b) 1200s, respectively

Area \ Location		Front (5m from walls)	Center	Back (5m from walls)
Simulation	Tunnel	<5% in area for smoke height<1.8m	Higher than 1.8m	Higher than 1.8m
	Station	<5% in area for smoke height<1.8m	Higher than 1.8m	<5% in area for smoke height<1.8m
Experiment [27]	Tunnel	Higher than 1.8m	Higher than 1.8m	Higher than 1.8m
	Station	Higher than 1.8m	Higher than 1.8m	Higher than 1.8m

Table 2. Comparison of simulation results and experimental observations.

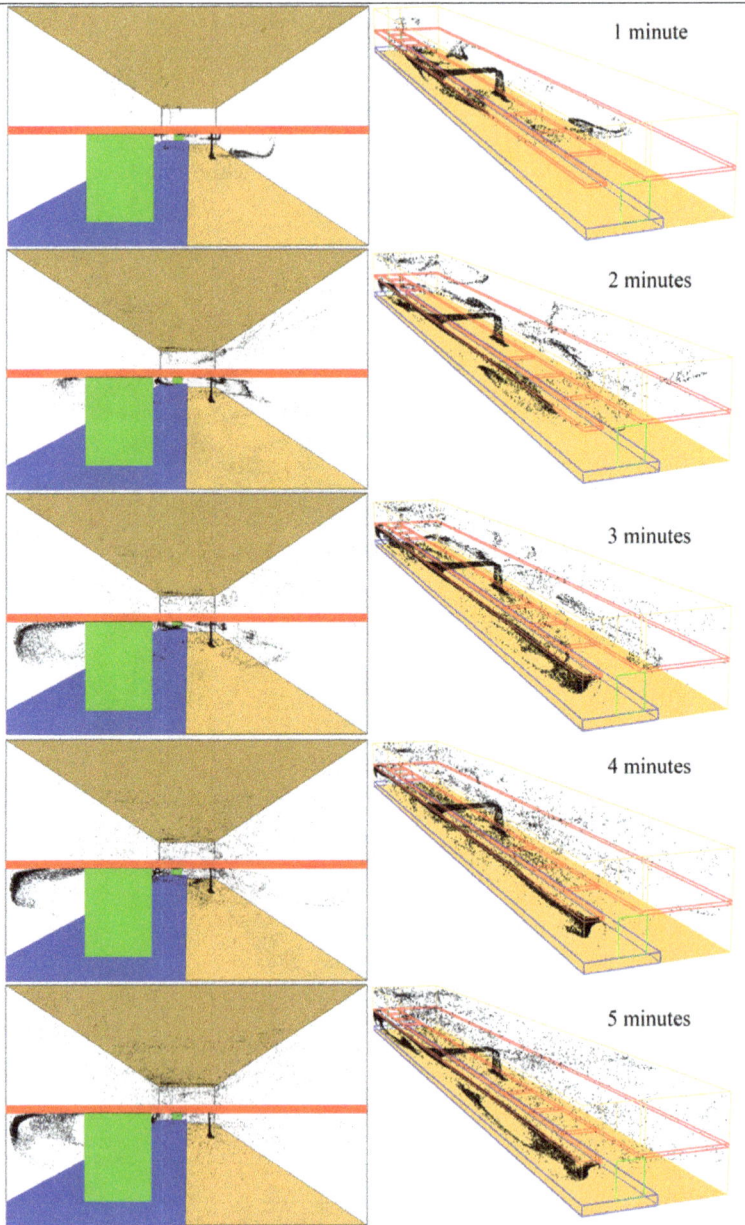

Figure 9. Two mechanical fans located on both sides of underground station B2 as indicated in green colour. Fire load is individually 1MW and the time is 1~6 minutes. Both front (left) and parametric (right) views are presented for comparison of smoke management effects.

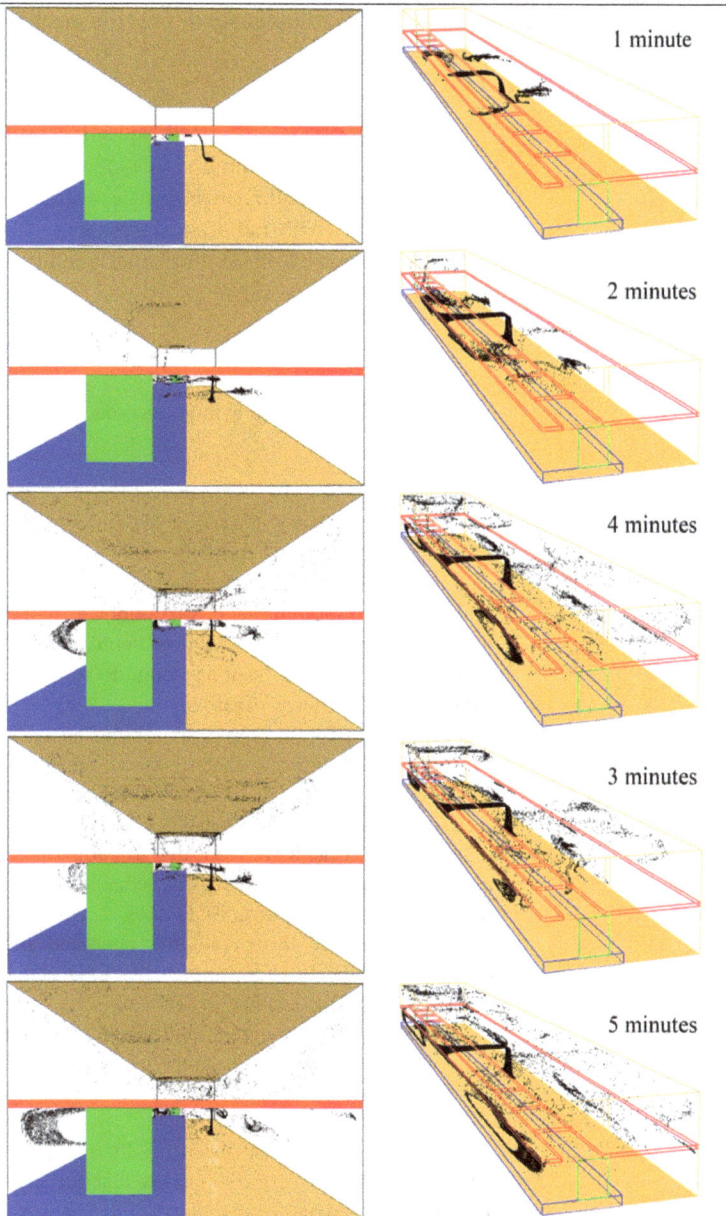

Figure 10. Two mechanical fans located on both sides of underground station B2 as indicated in green colour. Fire load is individually 5MW and the time is 1~6 minutes. Both front (left) and parametric (right) views are presented for comparison of smoke management effects.

4.2. Operation mode 2, two fans operate simultaneously to provide mechanical ventilations

Two mechanical fans (volume flow rate at 525m³/sec) located on both sides of underground station B2 as indicated in green colour in Fig.9-10. Fire load are 1MW and 5MW and the time is 1~6 minutes. It is simulated that smoke can be greatly diffused due to fan-driven flows, resulting in less than 1% area of 1.8m smoke height and relatively safe egress mode after 6 minutes of 1MW fire. Simulation results agree with full scale observations [27] where located fans near escalator area would effectively re-distribute the thermally-induced smoke to the overall ceilings before accumulating in the escape route.

5. Conclusion

From the simulations on mechanical ventilation of previous sections, the following conclusions can be drawn:

1. The first case numerically studies the effect of ventilation rates and bathroom equipment locations on the odor removal efficiency in a modeled bathroom. Our study agrees well with the results of a previously published study [9].
2. Qualitatively, numerical calculations result in similar trend such that the influence of ventilation rates and toilet locations on the odor removal is consistent with test data in a mock-up bathroom. Similar conclusion is drawn that higher ventilation rates and shorter distances between the toilet and exhaust air vent are found to be more capable of raising local air quality and the removal efficiency of odors.
3. Quantitatively, two indexes of QI and ORE, are employed for the analysis of the ventilation effect. The experimental results indicate that doubling the flow rate of supply air from 8.5 to 17 ACH leads to a moderate 4% increase in ORE, while the result of our numerical study is found that doubling area of exhaust vents leads to a moderate 30% increase in ORE in 8.5 ACH rates. That is, if we want design bathroom from the point of view's energy-saving, the new idea is that increasing exhaust-vent area also arise more efficient of ventilation system than enhancing ACH rates.
4. The second case study aims to numerically investigate the operation modes of mechanical ventilation in underground tunnels and compare the CFD effectiveness with full-scale experiments and existing fire codes in Taiwan. We numerically investigate the effectiveness of the smoke control systems via mechanical ventilation. In particular, focus is placed on safe egress time in different operation modes and comparison with full scale experiments as well as fire codes in Taiwan. The following conclusions can be drawn from simulation results.
5. Fire scenarios are very complex phenomena and these are especially true for the cases of underground channels and subway stations. It is simulated that for the case of Nangkang North Tunnel in Taipei area, the worst case scenario of mechanical failures in ventilation systems will still hold the egress time of six minutes or more under fire loads of 1 and 5MW, respectively. The conservative design criterion is in compliance with NFPA 130 and domestic fire codes in Taiwan. Moreover, it is imperative to

perform full-scale experiment for special structures such as underground station, i.e., performance-based fire codes and our simulation results agree well experiments.

6. Current design of underground station in Taiwan relies heavily on the smoke storage space to extend the egress time while the mechanical ventilation systems have only a minor contribution in smoke management. On the other hand, mechanical ventilation systems can play a more significant role in rescue mission after successful personnel evacuation. For example, numerical results indicate that smoke chimney effect can be effectively dispersed through mechanical ventilation to both safeguard firefighters and rescue members.

7. In terms of smoke control strategy, transient simulations are imperative in order to test the fire codes of various parametric values and different scenarios. Transient simulations allow one to check not only the time required to confine the smoke front, but also the extension of all the surrounding environment which can be potentially occupied by smoke. Educationally, it is also a very good teaching tool to train the firefighters and managers in order to construct a better rescue strategy.

Author details

Yiin-Kuen Fuh, Wei-Chi Huang and Jia-Cheng Ye
National Central University, Department of Mechanical Engineering & Institute of Energy, Taiwan, ROC

Acknowledgement

Thanks for the National Institute of Standards and Technology develop Fire Dynamics Simulator and Smokeview (FDS-SMV).

6. References

[1] NIST Special Publication 1018-5, Fire Dynamics Simulator (Version 5) Technical Reference Guide.

[2] W.S. Dols, A.K. Persily, S.J. Nabinger: IAQ '92, Environments for People (1992), p.85.

[3] K.C. Chung, S.K. Wang: Indoor Environment Vol. 3 (1994), p.149.

[4] M. Sandberg, C. Blomqvist: ASHRAE Transaction Vol. 95 (1989), p.1041.

[5] ASHRAE: fundamental handbooks. Atlanta, GA: *American Society of Heating, Refrigerating, Air-Conditioning*, Inc.(2005).

[6] K.C. Chung, C.M. Chiang, W.A. Wang: Building and Environment Vol.32, No. 2(1997), p.149.

[7] Sandberg M. What is ventilation e1ciency? Building and Environment 1981;16(2):123–35.

[8] K.C. Chung, S.P. Hsu: Building and Environment Vol.36 (2001), p.989.

[9] Y.C. Tung, S.C. Hu, T.Y. Tsai: Building and Environment Vol. 44 (2009), p.1810.

[10] Fire dynamics simulators, http://fire.nist.gov/fds/, accessed in Feb. 14, 2010

[11] K. Hagstrom, A.M. Zhivov, K. Siren, L.L. Christianson: Building and Environment Vol.37 (2002), p.55

[12] Ministry of Transportation, Taiwan, National fire codes for fire prevention equipments and egress facilities in railway tunnels and underground stations. (2008) (In Chinese)

[13] B. Merci: Fire Safety Journal, Vol. 43 (2008), p.376.

[14] AS 4391 (1999), Smoke Management Systems Hot Smoke Test, Australian Standard

[15] NFPA 130 (2003), Standard for Fixed Guideway Transit and Passenger Rail Systems

[16] E. Kim, J. P. Woycheese and N. A. Dembsey: Fire Technology Vol. 44 (2008), p.137.

[17] K. McGrattan, A. Hamins: Fire Technology Vol. 42 (2006), p.273

[18] J. Tan; Y. Xie and T. Wang: IEEE 2010 International Conference on Mechanic Automation and Control Engineering (MACE) Issue. 26-28 (2010), p.1028 - 1032

[19] M. Jurij: Tunnelling and Underground Space Technology Vol. 18 (2003), p.525

[20] F. Chen, S.-C. Guo, H.-Y. Chuay and S.-W. Chien: Theoretical and Computational Fluid Dynamics Vol. 16 (2003), p349

[21] F. Chen, S.-W. Chien, H.-M. Jang and W.-J. Chang: Continuum Mechanics and Thermodynamics Vol. 15 (2003), p.425

[22] R. Borchiellini, V. Verda: Fire Safety Journal Vol.44 (2009), p.612

[23] C.C. Hwang, J.C. Edwards: Fire Safety Journal Vol. 40 (2005), p213.

[24] Taipei municipal department of underground construction, Ministry of Transportation, Taiwan, Banciao station full-scale tests on fire and smoke management for emergency operation. (1999) (In Chinese)

[25] Underground construction department, Ministry of Transportation, Taiwan, Nangkang North Tunnel demonstration of mechanical ventilation. (2005) (In Chinese)

[26] S. P. Su, Y.H. Tsai: *Underground tunnel integration with ventilation- case study*，7th across Taiwan Strait tunnel and underground conference. (2008) (In Chinese)

[27] S. P. Su, Y.H. Tsai: *Full-scale experiment on operation modes for underground tunnel ventilation*，8th across Taiwan Strait tunnel and underground conference. (2009) (In Chinese)

[28] D. A. Anderson, J.C. Tannehill, and R.H. Pletcher. Computational Fluid Mechanics and Heat Transfer. Hemisphere Publishing Corporation, Philadelphia, Pennsylvania, 1984. 11, 15

[29] H. Qian, Y. Li, P.V. Nielsen, C.E. Hyldgaard, T.W. Wong, A.T.Y. Chwang: Indoor Air Vol.16 (2006), p.111.

[30] H.Brohus: *Personal Exposure to Contaminant Sources in Ventilated Rooms.* Ph.D. thesis, Aalborg, Denmark: Department of Building and Structural Engineering, Aalborg University (1997).

[31] W.J. Fisk, D. Faulkner, D. Sullivan, F. Bauman: Indoor Air Vol.7 (1997), p.55.

Chapter 7

Identification from Flight Data of the Aerodynamics of an Experimental Re-Entry Vehicle

Antonio Vitale, Federico Corraro, Guido De Matteis and Nicola de Divitiis

Additional information is available at the end of the chapter

1. Introduction

Post flight data analyses are essential activities in aerospace projects. In particular, there is a specific interest in obtaining vehicle aerodynamic characteristics from flight data, especially for re-entry vehicle, in order to better understand theoretical predictions, to validate wind-tunnel test results and to get more accurate and reliable mathematical models for the purpose of simulation, stability analysis, and control system design and evaluation. Indeed, due to atmospheric re-entry specificity in terms of environment and phenomena, ground based experiments are not fully exhaustive and in-flight experimentation is mandatory. Moreover pre-flight models are usually characterised by wide uncertainty ranges, which should be reduced. These objectives can be reached by performing vehicle's model identification from flight data.

The Italian Aerospace Research Centre (CIRA) has faced the problem of re-entry vehicle model identification from flight data within the framework of its Unmanned Space Vehicle (USV) program. The main objective of the USV program is designing and manufacturing unmanned Flying Test Beds (FTBs), conceived as multi-mission flying laboratories, in order to test and verify innovative materials, aerodynamic behaviour, advanced guidance, navigation and control functionalities as well as critical operational aspects peculiar of the future Reusable Launch Vehicle. Based on the velocity range under investigation, the whole USV program has been divided into several parts, the first of them, named USV_1 project, is aimed at investigating the terminal phase of re-entry mission, that is, subsonic, transonic and low supersonic flight regimes. Two identical autonomous Flying Test Beds (called FTB_1 but nicknamed Castore and Polluce) were designed and produced to support the execution of the USV_1 project. The FTB_1 vehicles are unmanned and un-powered. They are winged slender configurations, with two sets of aerodynamic effectors: the elevons, that provide both pitch control when deflected symmetrically and roll control when deflected asymmetrically, and the rudders, that deflect only symmetrically to allow yaw control. Lateral-directional stability is enhanced by means of two ventral fins. A Hydraulic Actuator System (HYSY) controls the aerodynamic effectors. The on-board computers host the software that implements the guidance, navigation and control algorithms and manages subsystems and experimental payloads. One of the FTB_1 vehicles is shown in Figure 1, while Figure 2 presents its three-view.

Figure 1. FTB_1 vehicle

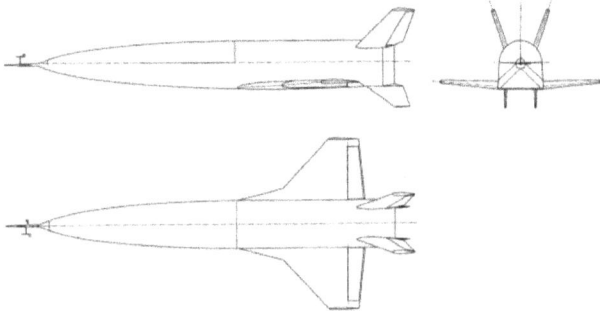

Figure 2. FTB_1 three-view

Two transonic flight missions, named Dropped Transonic Flight Test 1 and 2 (DTFT1 and DTFT2), were already carried out. They were aimed at evaluating the transonic flight of a re-entry vehicle. Data gathered during the FTB_1 missions have been analysed by the proposed identification techniques, in order to increase the accuracy of the vehicle model. Model identification of a re-entry vehicle is a very challenging task for the following main reasons:

1. The aerodynamic behaviour of a re-entry vehicle is characterised by a complex flow structure that produces significant variations of all the aerodynamic coefficients depending on Mach number and angle of attack. It makes it difficult to model the vehicle aerodynamics, particularly in transonic regime.

2. Experimental re-entry missions are typically performed once, providing a limited number of suitable data, and the experiment cannot be repeated in the short term. Therefore, it is difficult to refine the vehicle model in the whole flight envelope.

3.   Due to safety constraints, manoeuvres specifically suited to the purpose of model identification are minimised.

The first two issues call for structured parametric models based on physical considerations, where the flow field characteristics in the regimes of interest are represented with adequate accuracy. As a major advantage, such a model would extend the results obtained from the analysis of a single trajectory to the whole flight envelope. On the other hand, the third topic above requires that as much as possible information is extracted from low excitation inputs, and it is thus related to the effectiveness of the adopted identification methodology.

In this chapter the parametric aerodynamic model is discussed first, the structure of which is based on first principles and specifically accounts for the peculiarities of a slender winged body configuration. The definition of this model has to face several challenging problems. The first of them is of a physical nature and arises from the variations of the flow structure about the aircraft, which depends on the current vehicle state variables and on some of their time derivatives. The simultaneous effect of all these quantities produces a pressure distribution on the aircraft surface, which depends on such variables in a complex fashion (Lamb, 1945). Because of this complexity, the determination of reasonable expressions of the aerodynamic coefficients, in terms of the state variables, can be very difficult. Although the aerodynamic performances of several lifting vehicles, such as HL-10, HL-20, X-33, and X-38, have only recently been analysed (Brauckmann, 1999; Kawato et al., 2005), the methodologies for calculating the aerodynamic characteristics of lifting bodies in subsonic, transonic, and supersonic regimes do not provide the same level of accuracy that is obtained for the classical wing-body configurations. This is apparent, in particular, for what concerns the variations of the lateral and directional coefficients with respect to aerodynamic angles and Mach number (Rayme, 1996). In fact, the simultaneous effects of lateral flow, body angular rates, and fluid compressibility can determine complex situations, where these coefficients exhibit nontrivial, non-monotonic variations (Kawato et al., 2005). The second problem is of a mathematical nature and regards the use of a tabular aerodynamic coefficients database. If the aerodynamic coefficients are known for assigned values of the state variables, the accuracy of the coefficient values out of the data points (calculated through an interpolation procedure) depends on the adopted interpolation method and on the number of independent variables. Because these coefficients depend on quite a large number of state variables, the interpolation provides in general poor accuracy (Hildebrand, 1987), especially for the transonic variations of the lateral and directional coefficients at null sideslip angle, roll and yaw rates. Nevertheless, structured models, where the aerodynamic coefficients are expressed using some interpolation technique as functions of Mach number, aerodynamic angles and control surfaces deflection, are usually proposed in the literature for the purpose of identification (Gupta & Hall, 1979; Trankle & Bachner, 1995). Since these models are not based upon first principles, they cannot, in general, be applied outside of the region of the flight envelope where flight trials are undertaken. Last, but not least, the aerodynamic controls, which influence the aerodynamic coefficients in conjunction with all the variables, determine a further difficulty for the determination of the aerodynamic coefficients of a lifting body.

The model proposed in the present work provides a continuous and regular analytical representation of non-dimensional aerodynamic force and moment coefficients acting on the vehicle in the three regimes of subsonic, transonic and supersonic flow. It is based on the Kirchoff theorem, which in origin was formulated for incompressible streams and is based on the linear property of the continuity equation. This theorem states that, for an incompressible flow, the local fluid velocity around an obstacle is a linear function of the characteristic velocities of the problem. To study the vehicle aerodynamics in the compressible regimes, the Kirchoff theorem is properly extended to the compressible streams, taking into account that the local velocity depends on the fluid compressibility through the von Kármán equation (de Divitiis & Vitale, 2010). The model allows expressing each aerodynamic coefficient as nonlinear function of Mach number, aerodynamic angles, control effectors deflections, angular rates, and a set of constant aerodynamic parameters. The nonlinear behaviour stems from the effect of Mach number in the transonic regime and from the aerodynamic characteristics of the FTB_1 low aspect ratio, lifting-body configuration. The parameters of the aerodynamic model are firstly determined before flight, fitting a pre-flight aerodynamic database, built upon wind-tunnel test data and computational fluid dynamics analysis (Rufolo et al., 2006). Next, a subset of the model parameters is identified from flight data analysis, in order to update their pre-flight values and to reduce the related uncertainty level.

Next, an original methodology for model identification from flight data is presented, which is applied in the framework of a two-step strategy called Estimation Before Modelling (EBM) (Hoff & Cook, 1996). This strategy is based on the classical decomposition principle, that is, it decomposes the starting identification problem in sub-problems which are easier to be solved. The EBM is introduced to manage the complex nonlinear structure of the vehicle dynamic equations and, above all, of the proposed aerodynamic model. The methodology allows to deal independently with the mission flight path reconstruction, that is the estimation of vehicle state vector and global aerodynamic coefficients, and the evaluation of aerodynamic model parameters. As for the latter sub-problem, the estimation process is carried out independently for each aerodynamic coefficient and for each flight regime (that is, subsonic and supersonic). The multi-step approach also permits to select a suitable estimation methodology to solve each sub-problem, exploiting in such a way the advantages of several identification techniques. Finally, it is specifically suited to deal with problems where identification manoeuvres are minimised and dynamic excitation is poor. In particular, the identifiable parameters are easily selected, and the identification (and related validation) can be carried out only for the model of the aerodynamic coefficients the parameters of which are in fact identifiable.

The proposed identification strategy is illustrated in Figure 3. In the first step of EBM, vehicle state vector, aerodynamic coefficients and some atmospheric properties (such as local wind experienced during the mission) are estimated. This step is formulated as a nonlinear filtering problem and solved using the Unscented Kalman Filter (UKF). In recent times, UKF has been proposed as a valid alternative to the Extended Kalman Filter (EKF) for nonlinear filtering, receiving great attention in navigation, parameter estimation, and dual estimation problems (Chowdhary & Jategaonkar, 2006). The UKF is based on the concept of Unscented Transformation (UT), introduced by Julier and Uhlmann, and, unlike EKF, provides at least second order accurate estimates of the first two statistical moments, not requiring approximations for state and output functions (Julier & Uhlmann, 1995). It enables a complete and structured statistical characterization of the estimated variables, leading to a reliable evaluation of uncertainties on the unknowns. The availability of the aerodynamic coefficients with related estimation uncertainty allows validating pre-flight aerodynamic databases and models. The second step receives in input the aerodynamic coefficients and related uncertainties calculated in the previous step, and provides an estimation for a subset of the aerodynamic model parameters that, as said before, is valid throughout the whole flight envelope of interest. This subset of parameters is selected using a sensitivity analysis based on the evaluation of the Cramer Rao Bounds. The parameters estimation could be performed using the UKF again as well as the simpler Least Mean Squares techniques. With respect to UKF, the LMS technique has the advantage that it does not require the tuning of the filter gains, neither the definition of an initial guess for the unknowns, which could eventually influence the estimation. When the estimation is carried out, the uncertainties on the aerodynamic coefficients identified in the first step are treated as measurement noise and they are rigorously propagated through the second step, whatever the applied estimation methodology is. Therefore, the identification process provides the nominal value and the related estimation uncertainty of the aerodynamic parameters, and guarantees an accurate and reliable characterisation of the identified aerodynamic model, by using all the available pre-flight information and in-flight gathered data. In this way the identified model is completely defined and the values of the estimated aerodynamic uncertainties are generally lower than the pre-flight ones.

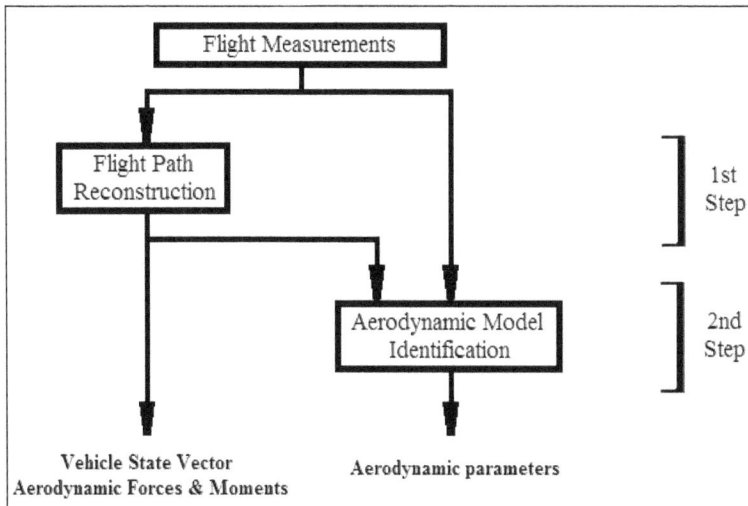

Figure 3. Identification strategy

The application of the above described aerodynamic modelling and identification methodology to the flight data of the first two missions of the FTB_1 vehicle has provided interesting results in terms of estimation convergence, reduction of uncertainty with respect to pre-flight model and capability of extracting useful information on the vehicle aerodynamics from a rather limited set of flight data.

2. DTFT missions profile

As said, the FTB_1 vehicle already performed two test missions, in winter 2007 (DTFT1) and in spring 2010 (DTFT2). Both mission profiles were based on the release of the vehicle from a high altitude scientific balloon at nominal mission altitude (about 20 km for the first mission and 24 km for the second one), followed by a controlled gliding flight down to the deployment of a recovery parachute. Key mission phases of DTFT missions are shown in Figure 4.

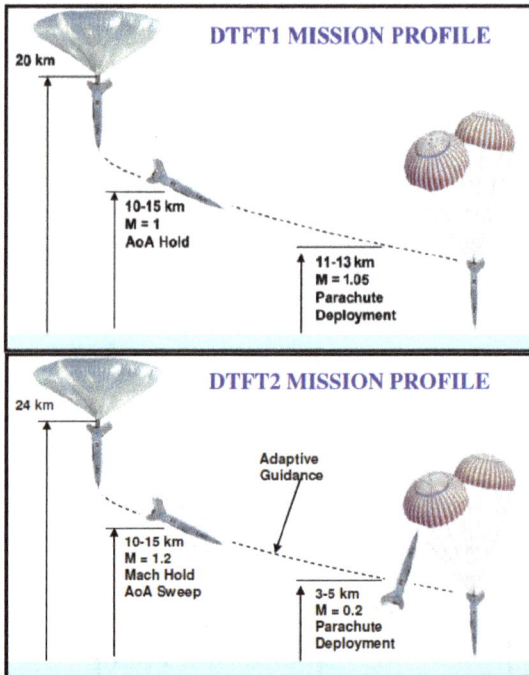

Figure 4. Pictorial representation of the DTFT Missions Profile

In the first mission the transonic regime of flight was reached (Mach ~1.08) while holding the angle of attack at a constant value. No lateral directional manoeuvres were foreseen and the flight was very short, lasting only about 44 seconds. Based on first mission experience, second mission was more complex. After release, the vehicle performed a pitch-up manoeuvre to reach and hold a specified value of the angle of attack while accelerating up to Mach 1.2 at about 15 km altitude; then a pull down manoeuvre was performed to keep the Mach number constant while a sweep in angle of attack was executed. The manoeuvre allowed the verification of the aerodynamic behaviour of the vehicle at constant Mach and variable angle of attack in full transonic regime as it would happen in a wind tunnel facility. At the end of this manoeuvre the vehicle began a pull up manoeuvre to decelerate to very low speeds (below Mach 0.2) and reached an altitude lower than 5 km where a subsonic parachute was opened, allowing a safe splashdown of the vehicle. Figure 5 shows the in-flight measured barometric altitude versus Mach profile for DTFT2, and also highlights the most relevant phases of flight.

In both missions, the on board navigation sensors suite was composed of Inertial Measurement Unit (INS), magnetometer and Air Data System (ADS). Flight measurements of load factors, centre of mass (CoM) velocity and position, angular velocity, Euler angles, aerodynamic angles, Mach number, total and static pressure, total temperature and aerodynamic effectors deflections are required in input by the parameter identification process. During DTFT1, these data were recorded at different sampling rates (10Hz and 100Hz). They were re-sampled and synchronized at 100Hz prior to perform further analyses. In the DTFT2 mission all the data were gathered at 100Hz. Post-flight meteorological data, namely, static pressure, static temperature and mean wind velocity, provided by the European Centre for Medium-Range Weather Forecasts (ECMWF) were also collected for identification purpose.

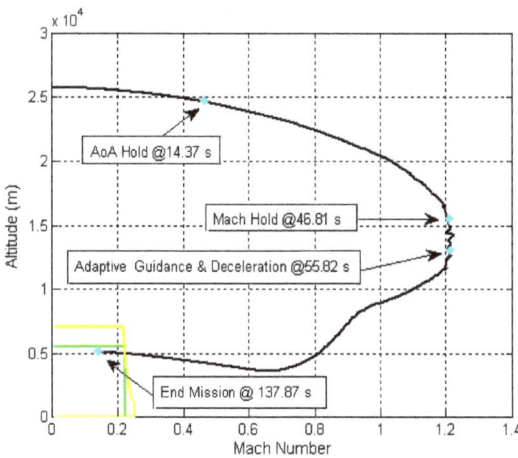

Figure 5. DTFT2 Altitude vs. Mach profile

3. Aerodynamic model

In this section the available pre-flight aerodynamic database is first described. Next the theoretical derivation and the final formulation of the analytical model, proposed for system identification purpose, is presented (de Divitiis & Vitale, 2010).

3.1. Pre-flight aerodynamic data base

The pre-flight Aerodynamic Data Base (ADB) was developed at CIRA in the framework of studies on transonic aerodynamics for the FTB_1 vehicle. Aerodynamic coefficients in ADB account for several inputs, that is, Mach number, aerodynamic angles, Reynolds number, rotary and unsteady effects along with the action of controls. The ADB is described in detail in (Rufolo et al., 2006). The primary sources of data were represented by the tests carried out at CIRA wind tunnel PT-1 and at the DLR-DNW Transonic Wind tunnel Gottingen (TWG). The experiments mainly addressed the transonic regime, according to its particular interest for the DTFT missions. Computational Fluid Dynamics (CFD) and simplified engineering methods were used to cross-check wind tunnel data and to analyse in detail flow conditions where measurements were not complete. Simplified methods like Vortex Lattice Method, Boundary Element Method and DATCOM were also employed to fill gaps in wind tunnel data, and allowed the extension of the database to low subsonic regime (Mach < 0.5), also including the effects of Reynolds number. The resulting ADB covers a wide envelope of flight conditions and provides aerodynamic coefficients in tabular form. Uncertainty of predictions was also estimated, taking into consideration random experimental errors (repeatability), systematic experimental errors (known and not removable errors) and CFD errors (effect of computational grid, convergence, level of turbulence modelling, boundary conditions, etc.). The ADB is implemented in the form of look-up tables for the purpose of simulation, control system design and validation.

3.2. Analytical aerodynamic model for identification purpose

The proposed analytical aerodynamic model provides a continuous and regular analytical representation of the aerodynamic force and moment coefficients of the FTB_1 in the form of parametric functions, based on first principles and valid for winged slender configuration across the three regimes of subsonic, transonic and supersonic flow (de Divitiis & Vitale, 2010). Its formulation is derived starting from the continuity equation. Under the hypothesis of small perturbations, that is, small angle of attack (α), sideslip angle (β) and body thickness, the perturbed velocity $\mathbf{v}$ in the proximity of the vehicle is described by the local perturbation $\tilde{\mathbf{v}}$, which in wind frame is

$$\tilde{\mathbf{v}}(\mathbf{r}) \equiv \left(\tilde{u}, \tilde{v}, \tilde{w} \right) \tag{1}$$

where $\tilde{u}$, $\tilde{v}$ and $\tilde{w} \ll V_\infty$, being V_∞ the stream velocity and $\mathbf{r}$ the position vector in body frame; thus

$$|\mathbf{v}| = \sqrt{\left(V_\infty + \tilde{u} \right)^2 + \tilde{v}^2 + \tilde{w}^2} \approx V_\infty + \tilde{u} \tag{2}$$

The component $\tilde{u}$ corresponds to the direction $\tilde{x}$ parallel to the flight velocity $\mathbf{V}$, whereas $\tilde{v}$ and $\tilde{w}$ are the lateral components of the perturbed velocity, corresponding to the normal coordinates to $\mathbf{V}$, that is, $\tilde{y}$ and $\tilde{z}$. In the small perturbations hypothesis, $\mathbf{v}$ satisfies the continuity equation in the von Kármán-Guderley form, which, in the wind frame, is written as follows (Cole & Cook, 1986)

$$B\frac{\partial \tilde{u}}{\partial \tilde{x}}+\frac{\partial \tilde{v}}{\partial \tilde{y}}+\frac{\partial \tilde{w}}{\partial \tilde{z}}=0 \tag{3}$$

$$B=1-M_\infty^2\left[1+(1+\gamma)\tilde{u}/V_\infty\right] \tag{4}$$

where γ is the air specific heat ratio and M_∞ is the flight Mach number. For a small enough M_∞, all the points around the aircraft are subsonic, B > 0 in any case, and (3) is an elliptical equation. On the contrary, when each point is supersonic, B < 0 everywhere and (3) is a hyperbolic equation. In both cases, B can be approximated by the expression

$$B=1-M_\infty^2 \tag{5}$$

and (3) can be reduced to a linear equation. Due to this linearity, the local velocity **v** is also a linear function of the characteristic velocities of the problem. This result is an extension to the compressible stream of the Kirchoff theorem (Lamb, 1945). With reference to Figure 6, the characteristic velocities for a rigid vehicle moving in a fluid are **V** and the angular velocity ω that, written in the body frame are:

$$\mathbf{V}=V\,\hat{\mathbf{V}} \tag{6}$$

$$\hat{\mathbf{V}}=\begin{bmatrix}\hat{u}\\\hat{v}\\\hat{w}\end{bmatrix}\equiv\begin{bmatrix}\cos\alpha\cos\beta\\\sin\beta\\\sin\alpha\cos\beta\end{bmatrix} \tag{7}$$

$$\omega=\begin{bmatrix}p\\q\\r\end{bmatrix} \tag{8}$$

where $\hat{u}$, $\hat{v}$ and $\hat{w}$ are the direction cosines of **V** and p, q, r are the angular rate components, both in body frame. The Kirchoff theorem allows to express the local fluid velocity as

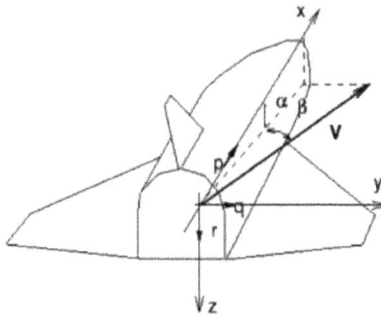

Figure 6. Characteristic velocities and reference frame

$$\mathbf{v}(\mathbf{r}) = \frac{\partial \mathbf{v}}{\partial \mathbf{V}} \mathbf{V} + \frac{\partial \mathbf{v}}{\partial \boldsymbol{\omega}} \boldsymbol{\omega} = \mathbf{A}(\mathbf{r}, M_\infty) \mathbf{V} + \mathbf{B}(\mathbf{r}, M_\infty) \boldsymbol{\omega} \tag{9}$$

The Jacobian matrices $\mathbf{A}$ and $\mathbf{B}$ are the influence functions that, for $M_\infty < 1$ or for $M_\infty > 1$, only depend on M_∞ and $\mathbf{r}$.

In the transonic regime, the nonlinear term of equation (3) is non-negligible with regard to the others, and the von Kármán equation is locally elliptic or hyperbolic, following the sign of $\mathbf{B}$. As a result, the influence functions will also depend on $\mathbf{V}$ and $\boldsymbol{\omega}$. The solutions of (3) are formally expressed by the continuation method in the form of (Guckenheimer & Holmes, 1990)

$$\mathbf{v}(\mathbf{r}) = \int_0^{\mathbf{v}} \frac{\partial \mathbf{v}}{\partial \mathbf{V}} (\mathbf{r}, M_\infty, \mathbf{V}, \boldsymbol{\omega}) d\mathbf{V} + \int_0^{\boldsymbol{\omega}} \frac{\partial \mathbf{v}}{\partial \boldsymbol{\omega}} (\mathbf{r}, M_\infty, \mathbf{V}, \boldsymbol{\omega}) d\boldsymbol{\omega} \tag{10}$$

This velocity, that accounts for the variations of the flow structure about the vehicle, depends on the path integrals of (10), which are described by the time histories of $\mathbf{V}$ and $\boldsymbol{\omega}$. For steady-state aerodynamics, the local fluid velocity depends only on the current state variables, so that (10) reads as follows

$$\mathbf{v}(\mathbf{r}) = \mathbf{A}(\mathbf{r}, M_\infty, \mathbf{V}, \boldsymbol{\omega}) \mathbf{V} + \mathbf{B}(\mathbf{r}, M_\infty, \mathbf{V}, \boldsymbol{\omega}) \boldsymbol{\omega} \tag{11}$$

It is worthy to remark that this analysis only holds if the variations of the flow structure around the vehicle are considered to be known when α, β, and $\boldsymbol{\omega}$ change. The flow structure is supposed to be assigned, and this implies that the solutions of (3) do not modify their analytical forms with respect to (11). Starting from (11), let us now detail the formulation of the aerodynamic coefficients, recognizing three distinct contributions: steady aerodynamics, unsteady aerodynamics, effect of the controls (de Divitiis & Vitale, 2010).

3.2.1. Steady aerodynamic coefficients

The aerodynamic force and moment are calculated as surface integrals of the pressure P over the vehicle wetted surface S_w:

$$F = -\int_{S_w} P\mathbf{n}dS, \qquad M = -\int_{S_w} P(\mathbf{r} - \mathbf{r}_{cg}) \times \mathbf{n}dS \tag{12}$$

with $\mathbf{n}$ normal unit vector to the wetted surface and $\mathbf{r}_{cg}$ vehicle centre of gravity location in the body frame sketched in Figure 6. The contribution of the skin friction does not appear in (12), and its effect is considered to be caused by a proper pressure reduction (Lamb, 1945).

Equation (11) can be reformulated in terms of dimensionless angular velocity, $\tilde{\boldsymbol{\omega}} \equiv (\hat{p}, \hat{q}, \hat{r}) = \boldsymbol{\omega} L / V$

$$\mathbf{v}(\mathbf{r}) = V \left(\mathbf{A}\hat{\mathbf{V}} + \frac{1}{L} \mathbf{B}\tilde{\boldsymbol{\omega}} \right) \tag{13}$$

Pressure P in equation (12) is determined using the steady Bernoulli theorem

$$\frac{\gamma}{\gamma-1}\frac{P}{\rho}+\frac{\mathbf{v}\cdot\mathbf{v}}{2}=\frac{\gamma}{\gamma-1}\frac{P_{\infty}}{\rho_{\infty}}+\frac{V_{\infty}^{2}}{2} \tag{14}$$

in which ρ is the air density and the square of $\mathbf{v}$ is provided by equation (13)

$$\mathbf{v}\cdot\mathbf{v}=\left|\mathbf{v}(\mathbf{r})\right|^{2}=V^{2}\sum_{i=1}^{3}\sum_{j=1}^{3}\sum_{k=1}^{3}\left(A_{ij}A_{ik}\bar{v}_{j}\bar{v}_{k}+A_{ij}B_{ik}\bar{v}_{j}\bar{\omega}_{k}+B_{ij}B_{ik}\bar{\omega}_{j}\bar{\omega}_{k}\right) \tag{15}$$

The aerodynamic force coefficients in the body frame exhibit more oscillating variations with regard to α than those in the wind axes (Lamb, 1945). On the contrary, the moment coefficients exhibit quite smooth variations in body axes (Lamb, 1945). For this reason, the aerodynamic force and moment are calculated in the wind frame and body axes, respectively. They are expressed through drag (C_D), lateral (C_S) and lift (C_L) force coefficients and roll (C_l), pitch (C_m) and yaw (C_n) moment coefficients, respectively:

$$F=-\frac{1}{2}\rho V^{2}S\begin{bmatrix}C_{D}\\C_{S}\\C_{L}\end{bmatrix}, \qquad M=-\frac{1}{2}\rho V^{2}SL\begin{bmatrix}C_{l}\\C_{m}\\C_{n}\end{bmatrix} \tag{16}$$

where L and S are vehicle characteristic length and surface. The generic aerodynamic coefficient C_i (i = D, S, L, l, m, n) is computed integrating equations (12). We get

$$C_{i}(\alpha,\beta,p,q,r)=\sum_{h=1}^{3}\sum_{k=1}^{3}\left(F_{hk}^{i}\bar{v}_{h}\bar{v}_{k}+G_{hk}^{i}\bar{v}_{j}\bar{\omega}_{k}+H_{hk}^{i}\bar{\omega}_{j}\bar{\omega}_{k}\right) \tag{17}$$

where $(\bar{v}_{1},\bar{v}_{2},\bar{v}_{3})\equiv(\hat{u},\hat{v},\hat{w})$, $(\bar{\omega}_{1},\bar{\omega}_{2},\bar{\omega}_{3})\equiv(\hat{p},\hat{q},\hat{r})$ and $F_{hk}^{i},G_{hk}^{i},H_{hk}^{i}$ are functions of $\mathbf{r}$ evaluated as surface integrals over S_{w} of $\sum_{k=1}^{3}A_{ij}A_{ik}$, $\sum_{k=1}^{3}A_{ij}B_{ik}$ and $\sum_{k=1}^{3}B_{ij}B_{ik}$, respectively.

Although $\mathbf{A}$ and $\mathbf{B}$ are functions of V and ω, the quantities $F_{hk}^{i},G_{hk}^{i},H_{hk}^{i}$, which represent the aerodynamic derivatives, for thin obstacles vary with M_{∞}, and exhibit quite small variations with respect to α, β, and ω (Ashley & Landahl, 1965). Hence, according to the literature, these integrals are supposed to be functions of M_{∞} alone. They show rather smooth variations with respect to M_{∞} in the subsonic and supersonic regions, whereas for $M_{\infty}\approx 1$, sizable variations, caused by the fluid transonic regime, are observed.

C_i incorporates three addends. The first addend is the static aerodynamic coefficient, whereas the second one, which provides the simultaneous effect of V and ω, represents the contribution of the rotational derivatives. The last term is a quadratic form of ω that, in the aerospace applications, is negligible with respect to the others. Therefore, C_i is expressed as follows

$$C_i\left(\alpha,\beta,p,q,r\right)=\sum_{h=1}^{3}\sum_{k=1}^{3}\left(F_{hk}^i\hat{v}_h\hat{v}_k+G_{hk}^i\hat{v}_j\hat{\omega}_k\right) \tag{18}$$

where F_{hk}^i and G_{hk}^i are called static and rotational characteristic functions, respectively. They are the second-order derivatives of the generic aerodynamic coefficient with respect to the direction cosines of **V** and to the dimensionless angular velocity. The structure of these derivatives is supposed to be

$$F\left(M_\infty\right)=F_{sub}\frac{H_{sub}\sqrt{m+\varepsilon}\left(1+h_1M_\infty^2+h_2M_\infty^3\right)}{\sqrt{\varepsilon+m\left|1-M_\infty^\xi\right|^2}}+F_{sup}H_{sup}\frac{1+g_1M_\infty^{y_1}}{1+g_2M_\infty^{y_2}} \tag{19}$$

where the indexes i, h and k have been omitted, and the same structure holds for G, too. $F_{sub},m,\varepsilon,h_1,h_2,\xi,F_{sup},g_1,y_1,g_2,y_2$ are aerodynamic constant parameters to be identified prior to flight, using the available tabular aerodynamic database, or after the flight by analysing the flight data. The quantities H_{sub} and H_{sup} are two sigmoidal functions of M_∞, which are chosen as follows:

$$H_{sub}\left(M_\infty\right)=\frac{\tanh 50\left(1-M_\infty\right)+1}{2} \tag{20}$$

$$H_{sup}\left(M_\infty\right)=\frac{\tanh 50\left(M_\infty-1\right)+1}{2} \tag{21}$$

Equation (19) incorporates two addends: the first one gives the variation of the aerodynamic coefficients in the subsonic regime, whereas the second one describes the supersonic region. Indeed, H_{sub} is about 1 if $M_\infty \leq 0.95$ and about 0 if $M_\infty \geq 1.05$, whereas H_{sup} is about 0 if $M_\infty \leq 0.95$ and about 1 if $M_\infty \geq 1.05$ In transonic regime both the sigmoidal functions assume values between 0 and 1 and the combination of the subsonic and the supersonic contributions provides the aerodynamic coefficients in the transonic regime. Substituting equations (7) and (8) in (18) and considering some simplifications due to the symmetry of the vehicle (de Divitiis & Vitale, 2010), we get the expressions for steady aerodynamic force coefficients in wind axes and the moment coefficients in the body frame. In particular, since the vehicle is symmetric with respect to the longitudinal plane, each longitudinal aerodynamic coefficient is an even function of β and results in an odd function of the products $\hat{v}\hat{r}$ and $\hat{v}\hat{p}$, whereas the lateral-directional coefficients are odd functions of β and the products $\hat{u}\hat{p}$, $\hat{u}\hat{r}$, $\hat{v}\hat{q}$, $\hat{w}\hat{p}$, and $\hat{w}\hat{r}$. The steady aerodynamic coefficients are

$$C_D=\left(F_{uu}^D\cos^2\alpha+F_{uw}^D\cos\alpha\sin\alpha+F_{ww}^D\sin^2\alpha\right)\cos^2\beta+F_{vv}^D\sin^2\beta+$$
$$+\left(G_{uq}^D\cos\alpha\cos\beta+G_{wq}^D\sin\alpha\cos\beta\right)\hat{q}+\left(G_{vr}^D\hat{r}+G_{vp}^D\hat{p}\right)\sin\beta \tag{22}$$

$$C_S = \left(F_{uv}^Y \cos\alpha + F_{vw}^Y \sin\alpha \right) \sin\beta \cos\beta +$$
$$+ \left(G_{up}^Y \hat{p} + G_{ur}^Y \hat{r} \right) \cos\alpha \cos\beta + G_{vq}^Y \hat{r} \sin\beta \hat{q} + \left(G_{wp}^Y \hat{p} + G_{wr}^Y \hat{r} \right) \sin\alpha \cos\beta \tag{23}$$

$$C_L = \left(F_{uu}^L \cos^2\alpha + F_{uw}^L \cos\alpha \sin\alpha + F_{ww}^L \sin^2\alpha \right) \cos^2\beta + F_{vv}^L \sin^2\beta +$$
$$+ \left(G_{uq}^L \cos\alpha \cos\beta + G_{wq}^L \sin\alpha \cos\beta \right) \hat{q} + \left(G_{vr}^L \hat{r} + G_{vp}^L \hat{p} \right) \sin\beta \tag{24}$$

$$C_l = \left(F_{uv}^l \cos\alpha + F_{vw}^l \sin\alpha \right) \sin\beta \cos\beta +$$
$$+ \left(G_{vp}^l \hat{p} + G_{vr}^l \hat{r} \right) \cos\alpha \cos\beta + G_{vq}^l \sin\beta \hat{q} + \left(G_{wp}^l \hat{p} + G_{wr}^l \hat{r} \right) \sin\alpha \cos\beta \tag{25}$$

$$C_m = \left(F_{uu}^m \cos^2\alpha + F_{uw}^m \cos\alpha \sin\alpha + F_{ww}^m \sin^2\alpha \right) \cos^2\beta + F_{vv}^m \sin^2\beta +$$
$$+ \left(G_{uq}^m \cos\alpha \cos\beta + G_{wq}^m \sin\alpha \cos\beta \right) \hat{q} + \left(G_{vr}^m \hat{r} + G_{vp}^m \hat{p} \right) \sin\beta \tag{26}$$

$$C_n = \left(F_{uv}^n \cos\alpha + F_{vw}^n \sin\alpha \right) \sin\beta \cos\beta +$$
$$+ \left(G_{vp}^n \hat{p} + G_{vr}^n \hat{r} \right) \cos\alpha \cos\beta + G_{vq}^n \sin\beta \hat{q} + \left(G_{wp}^n \hat{p} + G_{wr}^n \hat{r} \right) \sin\alpha \cos\beta \tag{27}$$

3.2.2. Unsteady aerodynamic coefficients

The unsteady effects are of two kinds (Ashley & Landahl, 1965): the first effect is directly related to the pressure forces through the Bernoulli theorem, it is instantaneous and depends only on the current value of the state variables; the second effect is caused by the unsteady motion of the wakes and it represents the story of this motion from the initial condition until the current time. The proposed model only takes into account the first effect. It is caused by the term $\Delta p_u = \rho \left(\partial\phi/\partial t \right)$ which appears in the Bernoulli equation in the case of unsteady flow, where ϕ is the velocity potential. For assigned velocity variations, this increment is function of the time derivatives of the aerodynamic angles and the flight speed, whereas the contribution produced by the time derivatives of the angular velocity is not taken into account, that is

$$\Delta p_u = \rho \left(\frac{\partial\phi}{\partial t} \right) = \rho \left(\frac{\partial\phi}{\partial\alpha} \frac{d\alpha}{dt} + \frac{\partial\phi}{\partial\beta} \frac{d\beta}{dt} + \frac{\partial\phi}{\partial V} \frac{dV}{dt} \right) \tag{28}$$

The pressure increment is the sum of three terms; for thin vehicles, the last addend, which is related to the variation of the velocity, is negligible with respect to the first one. Thus, it is not considered in the present analysis. Δp_u can also be written in terms of the derivatives of the velocity potential $\partial\phi/\partial\hat{u}$, $\partial\phi/\partial\hat{v}$ and $\partial\phi/\partial\hat{w}$. Using equation (7) we get

$$\frac{\Delta p_u}{\rho} = \frac{\partial\phi}{\partial\hat{u}} \left(-\sin\alpha \cos\beta \frac{d\alpha}{dt} - \cos\alpha \sin\beta \frac{d\beta}{dt} \right) + \frac{\partial\phi}{\partial\hat{v}} \cos\beta \frac{d\beta}{dt} +$$
$$+ \frac{\partial\phi}{\partial\hat{w}} \left(\cos\alpha \cos\beta \frac{d\alpha}{dt} - \sin\alpha \sin\beta \frac{d\beta}{dt} \right) \tag{29}$$

This increment induces aerodynamic force and moment, which are obtained by integrating Δp_u on Sw. Therefore, the unsteady increment ΔC_i of each aerodynamic coefficient varies linearly with the aerodynamic angles derivatives, resulting in

$$\Delta C_i = \left(A_1^i \sin\alpha + A_2^i \cos\alpha \right) \cos\beta \,\dot{\alpha} + \left(B_1^i \cos\alpha\sin\beta + B_2^i \cos\beta + B_3^i \sin\alpha\sin\beta \right) \dot{\beta} \qquad (30)$$

where A_h^i and B_h^i (i = D, S, L, l, m, n) are appropriate surface integrals over Sw of $\partial\phi/\partial\hat{u}$, $\partial\phi/\partial\hat{v}$ and $\partial\phi/\partial\hat{w}$, that are functions of M_∞. These integrals are supposed to be described by expressions such as (19), whereas $\dot{\alpha}$ and $\dot{\beta}$ are the dimensionless time derivatives of the aerodynamic angles, defined as

$$\dot{\alpha} = \frac{d\alpha}{dt}\frac{L}{V}, \qquad\qquad \dot{\beta} = \frac{d\beta}{dt}\frac{L}{V} \qquad (31)$$

The terms $\left(A_1^i \sin\alpha + A_2^i \cos\alpha \right)\cos\beta$ and $\left(B_1^i \cos\alpha\sin\beta + B_2^i \cos\beta + B_3^i \sin\alpha\sin\beta \right)$ represent the unsteady aerodynamic derivatives. Due to the vehicle symmetry with respect to the longitudinal plane, the derivatives with respect to $\dot{\alpha}$ of the longitudinal coefficients are even functions of β, while the analogous derivatives of the lateral-directional coefficients are identically equal to zero. The aerodynamic derivatives with respect to $\dot{\beta}$ are even functions of β for the longitudinal coefficients and odd functions of β for the lateral-directional coefficients.

3.2.3. Effects of the controls

The FTB_1 vehicles have two sets of aerodynamic effectors: the elevons, that provide both pitch control when deflected symmetrically (δ_e) and roll control when deflected asymmetrically (δ_a), and the rudders, that deflect only symmetrically (δ_r) to allow yaw control. The rotation of the aerodynamic control surfaces modifies the vehicle geometry, which in turn determines a variation of the aerodynamic force and moment coefficients. These coefficients are also expressed by equations (22) - (27), because the analytical structure of these equations holds also when the control surfaces are deflected. Thus, it is reasonable that the increment of the aerodynamic coefficient caused by the effect of the controls is expressed by

$$\Delta C_e^i\left(M_\infty, \alpha, \delta_e \right) = F_{e1}^i\left(M_\infty \right)\delta_e + F_{e2}^i\left(M_\infty \right)\delta_e\alpha + F_{e3}^i\left(M_\infty \right)\delta_e^n \qquad (32)$$

$$\Delta C_a^j\left(M_\infty, \alpha, \delta_a \right) = F_{a1}^j\left(M_\infty \right)\delta_a + F_{a2}^j\left(M_\infty \right)\delta_a\alpha + F_{a3}^j\left(M_\infty \right)\delta_a^2 \qquad (33)$$

$$\Delta C_r^j\left(M_\infty, \alpha, \delta_r \right) = F_{r1}^j\left(M_\infty \right)\delta_r + F_{r2}^j\left(M_\infty \right)\delta_r\alpha + F_{r3}^j\left(M_\infty \right)\delta_r^2 \qquad (34)$$

where i = D, L, m, and j = S, l, n. Indeed, the effects of the elevator on the lateral aerodynamic coefficients, which can occur for $\beta \neq 0$, are not taken into account in the present analysis. Similarly, the effects of the ailerons and of the rudders on the longitudinal aerodynamic coefficients are considered negligible. In the above equations, the first and the second terms on the right hand sides represent, respectively, the linear effect of the control and the combined effect of control and angle of attack, whereas the third addend is the nonlinear term. In (32) the exponent n varies, depending on the coefficient: it is assumed equal to 2 for C_D, whereas it values 3 for C_L and C_m. The functions $F_{e1}^i(M_\infty)$, $F_{e2}^i(M_\infty)$, $F_{e3}^i(M_\infty)$, $F_{a1}^j(M_\infty)$, $F_{a2}^j(M_\infty)$, $F_{a3}^j(M_\infty)$, $F_{r1}^j(M_\infty)$, $F_{r2}^j(M_\infty)$, $F_{r3}^j(M_\infty)$ are called elevator, ailerons and rudder characteristic functions. They correspond to surface integrals over S_w, which can be obtained as the difference between the aerodynamic coefficients when the controls are deflected and those for clean configuration (null deflections). These integrals are functions of M_∞, and their analytical structure is assumed to be described by equation (19).

In conclusion, the aerodynamic coefficients are computed summing steady and unsteady contributions plus the effect of the controls, that are expressed by equations (22) - (27), equation (30), and equations (32) - (34), respectively. Each addendum in these equations contains a function of M_∞ expressed through (19), which also depends on a vector of free model parameters

$$\theta_1^i = \left[F_{sub}, m, \varepsilon, h_1, h_2, \xi, F_{sup}, g_1, y_1, g_2, y_2 \right] \tag{35}$$

with $l = \left[1, ..., Q(C_i) \right]$, being $Q(C_i)$ the total number of addends for the coefficient C_i $(i = D, S, L, l, m, n)$.

3.2.4. Pre-flight identification

All the constant parameters of the proposed model are estimated before flight, using the information provided by the pre-flight aerodynamic database. The pre-flight estimation is carried out through a least minimum square (LMS) method, which for each aerodynamic coefficient, is applied to the following optimization problem:

$$\min_{\vartheta_i} J_i = \min_{\vartheta_i} \sum_{i=1}^M \left[\left(C_k^i \right)_{ADB} - C_k^i \right]^2, \qquad i = D, S, L, l, m, n \tag{36}$$

where C_{ik} and $\left(C_{ik} \right)_{ADB}$ are the aerodynamic coefficients calculated in M points of the flight envelope, with the proposed model and the pre-flight aerodynamic database, respectively. J_i is the goal function, defined for each aerodynamic coefficient, for which the arguments are the free parameters $\theta^i = \left(\theta_1^i, \theta_2^i, ..., \theta_{Q(C_k)}^i \right)$ with the generic θ_l^i given by (35). To obtain the combined effects of all the vehicle state variables and those of the controls, the coefficients C_{ik} and $\left(C_{ik} \right)_{ADB}$ are calculated in a wide range of variation of these variables.

4. System identification methodology

In order to improve the reliability of the aerodynamic model, it is validated and refined using flight data. To this end, a suitable identification methodology is proposed in this section.

4.1. Problem formulation

Vehicle dynamics are represented as a stochastic process in continuous state space form, along with the measurements equations, as follows

$$\dot{x}_V(t) = f\big(\mathbf{x_V}(t), \mathbf{c}(t), \eta(t)\big), \qquad \mathbf{x_V}(t_0) = \mathbf{x_{V0}} \tag{37}$$

$$\mathbf{y}(t) = h\big(\mathbf{x_V}(t), \mathbf{c}(t), \mathbf{v}(t)\big) \tag{38}$$

$$\mathbf{c}(t) = l\big(\mathbf{x_V}(t), \mathbf{U}(t), \Theta\big) \tag{39}$$

where $\mathbf{x}_V$ is the state vector of the vehicle, t_0 is the initial mission time and f and h are generic nonlinear real-valued functions. Measurements are available for inputs $\mathbf{U}$ and outputs $\mathbf{y}$ of the model with a fixed sampling time. The vector of aerodynamic force and moment coefficients, denoted as $\mathbf{c}$, depends on vehicle state $\mathbf{x}_V$, input $\mathbf{U}$ and on a set of unknown aerodynamic parameters $\Theta = \left[\theta^D, \theta^S, \theta^L, \theta^l, \theta^m, \theta^n\right]$, through the aerodynamic model represented by the nonlinear real-valued function l (which translates the aerodynamic model defined in section 3). Finally, η and $\mathbf{v}$ are process and measurement noises, respectively. All noises are assumed zero mean and are characterized by covariance matrices.

We aim at estimating the parameter vector Θ, using flight data measurements. The identification process is solved according to the Estimation Before Modelling (EBM) approach (Vitale et al., 2009), where the time histories of state vector $\mathbf{x}_V$, some air properties (that is, wind velocity, air temperature and pressure) and global aerodynamic coefficients $\mathbf{c}$ are estimated first, using (37) and (38) and a set of measurements. Aerodynamic parameters identification, that is, the determination of Θ, is conducted in the second step using (39) and the values of $\mathbf{x}_V$ and $\mathbf{c}$ evaluated in the first step, with their covariance matrices. In this respect, computation of the covariance matrix of $\mathbf{x}_V$ and $\mathbf{c}$ provides information on the uncertainty of the inputs to the second step, where this uncertainty is regarded as measurements error on the inputs and is characterized by the computed covariance matrix. The two identification steps are described in detail in the following sub-sections.

4.2. First identification step

4.2.1. Identification methodology

The first identification step is formulated as nonlinear filtering problem and solved using the Unscented Kalman Filter (UKF). The nonlinearity stems from the vehicle nonlinear equations of motion.

The UKF is a nonlinear filtering technique based on the concept of Unscented Transformation (UT), an analytical method for propagating a probability distribution through a nonlinear transformation. In more details, the UT allows estimating the mean and the covariance of the nonlinear function by computing the weighted mean and covariance of a discrete set of function values, obtained propagating through the function a set of points (named sigma points) deterministically chosen in the domain of the function. The UKF provides at least second-order accurate evaluations of the first two statistical moments of the unknowns (Julier & Uhlmann, 1995), enabling a complete and structured statistical characterization of the estimated variables and leading to a reliable evaluation of the uncertainties on the estimations. Like all the Kalman filters, the UKF performs the estimation in two sequential phases: first a dynamic model, whose state vector is composed of the unknowns, is used for time propagation of the estimation (prediction phase); next, at each time step, the available flight measurements are compared with the prediction (that is, the dynamic model outputs) to refine the estimation (correction phase).

The UT is applied in the prediction phase of the filter. Several implementation of the UT, and consequently of the UKF are available in the literature (Wan & van der Merwe, 2000; Van Dyke et al. 2004), characterized by different number of sigma points, weights and free parameters. We adopted a non-augmented version of the UKF algorithm with additive process and measurements noises, in order to reduce the number of sigma points (Chowdhary & Jategaonkar, 2006). Different formulations are not expected to introduce significant improvements in the algorithm performance, while they could increase the computational effort. In order to avoid losing information on the effect of process noise on the outputs, two concatenated Unscented Transformations are performed during the prediction phase, to account for the propagation throughout the nonlinear process and measurement equations (Wu et al., 2005). Although the detailed mathematical formulation of the filter is not reported here for the sake of brevity, the main steps to be performed in each filtering phase are summarized. The prediction phase is composed of

1P. First generation of sigma points and related weights, based on the current estimate of the filter state vector and related covariance matrix.
2P. Propagation of the sigma point through the process equations.
3P. Prediction of the filter state vector, computed as weighted mean of the propagated sigma points.
4P. Prediction of the covariance matrix of the filter state. It is computed as summation of two terms: the first one is the weighted variance of the propagated sigma points (step 2P) with respect to the state vector prediction (step 3P); the second term is the process noise covariance matrix.
5P. Second generation of sigma points and related weights, based on the predicted filter state vector (step 3P) and covariance matrix (step 4P).
6P. Propagation of the sigma points through the measurement equations.
7P. Prediction of the filter outputs, computed as weighted mean of the propagated sigma points.

8P. Prediction of the covariance matrix of the filter outputs. It is computed as summation of two terms: the first one is the weighted variance of the propagated sigma points (step 6P) with respect to the filter outputs prediction (step 7P); the second term is the measurements noise covariance matrix.

9P. Prediction of the state-output correlation matrix. It is computed as the weighted deviation of the sigma points propagated through the process equation (step 2P) with respect to the predicted state vector (step 3P) times the deviation of the sigma points propagated through the measurement equation (step 6P) with respect to the predicted filter outputs (step 7P).

The correction phase is based on the following steps:

1C. Computation of the residual, that is, the difference between flight measurements and related filter outputs prediction (step 7P).

2C. Computation of the Kalman filter gain. It depends on filter outputs covariance matrix (step 8P) and state-output correlation matrix (step 9P).

3C. Correction of the predicted filter state. The corrected filter state is given by the summation of state prediction (step 2P) and Kalman gain (step 2C) times computed residual (step 1C).

4C. Correction of the predicted covariance matrix of the filter state.

4.2.2. Filter model

The UKF requires the definition of a dynamic model describing the behaviour of the unknowns, that represent the filter state vector. The adopted filter state is composed of the vehicle state vector, some local environment properties (wind velocity, temperature, and static pressure) and the aerodynamic coefficients. The filter model should be completed by the measurements equations, that is, algebraic equations for the evaluation of model outputs starting from the state variables. The model for the first identification step is sketched in Figure 7.

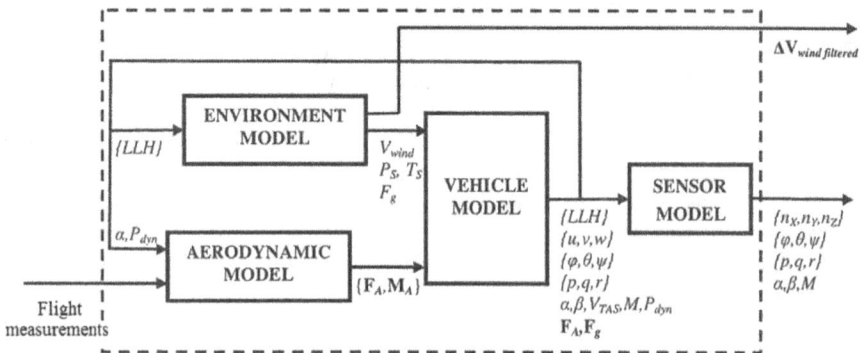

Figure 7. Filter model

It is composed of four main blocks: Vehicle model, Environment model, Aerodynamic model, Sensor model.

The Vehicle model is based on the classical rigid body nonlinear equations of motion (Stevens & Lewis, 2003). Vehicle state vector is composed of Centre of Mass (CoM) position and velocity components, attitude angles, and angular rates. Static algebraic expressions for the computation of aerodynamic angles, Mach number and dynamic pressure, are also included in the model (measurement equations).

The Aerodynamic force (C_F) and moment (C_M) coefficients are computed by the aerodynamic model. They are transformed in dimensional force and moment and sent in input to the vehicle model. More in detail, the aerodynamic coefficients are computed as summation of baseline deterministic components $(\bar{C}_F, \bar{C}_M)$ and corrections $(\Delta C_F, \Delta C_M)$ resulting from stochastic processes. The former are evaluated from the in-flight measurements of load factors $\mathbf{n}$, angular rates ω, and dynamic pressure P_{dyn}, namely

$$\bar{C}_F = -\frac{\left(m \cdot \mathbf{n}|\mathbf{g}|\right)}{P_{dyn}S}, \qquad \bar{C}_M = \frac{\mathbf{I}\dot{\omega} + \omega \times \mathbf{I}\omega}{P_{dyn}SL} \qquad (40)$$

where $\dot{\omega}$ is obtained by numerical differentiation of ω, m and $\mathbf{I}$ are mass and inertia matrix of the vehicle, respectively, $\mathbf{g}$ is gravitational acceleration, S and L are aerodynamic reference surface and length, respectively. The corrections to the baseline components are the unknowns to be estimated by the filter. They are modelled using Gauss-Markov (GM) stochastic models (Gelb, 1989), that require a suitable characterization.

The Environment model is composed of the WGS84 (World Geodetic System), for the computation of the gravitational acceleration as a function of vehicle position, and the atmospheric model. The latter is based on the meteorological data of the European Centre for Medium-Range Weather Forecasts (ECMWF), that provides baseline profiles for wind velocity, air temperature and pressure during the missions. High frequency corrections to these baseline trajectories are estimated by the filter and their dynamic behaviour is again modelled by means of Gauss-Markov models. Concerning the wind velocity, the high frequency corrections are low pass filtered in order to compute their low frequency content. Since we assume that the low frequency content is correctly provided by the ECMWF (that is, the low frequency component of wind velocity coincides with the baseline profile), the output of the low pass filter should be null, therefore it could be compared with a zero virtual measurement in the correction phase of the UKF.

Finally, the Sensor model is implemented to match the specifications of the actual on-board sensors. Globally, the filter models have 25 states to be estimated, that is, 12 states for the rigid vehicle, 6 from the aerodynamic coefficients (corrections to the six baseline trajectories) and 7 from the Environment model (corrections to the baseline trajectories of three wind components, atmospheric temperature and pressure, plus two states related to the low-pass filter).

4.2.3. Characterization of stochastic processes and uncertainties

The stochastic models used by the UKF are to be suitably characterized through the definition of some properties, such as model order, correlation time, process and measurements noises variance, that could affect the filter convergence. Most of them are specified in a rigorous way, as shown in this section. The remaining parameters are considered as free variables for the filter design, tuned when the identification procedure is preliminarily carried out on simulated data. The process noises related to the Vehicle model and to the low-pass filter applied to wind velocity correction are considered very low, due to high confidence in the pertinent models. The measurement noises of Sensor model are described by sensors datasheet, whereas the noise on filtered wind is characterized through the noise covariance matrix given by the ECMWF for the baseline, low-frequency profiles of wind velocity, air temperature and pressure.

The order and statistical characterization of the GM models adopted for the wind correction are assessed through the analysis of flight data collected during the ascent phase of the mission, when the vehicle is carried by a balloon at the release altitude. We assume that, in the ascent phase, the horizontal components of wind velocity in the North-East-Down (NED) reference frame are almost coincident with the corresponding components of the CoM measured velocity (balloon transported by the wind) and that the wind does not change in the time frame between ascent and descent phases. Under these hypotheses, the high frequency correction versus altitude is determined (and stored in a lookup table) as the difference between CoM velocity and wind speed given by the ECMWF in the ascent phase. Then the table is queried with the altitude trajectory of the mission descent phase to get the related correction, and the autocorrelation function of the correction is evaluated. The normalized autocorrelation of the North component of wind correction for DTFT1 is shown in Figure 8 (top plot); a similar plot is obtained for the East component, too.

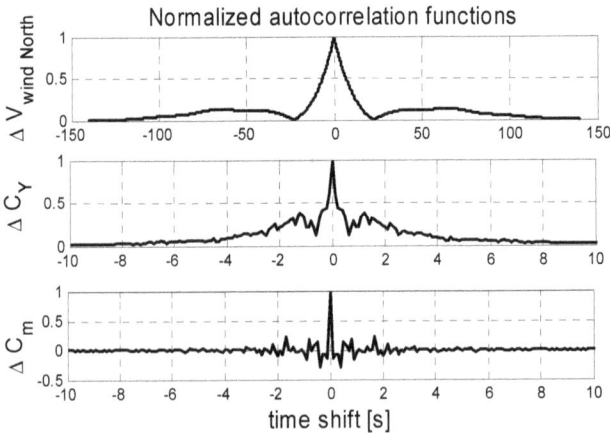

Figure 8. Normalized autocorrelation functions of corrections on North component of wind (top), lateral force (middle) and pitch moment (bottom)

The autocorrelation is typical of a first-order process (Gelb, 1989), described by the model

$$\Delta \dot{V}_{wind\,i} = -\frac{1}{\tau_{wnd\,i}} \Delta V_{wnd\,i} + \eta_{wnd\,i}, \quad i = \{East, North\} \tag{41}$$

where τ_{wind} and η_{wind} are correlation time and process noise, respectively. The correlation time is equal to 1/3 of the time delay, where the normalized autocorrelation function has a value of 0.05. The process noise, characterized by its variance, is a free parameter for the UKF design. The obtained model has also been applied to the Down component of wind correction, where no information can be extracted from the ascent phase data. Since no a priori information was available on the high frequency corrections of static temperature (T_S) and pressure (P_S) with respect to ECMWF, we assume they can be described by a zero-order GM model

$$\Delta \dot{T}_S = \eta_{Ts}, \quad \Delta \dot{P}_S = \eta_{Ps} \tag{42}$$

where the process noises η_{Ts} and η_{Ps} are again design parameter for the filter. The initial value of all the GM state is set to zero.

The characterization of GM models for the aerodynamic corrections is performed through simulation, taking advantage of the a priori information provided by the pre-flight aerodynamic database. As many as 2,000 Monte Carlo simulations of each mission were carried out before flight considering uncertainties on aerodynamics, inertia, initial state, sensors and actuators characteristics, and environmental disturbances. For each simulation, the aerodynamic corrections are evaluated as differences between true aerodynamics (known in simulation) and baseline aerodynamic terms, provided by (40). Then the autocorrelation functions related to the corrections are computed. Finally, for each aerodynamic coefficient a mean normalized autocorrelation function is evaluated, as shown in Figure 8 for the lateral force (middle plot) and pitching moment (bottom plot) corrections. The other force and moment corrections have similar behaviours.

A first-order GM model is selected for the force coefficients, with correlation time computed as described above for the wind corrections. The autocorrelation functions for moment coefficients corrections have an impulsive shape, typical of a zero-order GM processes. Accordingly, we get

$$\Delta \dot{C}_F = -\frac{1}{\tau} \Delta C_F + w_F, \quad \Delta \dot{C}_M = w_M$$
$$C_F = \bar{C}_F + \Delta C_F, \quad \quad C_M = \bar{C}_M + \Delta C_M \tag{43}$$

correlation time, w_F and w_M are the process noises. The variances of process noises σ_{WF} and σ_{WM} are related to the correlation time τ and to the variance σ_{MC} of the simulated trajectories for the aerodynamic coefficients in the aforementioned Monte Carlo analysis, namely

$$\sigma_{WF}^2 = 2\sigma_{MC}^2 / \tau, \quad \quad \sigma_{WM}^2 = \sigma_{MC}^2 \tag{44}$$

4.3. Second identification step

4.3.1. Identifiability analysis

The second identification step aims at estimating from flight data the identifiable subset $\bar{\vartheta} \subseteq \Theta$ of the parameters of the aerodynamic model defined in section 3. Indeed, this model presents many parameters and, taking into account the limited amount of available flight data, not all of them can be updated in post flight analysis. In particular, the attention is focused on the gains F_{sub} and F_{sup} which appear in the addends on the right hand side of equations (22) - (27), (30), (32) - (34). Some of these gains are identifiable and estimated from the flight data. The other gains, as well as all the other parameters of the model, are kept equal to the pre-flight identified values. The selection of the identifiable gains is performed considering the Cramer–Rao bounds (CRBs). The CRB related to the generic parameter $\vartheta_k \in \vartheta$, denoted as σ_{ϑ_k}, is computed through (Jategaonkar, 2006):

$$\mathbf{D} = \mathbf{F}^{-1} = \left[\sum_{i=1}^{N} \left[\left(\frac{\partial \mathbf{y}(t_i)}{\partial \vartheta} \right)^T \mathbf{R}^{-1} \left(\frac{\partial \mathbf{y}(t_i)}{\partial \vartheta} \right) \right] \right]^{-1} \tag{45}$$

$$\sigma_{\vartheta_k} = \sqrt{D_{kk}} \tag{46}$$

where $\mathbf{y}$ is the output vector of the system to be identified, recorded in N time instants, denoted as t_i. $\mathbf{R}$ is the covariance matrix of measurements error on $\mathbf{y}$. ϑ is the set of all the subsonic and supersonic gains of the aerodynamic model. $\mathbf{F}$ represents the information matrix (also named Fisher matrix), $\mathbf{D}$ is the dispersion matrix and D_{kk} is the k-th element on the main diagonal of $\mathbf{D}$.

The CRBs indicate the theoretically maximum achievable accuracy of the estimates and can be considered as a measurement of the sensitivity of system outputs with regard to parameter variations. If the CRB associated to a parameter is bigger than a suitable threshold, the parameter cannot be identified, because its variation has no relevant effect on system outputs and therefore on flight measurements. Concerning the computation of the information matrix, in our case the output $\mathbf{y}$ coincides with the vector composed by the aerodynamic coefficients. Since they are expressed by regular analytical functions, their derivatives with respect to each gain (that is, $\partial \mathbf{y}(t_i)/\partial \vartheta$) can be analytically computed. Finally these derivatives are evaluated along the flight trajectories of DTFT1 and DTFT2 using the flight measurements of Mach number, aerodynamic angles, control effectors deflections and vehicle angular rate. The matrix $\mathbf{R}$ is diagonal and its elements are the aerodynamic coefficients variances. Based on these considerations, the CRB for each gain can be computed and only the parameters having CRB less than 30% of their pre-flight nominal value are selected as identifiable and updated through the analysis of flight data.

4.3.2. Identification methodology

Two different estimation methodologies can be applied in this step. Due to the structure of the aerodynamic model, in both cases parameters estimation is performed independently for each global aerodynamic coefficient and for subsonic and supersonic regime.

The first approach is based on the UKF, already described in section 4.2. It was used for the analysis of DTFT1 flight data (Vitale et al., 2009). The UKF requires the definition of a dynamic model for the unknown parameters. Since they are constant, their dynamics are described by zero order GM processes

$$\dot{\vartheta}_k = \eta_{\vartheta k} \tag{47}$$

The initial condition of this equation is the pre-flight value of the parameter. Covariance matrices of initial condition and process noise are used as design parameters to tune the filter. The output equation is obtained from the analytical model. The first identification step provides a joint characterization of uncertainties on aerodynamic angles, Mach number, angular velocity, and aerodynamic coefficients. In order to properly manage the uncertainty characterization, these variables are all considered as inputs for the output equation of the second step, which is rearranged in term of residual on the aerodynamic coefficient, that is

$$res_i = C_{i-step1} - C_i\left(\mathbf{u}, \overline{\vartheta}^i\right), \quad i = \{D, S, L, l, m, n\} \tag{48}$$

where $C_{i-step1}$ is the i-th aerodynamic coefficient estimated in the first step and C_i is the analogous coefficient provided by the analytical model. The vector u includes Mach number, aerodynamic angles and angular rate estimated in the first step, plus the flight measurements of aerodynamic effectors. Finally $\overline{\vartheta}^i$ is the vector of identifiable parameters associated to coefficient C_i. Equations (47) and (48) are used in the prediction phase of the filter, whereas in the correction phase the residual (resi) is compared with a virtual null measurement.

The second estimation methodology is the Least Mean Square (LMS), that was used for the analysis of DTFT2 flight data. LMS only requires measurements equations, that is the analytical model, and does not need any initial guess or dynamic model describing the dynamics of the unknowns. Since the aerodynamic model is linear in the unknown parameters, in order to perform the estimation, the expression of the i-th aerodynamic coefficient is rearranged as

$$Y_i = \mathbf{A}_i \overline{\vartheta}_i \tag{49}$$

It can be easily demonstrated that Y_i is given by the difference between the global aerodynamic coefficient $C_{i-step1}$ (which is estimated in the first step) and the summation of all the additive terms on the right hand side of equations (22) - (27), (30), (32) - (34) that are related to non-identifiable gains. These additive terms are evaluated using Mach number, aerodynamic angles and angular rate estimated in the first step, and the flight measurements of aerodynamic effectors. $\mathbf{A}_i$ is the matrix of the regressors, which is composed of the additive terms on the right hand side of equations (22) - (27), (30), (32) - (34) related to the identifiable gains divided by the gains themselves, which are included in $\bar{\vartheta}^i$. The unknowns are given by

$$\bar{\vartheta}^i = inv(\mathbf{A}_i^T \mathbf{A}_i)\mathbf{A}_i^T Y_i \qquad (50)$$

Finally, for the LMS technique, the uncertainties on the estimated parameters are evaluated through a Monte Carlo analysis. To this end, many estimations of the same unknown parameters are carried out by using in input flight measurements and global aerodynamic coefficients randomly selected in their range of uncertainty. The statistics of the estimated parameters are then evaluated and used to define the estimation uncertainty on each of the evaluated aerodynamic parameters.

5. Flight data analysis

The analytical aerodynamic model and the identification methodology proposed in this chapter were applied to flight data gathered during the DTFT1 and DTFT2 missions, in order to identify the model of the FTB_1 vehicles. Post flight data analyses of these missions are described in the present section. The time histories of Mach number and angle of attack for the two missions are presented in Figure 9.

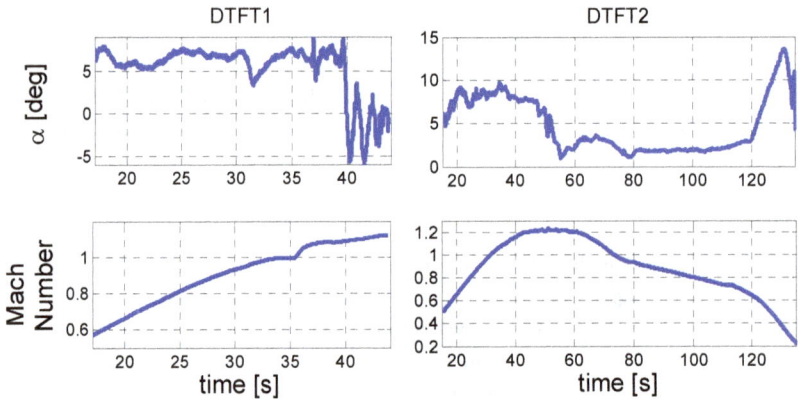

Figure 9. Angle of attack and Mach number time histories for DTFT1 and DTFT2

For both missions, the examined time frame starts 17 seconds after the vehicle drop, when the air data measurement noise is suitably low. For DTFT1 Mach number varied from 0.57 to about 1.08, whereas the angle of attack was held nearly constant at about 7 deg until 39 s. Transonic regime started about 31 s after the drop, where the displacement of the aerodynamic centre created a large perturbation in the pitch moment. At t = 39 s, due to a problem concerning the parachute deployment system, the flight control system switched into a safety mode. Consequently the aerodynamic control surfaces were brought to the neutral position, leading to the variations of α visible in the figure at t > 39 s that resulted from the excitation of the short period dynamics of the vehicle. In the DTFT2 mission Mach number varied from 0.2 to about 1.2. Transonic regime started about 30 s after the drop, while after 77 s the regime was again subsonic. The vehicle performed two sweeps in angle of attack: the first at maximum and constant Mach number, the second in subsonic regime at the end of the mission. In both flights, the sideslip angle was almost always close to 0 deg reference value. Before analyzing the flight data and starting the identification process, a compatibility check on the available measurements was performed, by using kinematic relations (Jategaonkar, 2006), in order to check the measurements consistency and the correctness of the measurement error characterization.

5.1. Results of DTFT1 data analysis

Aerodynamic force and moment coefficients, wind velocity, static temperature and pressure, and vehicle states were estimated in the first step of EBM procedure for the time interval [17, 44 s]. Figure 10 shows the identified longitudinal aerodynamic coefficients, that are compared with the values obtained using the pre-flight ADB and flight measurements required in input by the ADB. Although the coefficients returned by the pre-flight ADB are not far from the estimated values, an update of the pre-flight database appears necessary. In particular, C_L is over predicted as well as C_D in the first 10 seconds of the considered time frame. The estimated values of C_m are very close to zero up to 39 s flight time, whereas the same coefficient computed using the pre-flight ADB assumes negative values. The comparison between the horizontal components of wind velocity estimated by UKF and computed through ECMWF is shown in Figure 11. The UKF, extending the frequency content of wind velocity with respect to ECMWF, improves the evaluation of the wind field experienced by the vehicle which, in turn, has a positive effect on the filtering of the aerodynamic angles. Not shown for the sake of conciseness, the estimated values of Down component of wind velocity, static temperature and pressure are very close to the ECMWF predictions, whereas the filtered states of the vehicle are nearly indistinguishable from the in-flight measurements.

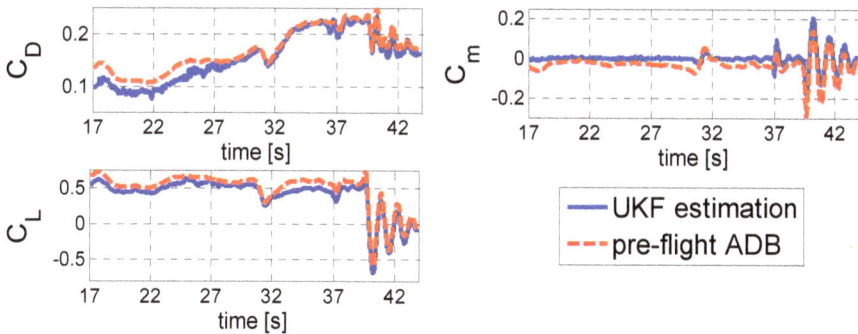

Figure 10. Pre-flight ADB and estimated longitudinal aerodynamic coefficients versus time

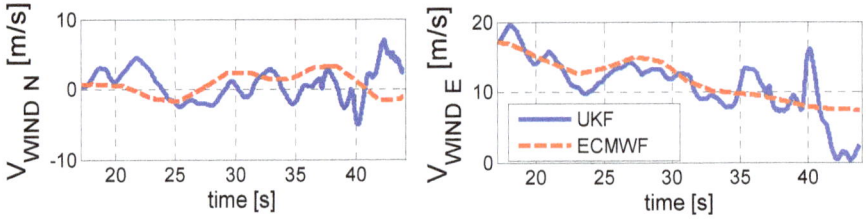

Figure 11. Wind horizontal velocity (in NED) estimated by UKF and provided by ECMWF

In the second identification step the analytical model was updated only for the longitudinal coefficients, because the flight trajectory was basically longitudinal and there was little excitation of lateral-directional dynamics. 6 aerodynamic parameters were estimated in subsonic regime, 3 related to drag coefficient and 3 to lift, by using the flight measurements gathered from 17 s through 36 s flight time. Cramer Rao bounds enhanced that, no parameters could be estimated for the pitch moment coefficient in the subsonic regime, due to the low excitation of attitude dynamics. In transonic regime, from 38 s to 44 s, 10 parameters were estimated, related to the supersonic drag coefficient (3 parameters), lift coefficient (3 parameters) and pitch moment coefficient (4 parameters). The estimated parameters are basically related to zero-order terms and to the aerodynamic derivatives with respect to α and δ_e. Figure 12 shows the convergence characteristics of the parameters related to the lift coefficient in subsonic regime. Similar plots were obtained for the other coefficients. The UKF also provided the uncertainties on the estimated parameters. Figure 13 presents the comparison between pre-flight and post flight uncertainties on main aerodynamic derivatives. The former are provided by the pre-flight ADB, whereas the latter are computed propagating the uncertainties on the estimated aerodynamic parameters through the analytical model. Model identification allowed to significantly reduce these uncertainties in most cases.

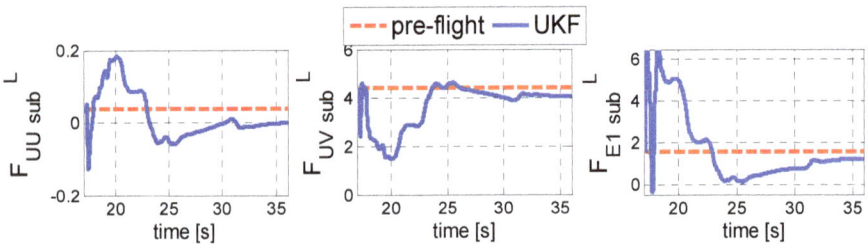

Figure 12. Estimation of the subsonic lift parameters

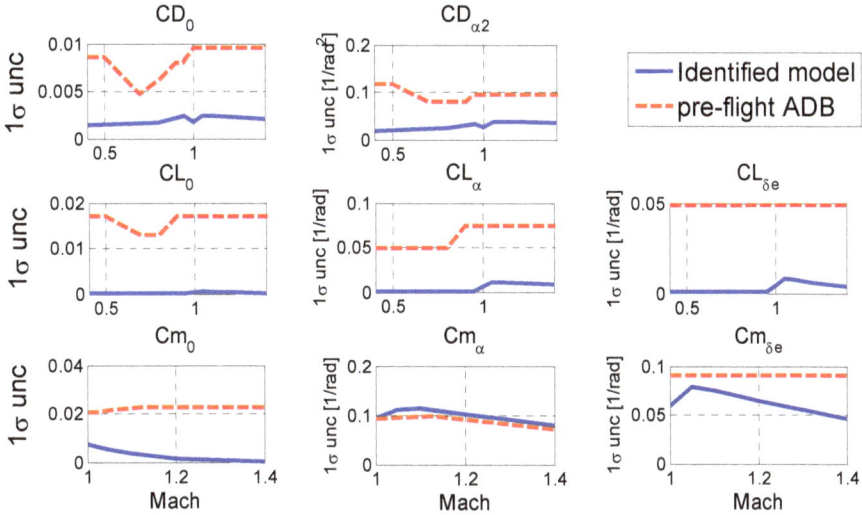

Figure 13. Pre-flight and post flight uncertainties on main aerodynamic derivatives

5.2. Results of DTFT2 data analysis

The DTFT2 mission allowed to identify also the lateral-directional aerodynamics. Figure 14 shows the comparison between the aerodynamic coefficients identified in the first step and the corresponding pre-flight behaviours, provided by the ADB. Matching between ADB and UKF is generally good, but for C_m in most of the trajectory, C_D in the very last part of the trajectory and lateral directional coefficients (C_s, C_l and C_n) in the time interval from 60 s to 80 s. Since in transonic regime the sideslip angle is always null except for the interval from 60 s to 80 s, where it varies between 2 deg and -2 deg (see Figure 17), it can be argued that ADB lateral directional coefficients seem to be too sensitive to sideslip angle variations in transonic regime. As for the pitching moment coefficient, the trajectory trends of the ADB is completely different from the UKF. The vehicle performed the mission in conditions very close to rotational equilibrium with respect to pitch, indeed the estimated pitch moment is about zero. On the contrary, the C_m profile provided by the ADB varies significantly and it is most of the time different from zero. Based on these considerations, a refinement of the model was performed in the second identification step, where 71 aerodynamic parameters were estimated (31 longitudinal and 40 lateral-directional).

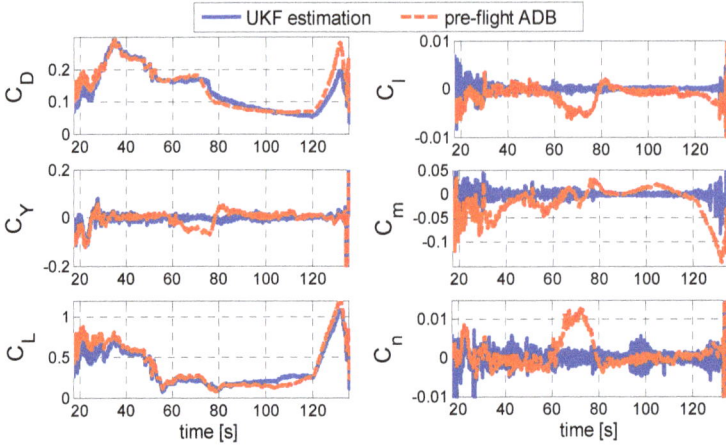

Figure 14. Pre-flight ADB and estimated aerodynamic coefficients versus time

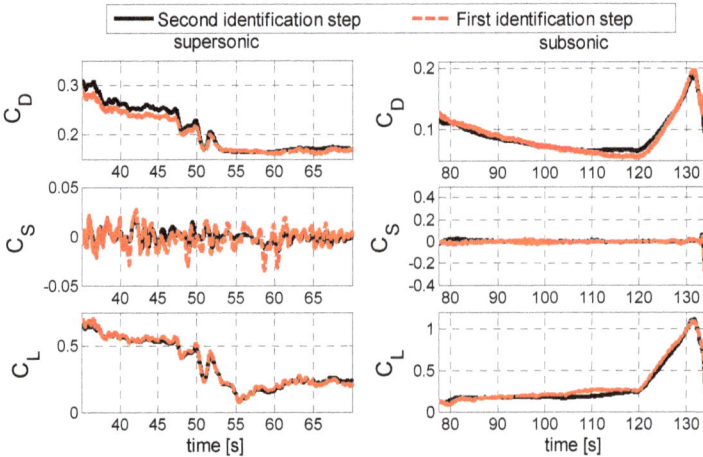

Figure 15. Comparison between aerodynamic force coefficients estimated in first identification step and provided by the identified model

The identified model was validated by using two different procedures. First, the aerodynamic coefficients provided by the model were compared (along the DTFT2 trajectory) with their time histories estimated by the UKF in the first identification step. Results are shown in Figure 15 (for the force coefficients) and Figure 16 (for the moment coefficients). The matching is generally very good, both in subsonic and in supersonic regimes, for all parameters but the pitching moment, the mean value of which is different from zero in some parts of the trajectory. This problem could be due to some of the parameters which were not updated using the flight data. However also for this coefficient the identified model works better than the pre-flight ADB.

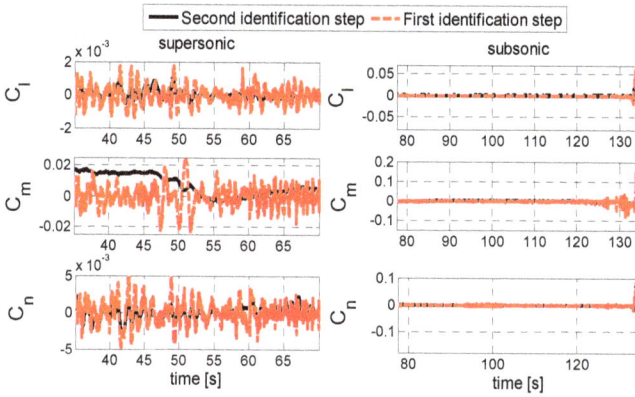

Figure 16. Comparison between aerodynamic moment coefficients estimated in first identification step and provided by the identified model

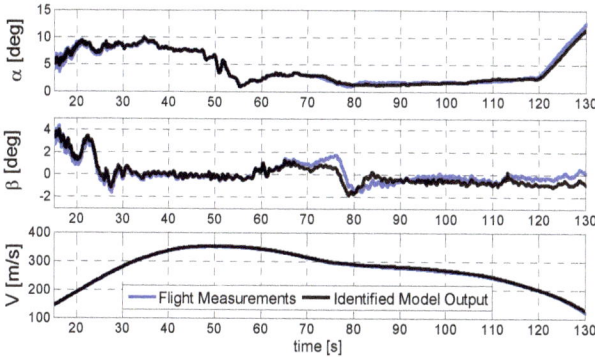

Figure 17. Validation of the identified model by open loop simulation

The second validation was performed through an open loop simulation of the DTFT2 mission (that is, without considering the action of the flight control system), where the identified model was used to simulate the aerodynamic behaviour of the vehicle. The measurements of aerodynamic effectors deflections were provided in input to the model and the outputs of the simulation were compared with the correspondent flight measurements. This test is very critical, because small errors in the identified model lead to the divergence of the simulation, due to the absence of a flight control system which allows tracking the reference trajectory. Indeed, if the aerodynamic force and moment coefficients computed by the identified model were used, the simulation diverged. On the other hand, if the simulation was carried out using the force coefficients provided by the identified model and the trajectory of the moment coefficients estimated by the UKF in the first step, then simulation results are very close to the flight measurements, as shown in Figure 17. This confirms the reliability of the estimated force model, whereas some more investigations are required on the aerodynamic moments model.

6. Conclusion

This chapter presented a novel analytical model for describing the aerodynamics of a re-entry vehicle in subsonic, transonic and supersonic regimes, and an innovative methodology for the estimation of model parameters from flight data.

The structure of the proposed aerodynamic model is based on first principles. As a major advantage, the model can extend the results obtained from the analysis of a single trajectory to the whole flight envelope. Model identification is performed in the framework of a multi-step approach, where the aerodynamic coefficients are identified first and, in a following phase, a set of model parameters is evaluated. In each step, a suitable estimation technique is used. This approach also provides the estimation of useful information on the environment conditions experienced by the vehicle during the flight, such as wind velocity and air temperature and pressure. Another relevant peculiarity of the identification method concerns the use of the Unscented Kalman Filter, the exploitation of all the available a priori information for the stochastic characterization of the filter models through Gauss-Markov processes, and the rigorous management of all the uncertainties involved in the system identification process. As a result, a reliable, complete, and structured statistical characterization of the identified model could be obtained.

The application of the proposed model and methodology to flight data of the first two missions of the Italian unmanned space vehicle provided very good results, in spite of the fact that flight maneuvers specifically designed for parameter estimation were not performed due to safety constraints. Furthermore, the applied estimation techniques did not present any convergence problem, not a trivial result for the considered field of application. Identification from flight data allowed to validate and refine the available pre-flight aerodynamic model in terms of nominal values update and significant reduction on model uncertainties. The availability of an updated aerodynamic model represents a fundamental step for the development of the upgraded version of the Guidance, Navigation and Control system for the next missions of the same configuration, where the accuracy of estimates and the reliability of the model over an expanded flight envelope will be carefully analyzed and assessed.

Author details

Antonio Vitale and Federico Corraro
Italian Aerospace Research Centre, Italy

Guido De Matteis and Nicola de Divitiis
University of Rome "Sapienza", Italy

7. References

Ashley, H., & Landahl, M. (1965). *Aerodynamics of Wings and Bodies* (6th Ed.), Dover, ISBN 0-486-64899-0 New York

Brauckmann, G. J. (1999). X-34 Vehicle Aerodynamic Characteristics. *Journal of Spacecraft and Rockets*, Vol. 36, No. 2, (March-April 1999), pp. 229-239, ISSN 0022-4650

Chowdhary, G., & Jategaonkar, R. (2006). Aerodynamic Parameter Estimation from Flight Data Applying Extended and Unscented Kalman Filter, *Proceedings of AIAA Atmospheric Flight Mechanics Conference*, Keystone (CO), August 2006

Cole, J. D., & Cook, L. P. (1986). *Transonic Aerodynamics*, North-Holland, ISBN 0444879587, New York

Divitiis, N., & Vitale, A. (2010). Fully Structured Aerodynamic Model for Parameter Identification of a Reentry Experimental Vehicle. *Journal of Spacecraft and Rockets*, Vol. 47, No. 1 (January-February 2010), pp. 113-124, ISSN 0022-4560

Gelb, A. (Ed.). (1989). *Applied Optimal Estimation*. M.I.T. Press, ISBN 0-262-20027-9, Cambridge, Massachusetts

Guckenheimer, J., & Holmes, P. (1990). *Nonlinear Oscillations, Dynamical Systems, and Bifurcations of Vector Fields*. Springer, ISBN 0387908196, New York

Gupta, N. K., & Hall, Jr. W. E. (1979). System Identification Technology for Estimating Reentry Vehicle Aerodynamic Coefficients. *Journal of Guidance and Control*, Vol. 2, No. 2 (March-April 1979), pp. 139-146, ISSN 0731-5090

Hildebrand, F. B. (1987). *Introduction to Numerical Analysis*, Dover, ISBN 0-486-65363-3, New York

Hoff, J. C., & Cook, M. V. (1996). Aircraft Parameter Identification Using an Estimation-Before-Modelling Technique. *Aeronautical Journal*, Vol. 100, No. 997 (August-September 1996), pp. 259-268, ISSN 0001-9240

Jategaonkar, R. (2006). *Flight Vehicle System Identification: A Time Domain Methodology*, AIAA, Reston, VA

Julier, S. J., & Uhlmann, J. K. (1995). New Extension of the Kalman Filter to Nonlinear Systems. *Proceedings of SPIE*, Vol. 3068, pp. 182-193 ISSN 0277786X

Kawato, H., Watanabe, S., Yamamoto, Y., & Fujii, K. (2005). Aerodynamic Performances of Lifting-Body Configurations for a Reentry Vehicle. *Journal of Spacecraft and Rockets*, Vol. 42, No. 2 (March-April 2005), pp. 232-239, ISSN 0022-4650

Lamb, H. (1945). *On the Motion of Solids Through a Liquid. Hydrodynamics* (6th Ed.), Dover, New York

Rayme, D. P. (1996). *Aircraft Design: A Conceptual Approach*. AIAA, ISBN 1563478293, Reston, VA

Rufolo, G. C., Roncioni, P., Marini, M., Votta, R., & Palazzo, S. (2006). Experimental and Numerical Aerodynamic Data Integration and Aerodatabase Development for the PRORA-USV-FTB_1 Reusable Vehicle, *Proceedings of 4th AIAA/AHI Space Plane and Hypersonic Systems and Technologies Conference*, Canberra, Australia, November 2006

Stevens, B. L., & Lewis, F. L. (2003). *Aircraft Control and Simulation* (2nd Ed.), John Wiley & Sons Inc., ISBN 0 471371459, Hoboken, New Jersey

Trankle, T. L., Bachner, S. D. (1995). Identification of a Nonlinear Aerodynamic Model of the F-14 Aircraft. *Journal of Guidance, Control, and Dynamics*, Vol. 18, No. 6 (November-December 1995), pp. 1292-1297, ISSN 0731-5090

Vitale, A., Corraro, F., Bernard, M., & De Matteis, G. (2009). Unscented Kalman Filtering for Identification of a Re-Entry Vehicle in Transonic Regime. *Journal of Aircraft*, Vol. 46, No. 5 (September–October 2009), pp. 1649–1659, ISSN 0021-8669

Wan, E. A., & van der Merwe, R. (2000). The Unscented Kalman Filter for Nonlinear Estimation, *Proceedings of IEEE 2000 Adaptive Systems for Signal Processing, Communication and Control Symposium (AS-SPCC)*, Lake Louise, Canada, October 2000.

Wu, Y., Hu, D., Wu, M., & Hu, X. (2005). Unscented Kalman Filtering for Additive Noise Case: Augmented vs. Non-augmented, *Proceedings of American Control Conference*, Portland, Oregon , June 2005

Van Dyke, M. C., Schwartz, J. L., & Hall, C. D. (2004). Unscented Kalman Filtering for Spacecraft Attitude State and Parameter Estimation, *Proceedings of AAS/AIAA Space Flight Mechanics Meeting*, Maui, Hawaii, February 2004

Chapter 8

Computer Modelling of Automobile Fires

Ladislav Halada, Peter Weisenpacher and Jan Glasa

Additional information is available at the end of the chapter

1. Introduction

The use of the *CFD (Computer Fluid Dynamics)* theory and its practical knowledge has become widespread in such academic disciplines as aerodynamics, fluid dynamics, combustion engineering and other fields. However, in the disciplines, which examine the ongoing processes in larger sizes, CFD was applied during the last decades only. One of such discipline is a spread of fire. Fire processes are a very complicated and complex phenomenon consisting of combustion, radiation, turbulence, fluid dynamics and other physical and chemical processes. A good knowledge about complex phenomena and processes occurring during fire in different environments is a significant component of fire safety. As fire itself is very complicated phenomenon, interdisciplinary approach to the problem is required.

This is also one of the reasons of a gradual development of programs in this area. In addition, large dynamics of burning processes requires high spatial resolution for sufficiently accurate calculation and also very small time step of calculation (thousandths of second) resulting in high demands on computer memory and high performance processors.

Knowledge of and experience with the combustion processes were achieved mostly by experiments in the past. This form of investigation is quite difficult not only with regard to financial aspects but also from the point of view of variability of the input data related to burning.

In recent years, computer simulation of fire is used as an economically least expensive method to obtain the knowledge about ongoing fire processes and their visualization. This approach is especially valid in the case of fires in tunnels, car parks and buildings because full-scale fire experiments in such structures could cause serious damages of material and technical equipment. Nowadays, monitoring the development of processes in fire environment allows to achieve relatively good knowledge about the dynamics of liquids and gases. Existing software tools provided by CFD simulation also enable to visualize their development.

In the literature, computer simulation of fire was firstly formulated in the seventies as *zone models* and later as *multi-zone models* [1, 2, 3, 4]. In these models, the fire area is divided into separate fire areas (zones) so that each of these ongoing processes has been settled. The theoretical basis of these methods comprises the laws of conservation of mass and energy. The whole space is divided into two spatially homogeneous zones: warmer upper volume containing heat and smoke, and lower part significantly less affected by heat and smoke. For each zone, mass and energy balances are enforced with additional models describing other physical processes such as fire plumes, flows through doors, windows and other vents, radiative and convective heat transfer and solid fuel pyrolysis. Relative physical and computational simplicity of the zone models has led to their widespread use in the analysis of compartment fires scenarios. However, for certain fire scenarios, more detailed spatial distributions of physical properties are required.

Computational fluid dynamics (CFD) models were introduced in the nineties and have reached significant development and relatively widespread use in various fields of human activity. The rapid growth of computing power and advances in CFD have led to the development of CFD based field models based on solving the Reynolds-averaged form of the Navier-Stokes equations. The use of CFD models has allowed the description of fires in complex geometries incorporating a wide variety of physical phenomena related to fire.

The simplified equations developed by Rehm and Baum [5], referred to in the combustion research community as low Mach number combustion equations, describe low speed motion of a gas driven by chemical heat release and buoyancy forces [6]. These equations are solved numerically by dividing the physical space where the fire is to be simulated into a large number of rectangular cells. Within each cell, gas velocity, temperature, etc., are assumed to be uniform changing only with time. The accuracy with which the fire dynamics can be simulated depends on the number of cells that can be incorporated into the simulation. This number is ultimately limited by the computing power available. Nowadays, single processor computers limit the number of such cells to at most a few millions. Parallel processing can be used to extend this range to some extent, but the range of length scales that need to be accounted for if all relevant fire processes are to be simulated is roughly 10^4 to 10^5 because combustion processes take place at length scales of 1 mm or less, while the length scales associated with building or automobile fires are of the order of tens of meters/centimeters.

Several advanced systems intended for simulation of combustion processes have been developed. CFX, SMARTFIRE and FDS [7, 8, 9] and others provide alternative models which may offer a good performance. We use the Fire Dynamics Simulator (FDS) system [9], whose first version was developed in 2000 by the NIST (National Institute of Standards and Technology, USA). At present, significantly better FDS version 5.5 is available already. The development of both CFD software systems and their program modules will certainly continue. It is obvious that a careful verification and validation of these systems will continue in the future to enhance their quality and reliability.

2. Basic equations of the FDS model

FDS solves a form of conservation equations for low speed, thermally driven flow. Smoke and heat transfer from fires is the main concern of this system, which also includes the thermal radiation, pyrolysis, combustion of pyrolysis products, flame spread and fire suppression by sprinklers. The basic set of the conservation mass, species, momentum and energy equations are as follows [9]:

$$\frac{\partial \rho}{\partial t} + \nabla.\rho \mathbf{u} = \dot{m}_b'''$$

$$\frac{\partial}{\partial t}(\rho Y_\alpha) + \nabla.\rho Y_\alpha \mathbf{u} = \nabla.\rho D_\alpha \nabla Y_\alpha + \dot{m}_\alpha''' + \dot{m}_{b,\alpha}'''$$

$$\frac{\partial}{\partial t}(\rho \mathbf{u}) + \nabla \rho \mathbf{uu} + \nabla p = \rho \mathbf{g} + \mathbf{f}_b + \nabla \tau_{ij}$$

$$\frac{\partial}{\partial t}(\rho h_s) + \nabla.\rho h_s \mathbf{u} = \frac{Dp}{Dt} + \dot{q}''' - \dot{q}_b''' - \nabla.\dot{\mathbf{q}}'' + \varepsilon,$$

where $\dot{m}_b''' = \sum_\alpha \dot{m}_{b,\alpha}'''$ is the production rate of species by evaporating droplets or particles; ϱ is the density; $\mathbf{u}=(u,v,w)$ is the velocity vector; Y_α, D_α, and $\dot{m}_{b,\alpha}'''$ are the mass fraction, the diffusion coefficient, and the mass production rate of α-th species per unit volume, respectively; p is the pressure; f_b is the external force vector; τ_{ij} is the viscous stress tensor; h_s is the sensible enthalpy; the term $\dot{q}'''$ is the heat release rate per unit volume from a chemical reaction and $\dot{q}_b'''$ is the energy transferred to the evaporating droplets; and the term $\dot{\mathbf{q}}''$ represents the conductive and radiative heat fluxes. Note that the use of the material derivative D( )/Dt = ∂ ( )/∂ t + u.grad ( ) holds in the last equation.

Other two equations, the pressure equation and the equation of state,

$$\nabla^2 H = -\frac{\partial}{\partial t}(\nabla \mathbf{u}) - \nabla.F \quad \text{and} \quad p = \frac{\rho RT}{W}$$

are added to the previous four equations. The pressure equation is obtained applying the divergence on the momentum equation. In this equation, the value H represents the total pressure divided by the density. R is the universal gas constant, T is the temperature and W is the molecular weight of the gas mixture.

Thus, we have the set of six equations for six unknowns, which are functions of three spatial dimensions and time: the density ρ, three components of $\mathbf{u} = (u, v, w)$, the temperature T and the pressure p. These equations must be simplified in order to filter out sound waves, which are much faster than typical flow speed. The final numerical scheme is an explicit predictor-corrector finite difference scheme, which is second order accurate in space and time. The flow variables are updated in time using an explicit second-order Runge-Kutta scheme.

Boundary conditions are prescribed on walls and vents. All input data for simulation are required in the form of a text file in prescribed format, which describes the coordinate system, geometry of domain and its location in given coordinates, mesh resolution obstacles, boundary conditions, material properties and other different simulation parameters. An important limitation of the system is that the domain should be rectilinear, conforming with the underlying grid. The domain is filled with rectangular obstructions representing real objects, which can burn, heat up, conduct heat, etc. Simulation outputs include quantities for gas phase (temperature, velocity, species concentration, visibility, pressure, heat release rate per unit volume, etc.), for solid surfaces (temperature, heat flux, burning rate, etc.), as well as global quantities (total heat release rate, mass and energy fluxes through openings, etc.). These outputs are saved during simulation with the desired format for visualization and can be visualized by the Smokeview program.

As it is mentioned in [9], the overall computation can either be treated as Direct Numerical Simulation (DNS), in which the dissipative terms are computed directly, or as Large Eddy Simulation (LES), in which large-scale eddies are computed directly and subgrid-scale dissipative processes are modelled. The numerical algorithm is designed so that LES becomes DNS as grid is refined. The numerical schemes used for the solution of all equations is completely described in [9].

All FDS calculations must be performed within a domain consisting of rectilinear meshes divided into rectangular cells. These cells can be either uniform in size or stretched, fulfilling the requirements of finite difference numerical scheme used in FDS. Their number depends on the desired resolution of fire scenario. As the FDS numerical scheme uses Fast Fourier Transforms (FFTs) in the y and z directions, the second and third mesh dimensions should each be of the form $2^l \times 3^m \times 5^n$ where l, m and n are integers.

In this chapter, we focus on computer simulation of:

- automobile engine compartment fire
- automobile passenger and/or luggage compartment fire
- fire spread from burning automobile onto a near standing automobile.

3. Automobile fire experiments

Automobile fires, causing enormous losses of property and lives as well as large environmental damages, have become a significant phenomenon, highly injurious to public. They occur in different surroundings, both in open air (on roads, on open parking areas, in vicinity to forest) and in closed or semi-closed structures (in underground car parks, in garages, in tunnels). Such different types of fire show own specific behaviour and the way of spread. Therefore, fire investigation has to take into account various conditions affecting the fire development and serious research is of great importance.

Numerous papers dealing with different aspects of automobile fire safety have appeared and several advanced computer fire simulation systems have been developed. Such systems are capable to model, simulate and visualize the spread of flames and smoke and the fire behaviour, and even to estimate environmental damages caused by fire in various conditions.

Full-scale automobile fire experiments belong to the important means which are necessary for improvement of automobile safety. They provide information about overall automobile fire behaviour which may be difficult to extrapolate from laboratory testing of the automobile components. They are usually expensive to carry out, leading to short experimental series consisting of a few or even only one experiment. Typically they are extensively instrumented in order to obtain as much information as possible from one experiment. They require special experimental facilities because of great size of automobile fires leading to damage of real structures. Several experiments on automobile fires had been carried out under different conditions (open, closed, or semi-closed car parks) and reported in the literature.

The measurements in older experiments [10, 11, 12] were restricted to gas and structure surface temperatures in the vicinity of burning car. During three full-scale experiments reported in [13], the heat release rate, mass loss and mass loss rate, heat flux, carbon monoxide, carbon dioxide and smoke production rate and gas temperatures above the automobile, and temperatures inside the automobile were determined as a function of time. Two full-scale fire experiments in a tunnel shuttle wagon [14] were carried out. A series of 10 experiments (5 with one and 5 with two automobiles) with a ceiling above the automobile and spread of fire from one automobile to another [15] were performed, collecting the combustion products and oxygen consumption where rate of heat release was measured using oxygen consumption calorimetry. A series of 10 experiments (3 experiments with a single automobile, 6 with two automobiles and 1 with three automobiles parked next to each other) were carried out [16, 17, 18], where heat release rate was measured using oxygen consumption calorimetry. Two experiments with three latest generation automobiles parked in an open car park were carried out [19], where gas and steel temperatures and vertical displacements of steel beams were reported. One experiment in car park (3 automobiles in 0.5 and 0.7 m distances) was performed [20]. An experiment with automobile fire in a four-storey car park was performed [21], measuring temperatures, displacement and strain of car park structures. Four full-scale automobile fire experiments were carried out [22], igniting in different parts of automobile and measuring mass loss, mass loss rate and temperature curves.

In 2009, a series of three full-scale automobile fire experiments with 4 cars in open air in the testing facilities of Fire Protection College of the Ministry of Interior of Slovak Republic in Povazsky Chlmec (Slovakia) and one full-scale automobile fire experiment with 2 cars in 0.5 m distance in the experimental tunnel of Scientific Research Coal Institute in Stramberk (Czech Republic) were carried out [23, 24, 25]. The primary objective of these experiments was to measure gas and surface temperatures (on the surface, above and inside the automobile and engine compartment) to gather data for validation of computer simulations of tested automobile fire scenarios. The main aim of the experiments was to obtain better knowledge about burning of a single automobile, to determine most combustible materials contributing to fire and to study the spread of fire from one vehicle to another. On the basis of these experiments we have consequently made computer simulations.

4. Automobile engine compartment fire experiment and simulations

This type of automobile fire has specific fire behaviour and nowadays it belongs to the most frequent vehicle fires [25, 26, 27, 28]. According to our knowledge, the spread of fire inside the engine compartment has not been analysed by computer fire simulator and published yet. We believe that such type of fire analysis and consequent computer simulation is very important. It provides more realistic fire spread in vicinity of the automobile and therefore more realistic input data for computer fire simulations in car parks, garages, or tunnels. We conjecture that ignition process for car fire simulations in such places simulated by a pool with burning gasoline (for example) is less realistic because it produces constant heat release rate.

The reported automobile fire experiments were carried out on automobiles of the same category. In the sequel, we focus on the first fire experiments from the reported series of full-scale automobile fire experiments. The first one was carried out on Audi 80 Quattro with just slightly distorted bodywork and missing front lights parked in open air on a concrete surface (Fig. 1). All doors and windows of the automobile were closed during the experiment. Gas temperatures inside the engine compartment (on the upper part of the engine block), inside the passenger compartment (in front part on the dashboard) and above the automobile (at 1.5 m height) were measured by thermocouples with 10 s recording interval. Temperature on the engine compartment lid surface was recorded by infra-red camera placed in front of the automobile at 3.5 m height and at 6 m distance from the automobile. The fire behaviour observations were recorded by digital cameras. Basic meteorological data (exterior temperature, wind speed, atmospheric pressure and others) were also collected.

Figure 1. Automobile with thermocouples prepared for the experiment

The fire was ignited by burning of a small amount of gasoline (about 5 ml) poured onto a small cloth, which was placed in the engine compartment under the rubber tube (Figures 2 and 3) imitating a frequent failure in the engine compartment. Immediately after the fire ignition, the engine compartment lid was closed. Sparse smoke above the lid appeared very soon and became denser gradually. During the first three minutes, smoke from the interstices at right edges of the lid and on the lid below the front window was prevailing. At the 4th minute, the smoke leaking out from the holes at the place of missing front lights became more intensive. At the 7th minute, weak flames appeared in the hole at the place of the missing left front light. Between the 7th and 8th minute of fire, flames appeared on the surface on the left side of the lid (varnish ignition). After 12 minutes, the fire was extinguished and inspection started to determine the components degraded most by the fire, and to detect the most flammable materials in the engine compartment.

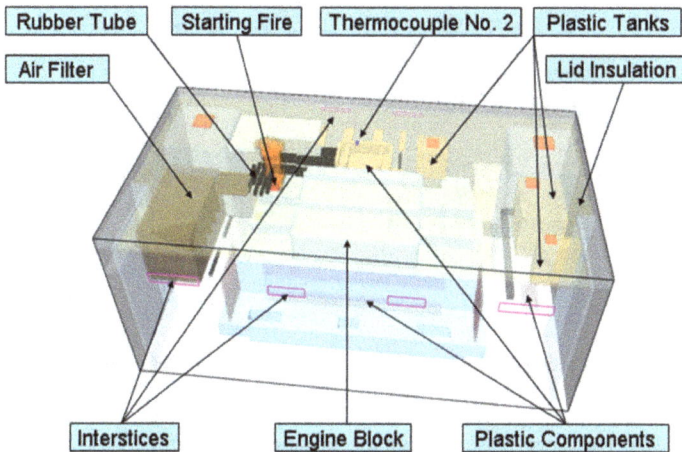

Figure 2. Scheme of automobile engine compartment

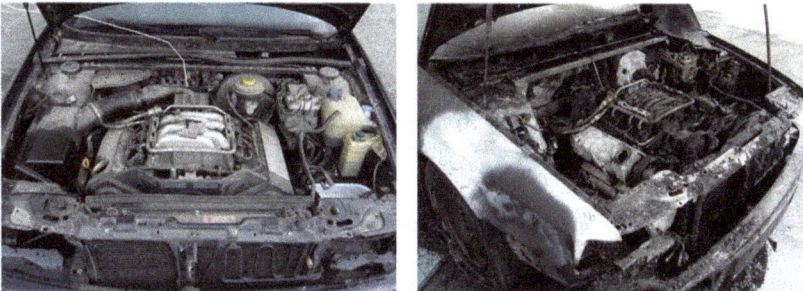

Figure 3. Engine compartment before (left) and after (right) the fire experiment

The most flammable objects in the engine compartment, completely or almost completely burnt during the first 12 minutes of burning, were: the rubber tube (above the fire ignition source), 3 plastic tanks (in the right back part of the engine compartment), air filter box with its paper content (on the left side of the engine compartment), and several other plastic components (Figures 2 and 3). It was confirmed that these components contribute most to fire in engine compartment. The inspection also showed significant degradation of several electric cables and other hoses, partially auto battery and partially molten parts of engine block, and other components afflicted by fire. However, the contribution of these components to fire seems to be small. Fig. 4 shows that the temperature recorded by the thermocouple in the automobile engine compartment reached as many as 900 °C.

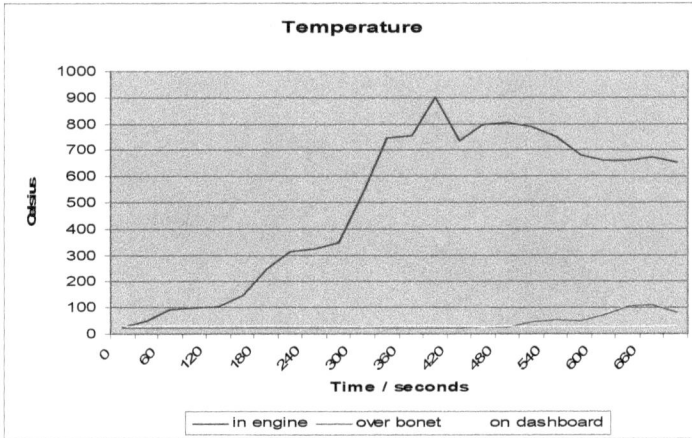

Figure 4. Temperature in the engine compartment, over the bonnet and on the dashboard

The next automobile fire experiment modelled a possible real situation of fire occured during car driving. The driver observed fuming, stopped the car after some hesitation, and decided to extinguish the fire in engine compartment through slightly ajar lid by portable car fire extinguisher. During this experiment, a technical failure was confirmed.

Fire fighters, who should have fought the fire, could not open the engine compartment due to the melting of the lock control mechanism of the engine compartment lid. It was even not possible to open the engine compartment by heaver and plate shears had to be used (see Fig. 5). For this reason, the fire fighting could start with a substantial delay only. Thus, it can be concluded that cars with such plastic control lock mechanism of engine compartment lid may not be extinguished by portable car fire extinguisher due to inaccessibility of the space where the combustion takes place.

The temperature peak over the bonnet comes earlier than that in engine compartment. Probable reason of that is insufficient oxygen supply in engine compartment, which suppressed the fire spread inside the engine compartment and caused that the thermocouple in the engine compartment was not exposed to direct plume. Instead, the sealing compound burnt down and gaseous fuel was created in the engine compartment.

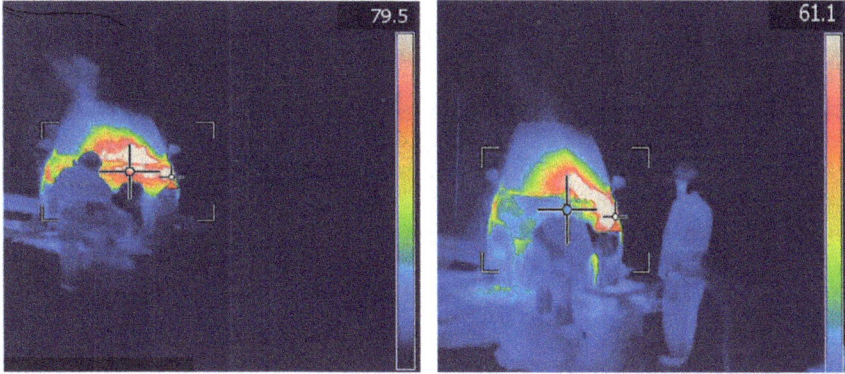

Figure 5. Destruction of the plastic control lock mechanism of engine compartment lid

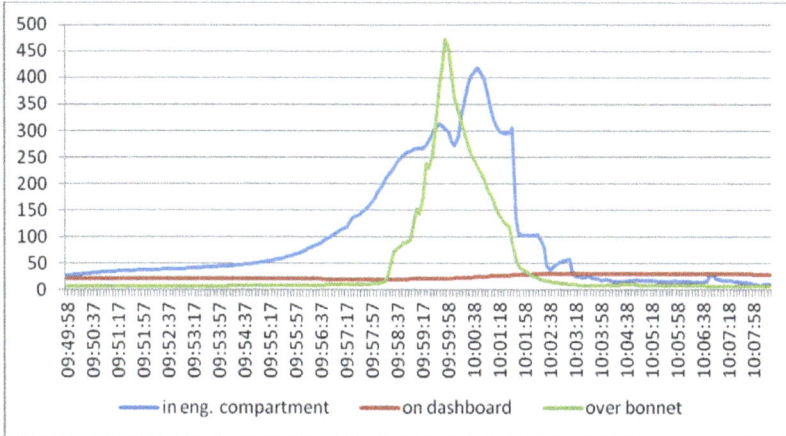

Figure 6. Temperature in the engine compartment, over the bonnet and on the dashboard

4.1. Computer simulation of engine compartment fire

Simulation of automobile engine compartment fire belongs to the most complex FDS simulation problems. Due to the extremely complex geometry of the burning space and objects inside the engine compartment, which affect the fire development and then have to be accurately captured, such a simulation requires very fine mesh resolution and therefore significant computational power for calculation. That is why the design of engine compartment geometry and of the components in the input FDS file is of great importance. In order to properly model the engine compartment as well as all relevant flammable components in its interior, corresponding input geometry of simulated space was elaborated using available 3D scans, as well as direct measurements of distances and proportions of detected flammable components (done before the start of experiment). The simulation domain includes all flammable components mentioned above and other plastic and metallic components which influence the direction of fire and smoke spread in time (see Fig. 2).

Engine compartment fire simulation is just the situation in which certain geometric features of engine compartment components do not conform to rectangular mesh, and have to be represented in a different way (electrical cables, tubes, hoses, etc.). The shape of plastic tanks and air filter can be captured almost realistically. However, the rubber tube (small thickness and cylindrical shape) has to be represented by a cluster of thin stripes where the total mass of stripes is equal to the mass of the tube itself. Moreover, the surface to volume ratio of the stripes is the same as that of the tube. Both these parameters, which are crucial for heating up and burning of material, were maintained. By this way some other components, such as paper interior of the air filter box, were modelled. Several other components, which did not appear to be critical for heat transfer and burning (e.g. small plastic caps of tanks too small to be properly captured), were represented by plastic or metallic slabs placed at appropriate places in the simulation space.

Proper determination of material properties (physical parameters describing burning properties of materials) in the engine compartment was another essential task to be solved. There were four prevailing types of materials identified: aluminium alloy (metallic components), rubber (rubber tube), paper (air filter interior) and mixture of polyethylene (plastic components). Some material parameters for automobile varnish were estimated (e.g. the heat release rate per unit area) and some of them were derived from observations (e.g. the ignition temperature was determined from recorded infra-red camera observations) during the experiment. Most material parameters were determined by specialists from University in Zilina (Slovakia).

In the simulation, the fire ignition source (the small burning cloth placed on the engine block under the rubber tube) is represented by a burning surface with the dimensions of 4 x 4 cm and total heat release rate set to 2.1 kW for the period of 60 s.

Several measurement devices were defined in the simulation to provide proper output parameters describing the fire behaviour in time, such as gas temperature and velocity, surface temperature, oxygen concentration, etc.

4.2. FDS simulation results

For the simulation of the fire scenarios described above, we used a PC with 4-core processor Intel Q9550 with the frequency of 2.83 GHz, 8 GB of RAM and 1TB hard disk.

Two types of the engine compartment fire simulation are briefly presented in this section. They differ from each other:

- in the manner in which the computational domain was chosen
- by their computational requirements, the range of consideration of the automobile environment in fire scenario

- by the number and precision of output quantities provided.

For each case, dozens of simulations were performed in order to test and find proper values of significant parameters and appropriate location of interstices affecting the oxygen supply.

4.2.1. Simulation A

In this simulation, the computational domain includes the interior of engine compartment and 3 cm space above the engine compartment lid to show also the temperature distribution on the upper surface of the lid. The domain boundary conditions defining the heat transfer are given by the material properties of the bodywork (aluminium alloy) and are set 'OPEN' for the upper boundary of the 3 cm space above the engine compartment. The interstices in the bodywork (Figures 2, 3 and 7) are represented by narrow vents with boundary conditions 'OPEN', which allow convective gas transfer between the engine compartment and the automobile environment, not explicitly included in this simulation. They represent the holes at the place of missing front lights and the corresponding interstices at the front and back sides of the engine compartment and on the lid edges.

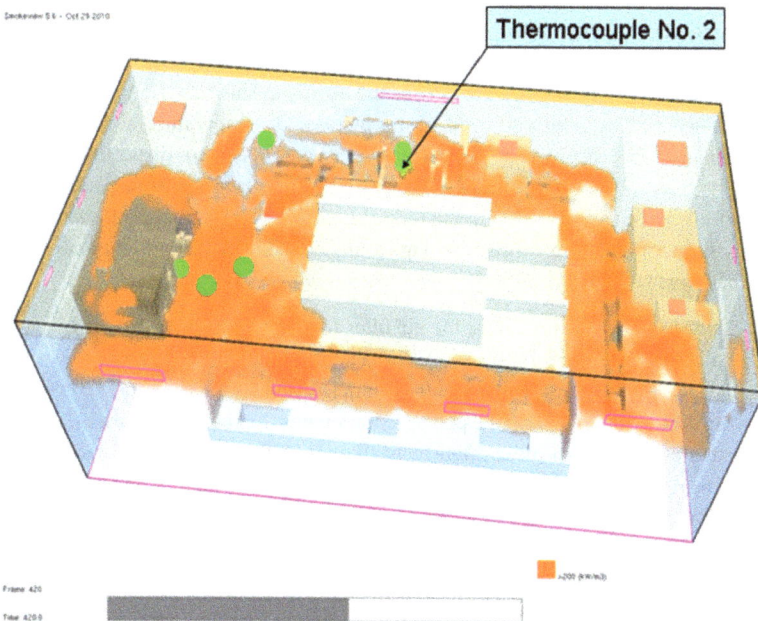

Figure 7. Simulation A in the 7th minute of burning. The orange colour volumes represent the volumes with HRR per unit volume values higher than 200 kW.m-3

The whole computational domain is represented by one computational mesh of 1 cm resolution with 503,712 cells which was assigned to one CPU core. Most of the dynamic processes of burning occurred in the left part of the mesh. The simulation of 720 s of fire required 207 hours of CPU time at Intel Q9550, 2.83 GHz CPU.

The main fire spread tendencies are similar to observations during the fire experiment. The simulated fire behaviour was as follows. At the 22nd second of fire, the rubber tube and the lid insulation layer above the burning cloth started to burn. At the 50th second of burning, flames were spread in the whole engine compartment. At that time, the total heat release rate (HRR) of about 15 - 20 kW was achieved. In rear part of the engine compartment, significant ascending air flow was created by the interstices located under the front window, which made the fire spread along the rubber tube to the rear part of the engine compartment.

This fire behaviour is in accordance with the observation of intensive smoke from the interstices below the front window in this phase of fire and with increase of the lid temperature detected by IR camera, as well as with increase of thermocouple No. 2 during the fire experiment (see Figures 4 and 8).

Meanwhile, the air filter was heated up and its paper content ignited, although its HRR was very low because of insufficient oxygen supply. Between the 200th and 300th second of simulation, the plastic components in the rear part of the engine compartment burnt away, which led to significant increase in air supply and caused temporary temperature increase at sensor No. 2 and to more intensive burning of the air filter. This behaviour is in accordance with the fire experiment, which showed strong increase of temperature of a part of the lid above the air filter in that phase of fire (see the surface temperature distribution on the lid shown in Fig. 8 as well as the peak in Fig. 4), as well as intensive smoke produced in the front part of the automobile. Although the range of colour schemes for surface temperatures in simulation and IR camera outputs in Fig. 8 are not the same, the temperatures show similar behaviour.

Heat and flames spread mainly along the air filter in the direction towards the hole at the place of missing left front light. At the 428th second of simulation (440th second in the fire experiment), the outer lid surface reached the temperature of 300 °C and the varnish ignited. After the air filter burnt away, the plastic components in the front part of the engine compartment started to burn more intensively (flames in the holes at the place of missing front lights were observed during the experiment too). The maximal value of total HRR achieved was about 40-50 kW.

The maximal surface temperature on the lid achieved by simulation is well comparable to the values recorded by the IR camera. It can be seen in Fig. 8 that the lid surface temperature distribution in time is in good spatial and temporal accordance with IR camera records (see also Fig. 9). The time of varnish ignition on the lid surface corresponds with observations during the experiment.

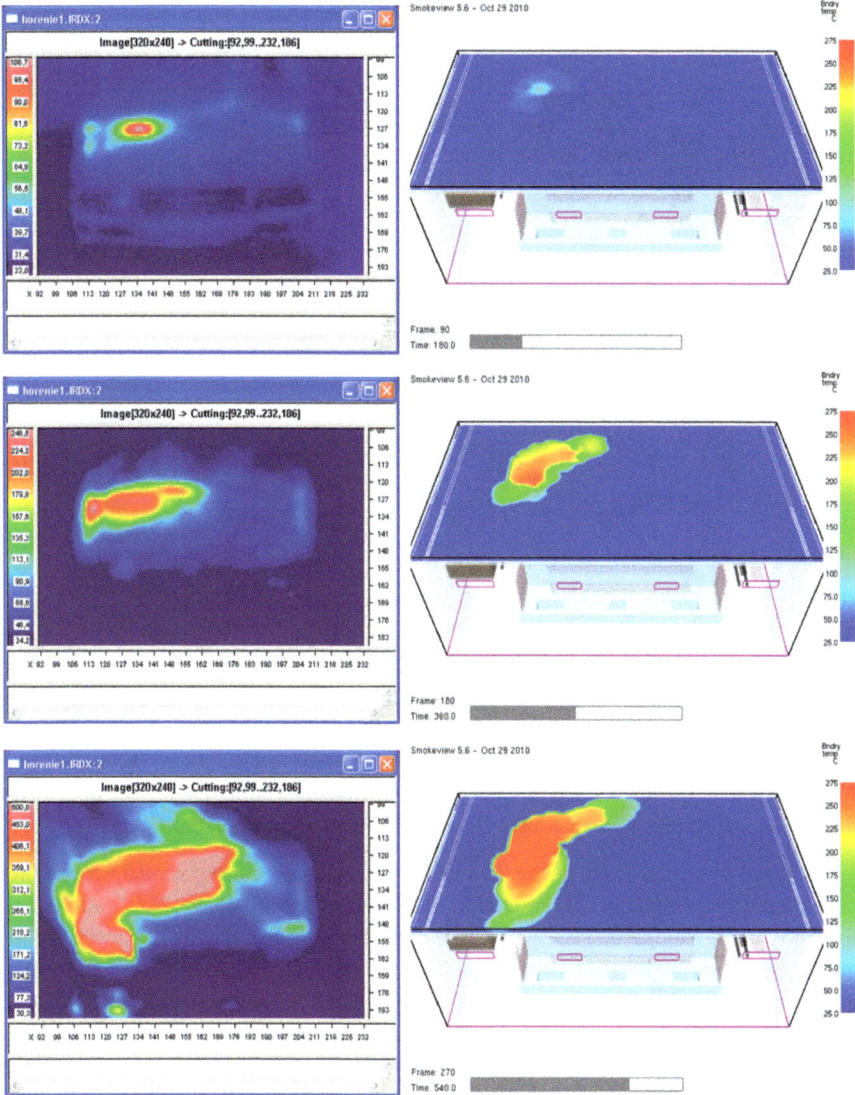

Figure 8. Surface temperature distribution on the engine lid: in IR camera output (left) and in simulation A (right) in the 3rd, 6th and 9th minute of burning

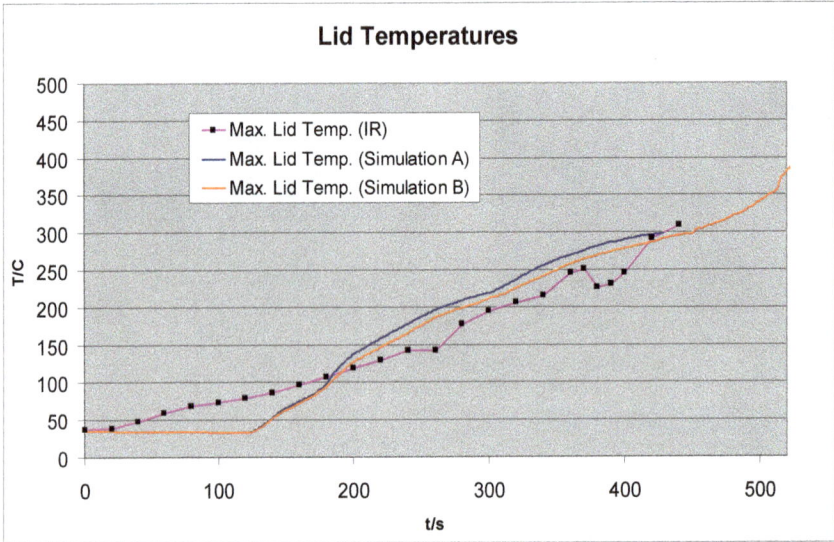

Figure 9. Maximal temperature of the lid surface in simulation and in the IR camera output

4.2.2. Simulation B

In this case, the computational domain includes the engine compartment interior as well as additional space above the engine compartment lid providing information about the fire spread above the engine compartment (see Fig. 10). The domain boundary conditions defining the heat transfer are again given by the material properties of the bodywork (aluminium alloy) and 'OPEN' for the space above the engine compartment. The interstices in the bodywork are represented as before by vents with boundary conditions 'OPEN'.

The additional mesh placed above the lid includes the space, in which potential varnish fire occurs during the simulation. Its resolution decreased to 2 cm having 192,640 additional cells. The engine compartment itself was assigned to one computational mesh. Most dynamic processes of burning occurred inside the engine compartment, therefore, the impact of such division on the accuracy of computation was small. The additional space above the engine compartment lid did not influence the simulation of fire in the engine compartment interior much (see Fig. 9). However, the simulation provides more realistic visualization of fire above the lid (see Fig. 10). The advantage of case B in comparison with case A is that it provides further additional output quantities of fire outside the engine compartment (e.g. the HRR of varnish fire, temperature above the lid, etc.) at the cost of growth of computational requirements. The simulation of 700 s of fire required 200 hours of CPU time at Intel Q9550, 2.83 GHz CPU.

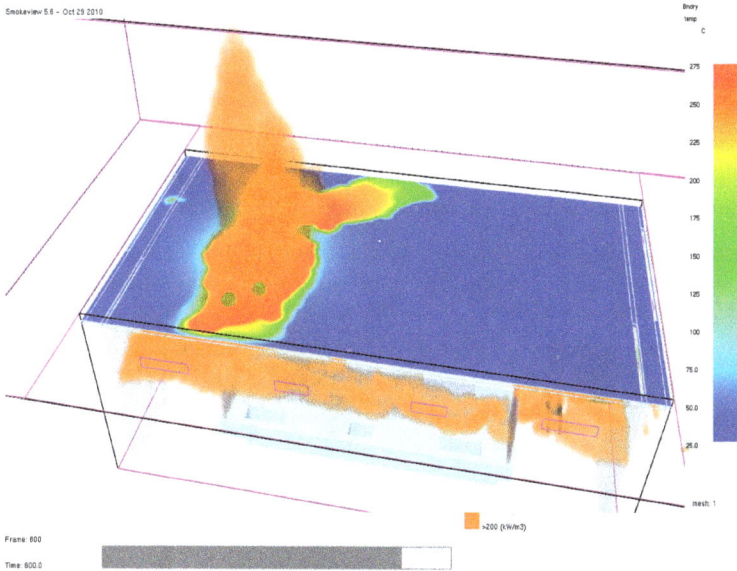

Figure 10. Simulation B in the 10th minute of burning. The orange volumes represent the ones with HRR per unit volume values higher than 200 kW.m⁻³

The main tendencies of the fire spread in this simulation are similar to the fire behaviour in simulation A and to observations during the fire experiment. The results of HRR of both simulation A and B are shown in Fig. 11.

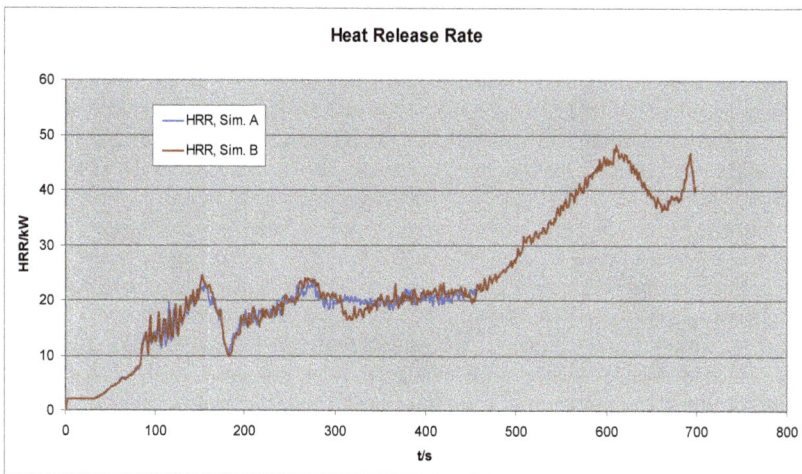

Figure 11. Heat release rate of fire in the simulations A and B

5. Automobile fires in passenger and luggage compartments

Fires of automobile passenger and/or luggage compartment are not very frequent. However, if they occur, they can be very dangerous particularly in urban conglomeration. They often accelerate fast especially in sooner manufactured automobiles with more flammable car seat upholstery materials and/or flammable vehicle load. Such fires can cause large damages especially in the case, when the interior of automobile is sufficiently supplied with oxygen during fire (for example, if it flows through a broken window or ajar door).

In this section, we will describe the FDS simulations of automobile fires in passenger and luggage compartment related to one of three full-scale fire tests conducted during the third fire experiment carried out in Slovakia in 2009 (see Fig. 12).

Figure 12. Full-scale fire experiment: passenger compartment fire

The fire was initiated on the back seat behind the driver seat. The left back window was open. The aim of the experiment was to monitor the course of passenger compartment fire and subsequent spread of fire from burning automobile onto a near-standing vehicle. Temperatures were measured by thermocouples on the engine block, over the engine compartment and in the middle of dashboard. Lateral air flow was imitated by a fan. The experiment showed a strong relation between the passenger compartment fire and the air supply via the open window. Substantial fire acceleration was observed in the moment of the windscreen destruction. High plumes and dense smoke appeared in the passenger compartment very soon in the same minute as the fire was ignited. The temperature in the passenger compartment reached the value of 750 °C in the second minute of fire. In the 3rd minute of burning, the windscreen destruction and airbags explosion occurred followed by strong plumes. At the 4th minute of fire, a driving mirror burning was observed caused by over-jumping a burning piece of flammable material from burning automobile in consequence of the airbag explosion. We performed a simulation of the fire scenario described above. The geometry of the used automobile was constructed and flammable components in the passenger compartment were modelled by the upholstery and plastic volumes (see Fig. 13).

The real material properties were estimated, or gathered from tables. Boundary conditions 'OPEN' were used for the domain boundaries. For the simulation, one computational mesh of 3 cm mesh resolution consisting of 1,049,760 cells was used. The simulation of 600 s of fire required 130 hours of CPU time at Intel i7-990X CPU, 3.46 GHz CPU. Some typical phases of the simulated passenger compartment fire are demonstrated in Figures 14 and 15.

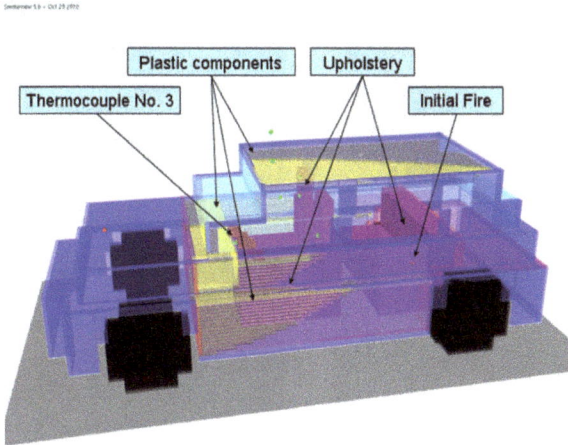

Figure 13. Scheme of automobile passenger compartment and used materials

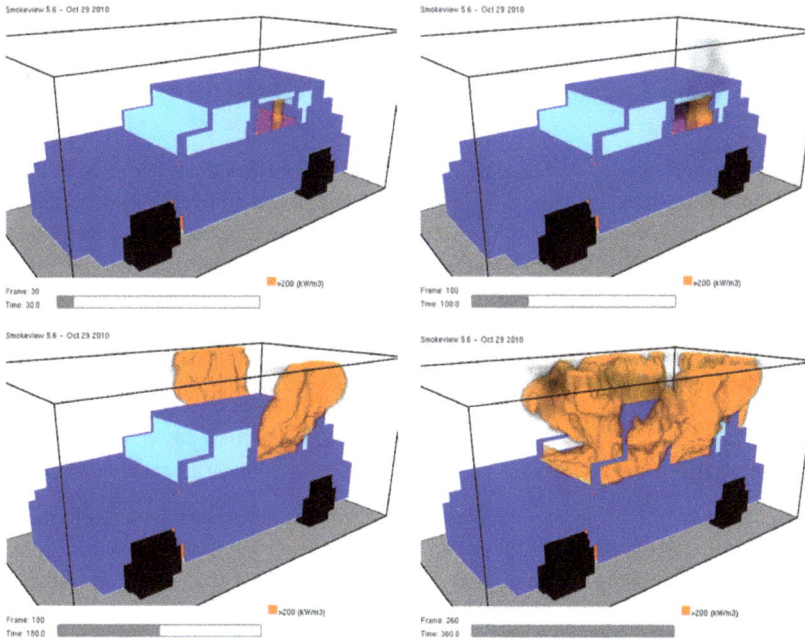

Figure 14. Simulation of passenger compartment fire in the 30th, 100th, 180th and 360th s of fire

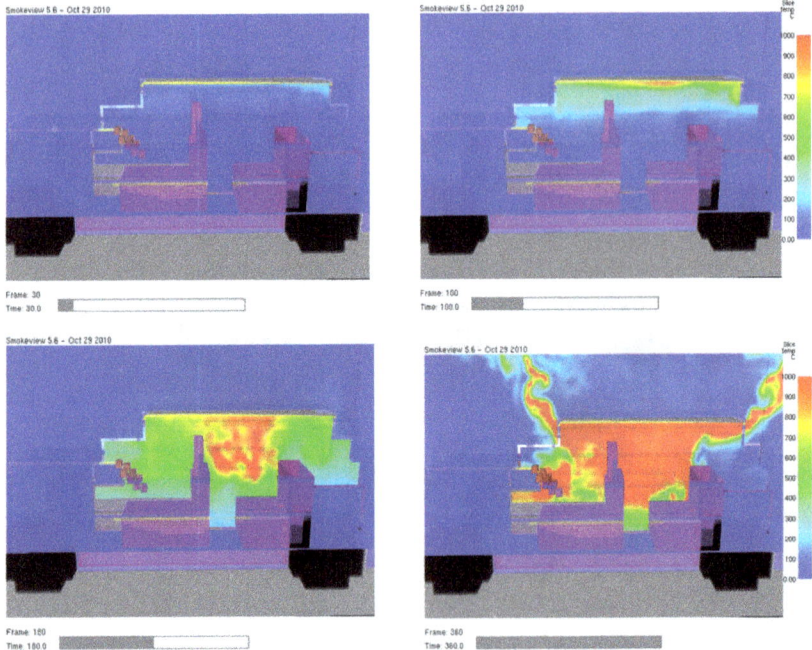

Figure 15. Simulation of passenger compartment fire: temperature cuts in the 30th, 100th, 180th and 360th second of fire

The fire behaviour was as follows. The fire was ignited in the same place as in the fire experiment, on the back seat behind the driver seat. Significant air supply through open left back window enabled rapid spread of fire throughout the whole back seat and significant increase of HRR. Other windows were removed in the simulation, step by step by observations from the fire experiment. Thus, the air supply increased even more in the next phases of burning. After 115 s of simulation, the whole back part of the passenger compartment was ignited and the right back window was removed. After 240 s, all remaining windows were removed and the HRR reached the maximal value of about 4 MW.

Fig. 16 shows the HRR behaviour during the fire, while temperature in passenger compartment and its comparison with the experimental values (thermocouple No. 3) are shown in Fig. 17.

The simulation results are in good agreement with experimental observations, although the value of the maximum temperature in simulation is higher by about 200 °C. In order to make this simulation more accurate, some additional research of real upholstery and plastic components material properties is required. The "step" behaviour of the temperature curve in the simulation, which can be seen in Fig. 17, was caused by the sudden removal of windows. In the real fire, this removal was more gradual. Therefore, no such sudden increase occurred. Thorough examination of material properties of glass and its incorporation into the simulation can solve this problem.

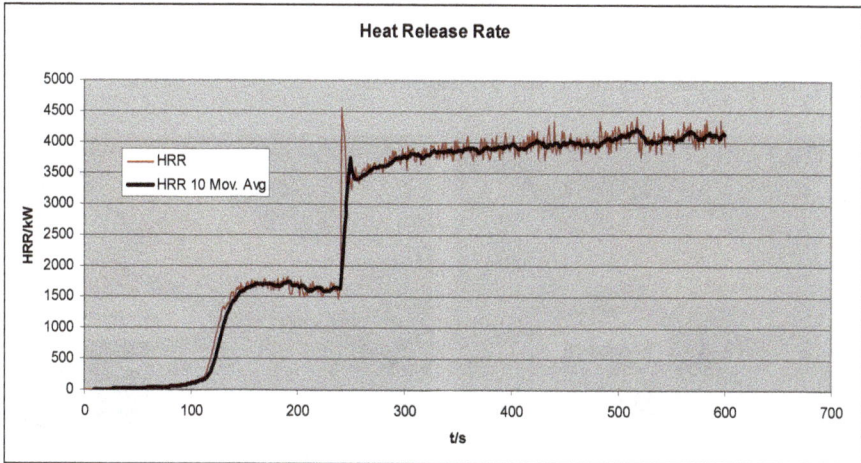

Figure 16. HRR of passenger compartment fire

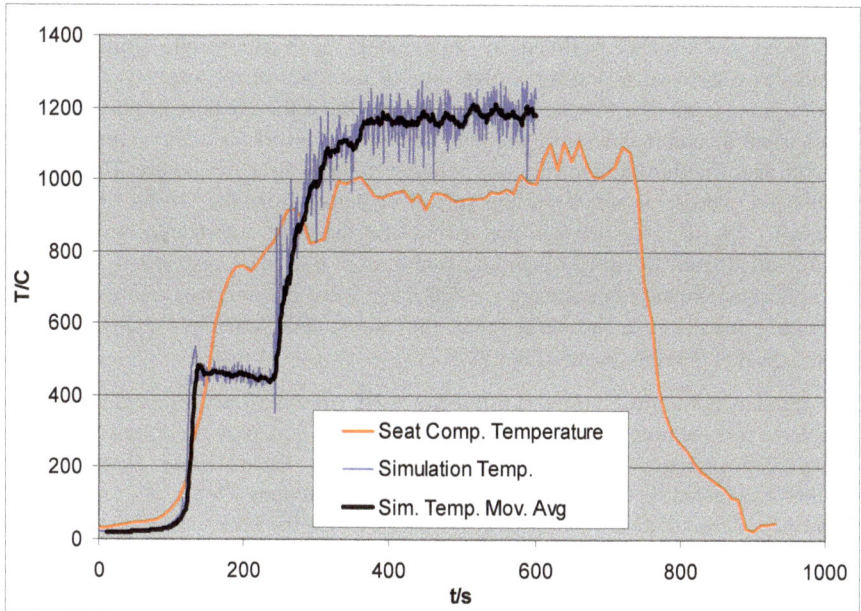

Figure 17. Temperature curves in the passenger compartment interior during the fire: simulation and experiment

6. FDS simulation of fire spread from burning automobile onto a near-standing vehicle

In this section, the FDS simulations of automobile fire spread onto a near-standing automobile will be described which are related to one of three full-scale fire tests conducted during the fourth fire experiment carried out in Slovakia in 2009 (see Fig. 18).

Figure 18. Full-scale fire experiment: spread of fire onto near-standing automobile

A simple case of the simulation of near-standing automobile ignition from burning vehicle is shown in Fig. 19. The aim of the simulation was to investigate the influence of wind speed and distance between the automobiles upon automobile ignition. Three automobiles were used in the simulation, two of them are of the same type as in the previous simulation. The first automobile (left) modified the air flow profile only. The second (central) automobile with slightly different bodywork geometry represented a source of fire. The third automobile was exposed to the central vehicle fire. The air flow direction was chosen to accelerate the spread of flames from the second to the third automobile. The distance between the second and the third automobiles and air flow velocity varied from 15 cm to 75 cm and from 0 m.s^{-1} to 3 m.s^{-1}, respectively. The total HRR of burning engine compartment was 787.5 kW (HRR per unit area was 500 kW.m^{-2}). In order to evaluate the impact of fire on the third automobile, a thermocouple was placed on the right side of its bodywork (Fig. 19).

The computational domain size was 810 x 540 x 225 cm with 5 cm computational mesh resolution. The total number of cells was equal to 787,320. Boundary conditions for the left and right domain boundaries (see the green surfaces in Fig. 19) are defined by the air flow velocities considered in the fire scenarios. Boundary conditions 'OPEN' are set for other domain boundaries. Intel i7-990X CPU, 3.46 GHz CPU was used for a series of simulations with different distance and velocity parameters. Duration of 100 s fire simulation was approximately 12 hours. Maximal simulation time was 600 s. By this way, it is possible to simulate and analyze the ignition time of near-standing automobile depending on different parameters in different conditions in open, semi-closed and closed areas for miscellaneous car categories.

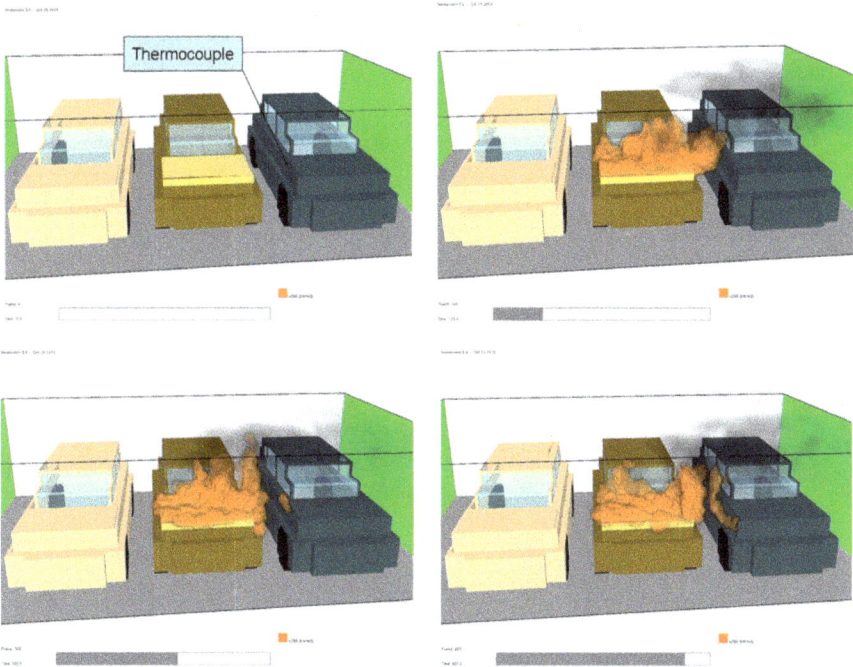

Figure 19. Simulation of near-standing automobile ignition: the case of 45 cm distance between automobiles and 2 m.s^{-1} air flow velocity in the 0th, 120th, 300th and 465th s of fire

Some typical phases of the simulation with 45 cm distance between automobiles and 2 m.s^{-1} air flow velocity are shown in Fig. 19. In the simulation, the ignition occurred at the 288th second of burning. In Table 1, the simulation results for different combinations of the chosen parameters are shown.

	0 m.s^{-1}	0.5 m.s^{-1}	1 m.s^{-1}	2 m.s^{-1}	3 m.s^{-1}
15 cm	304 s	251 s	185 s	137 s	107 s
30 cm	384 s	327 s	254 s	193 s	167 s
45 cm	494 s	441 s	348 s	288 s	268 s
60 cm	-	-	474 s	407 s	389 s
75 cm	-	-	-	575 s	-

Table 1. Time of the third automobile ignition for different distance and air flow velocity values

Fig. 20 shows the temperature behaviour at the thermocouple placed on the third automobile (see Fig. 19) for 2 m.s^{-1} air flow velocity and different distance values.

The simulation results confirmed the observations during the fire experiments conducted, that 60 cm distance between automobiles can be considered to be a reasonable safe distance between automobiles in car parks. For this value of the distance, ignition occurred only if the air flow velocity was above 1 m.s^{-1} and a relatively intensive fire lasted for almost 10 minutes. Such conditions are relatively rare in semi-closed or closed compartments.

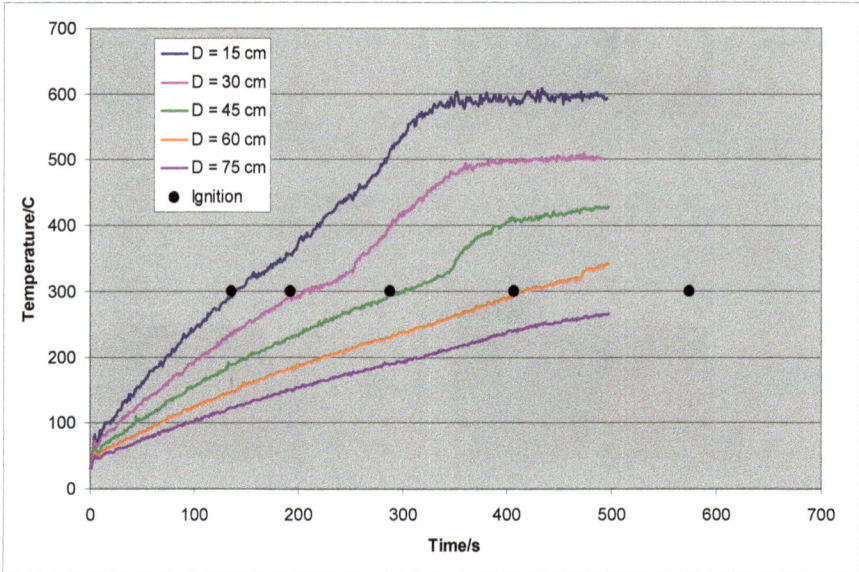

Figure 20. Temperature behaviour for 2 m.s^{-1} air flow velocity and different distance values. Black points represent the time of ignition of the third automobile observed in simulation. This time is not inevitably the same as the time when the temperature in control thermocouple reached the ignition temperature. The place of ignition was different in every simulation.

Another important observation is that a higher air flow velocity does not necessarily mean a higher probability of the ignition. For 75 cm distance and 3 m.s^{-1} air flow velocity, the ignition did not occur, although for 3 m.s^{-1} velocity the third automobile was ignited. In this case, a cooling effect of strong air flow and its influence on flame distraction probably prevailed and heating effect of burning vehicle was suppressed.

Simulations of this kind can be used for specific automobile fire scenarios expected in practice to evaluate the impact of fire, as scenarios parameters can be easily modified according to user requirements.

7. Conclusion

In this chapter, three simple cases of typical automobile fire safety problems arising in real traffic situations and their computer simulation are briefly described:

- automobile engine compartment fire
- automobile passenger and/or luggage compartment fire

- fire spread from burning automobile onto a near-standing automobile.

These simple fire scenarios were simulated by using advanced fire simulator FDS (Fire Dynamics Simulator) and validated by results of the full-scale fire experiments in the testing facilities of Fire Protection College of the Ministry of Interior of Slovak Republic performed in 2009 in Povazsky Chlmec (Slovakia). These fire types correspond to most frequent automobile fires associated with real situations on roads, parking areas, underground car parks, garages and road tunnels. During these experiments, several automobile fires have been performed to better understand the burning process itself and to collect proper fire parameters describing the fire behaviour for the purposes of verification of the computer FDS simulations of these fires.

The developed simulations indicate the ability of the used fire simulation system to model, simulate and visualize various practical fire scenarios and can help test the fire behaviour and various critical parameters which are important for automobile fire safety (such as e.g. critical distance between burning vehicle and near-standing automobile and/or other objects, ignition time of near-standing objects afflicted by automobile fire or vice-versa, tunnel ventilation parameters in the fire regime, etc.). Computer simulation allows also to test the influence of flammable material properties on the fire behaviour for certain fire scenario. The simulation can provide us with information about the main fire spread tendencies, as well as about its quantitative characteristics like fire heat release rate, temporal temperature curves, etc. It allows to evaluate fire risk for the automobile environment and to test possible fire scenarios for concrete situations, which can help evaluate fire suppression options in further fire phases.

General agreement between the simulations presented and the fire experiment observations have been confirmed. However, several specific features can be seen when analyzing the suggested simulations:

- Size of flames from engine compartment interstices in initial phase of the simulation. They seem to be longer than observed. A possible reason of that phenomenon can be the visualization of hot gases coming out from interstices (HRR per unit volume of which exceeds 1199 $kW.m^{-3}$), which are visualized as "flames" but are still not recognized as "flames" by human eye. Moreover, the real interstices were a bit narrower and longer than the simulated ones, maintaining the realistic total leakage area for the air flow, and they have more intricate form (of cavity under the lid). This imperfect "geometry" of interstices following from the mesh limitations could not entirely capture more complicated burning in these small, but intricate spaces. Hot gas could interact with metal surface of the lid in more complicated way than the simple open vent in the simulation. Although this phenomenon was manifested in the simulation B by a bit longer "flames" coming out from the interstices, the simulation of burning inside the engine compartment was reliable.
- Limitations of models implemented in FDS concerning heat conduction modelling in solids probably affect the surface temperature distribution in the lid. Since the FDS does not simulate 3D heat transfer in the lid, the shape of isotherms on the lid is slightly

different than in the IR pictures recorded during the fire experiment. However, this influence was of minor importance for the simulation reliability.

Despite the fact that the automobile engine compartment fire is a complex FDS simulation problem, the FDS simulator proved its applicability for modelling such complex fires. Although the exact geometry of objects in the simulated space is important, if the mesh resolution limitations prevent to capture them precisely, at least their surface to volume ratio and the total mass should be maintained. Air supply and proper size and position of interstices is crucial for the fire development; however, this information is usually not available and user should experiment with several simulations in order to find the proper solution. Next important problem is the determination of material properties where small differences in some values can change the fire behaviour qualitatively. Moreover, efficient realization and validation of fire simulations of large places (for example in a road tunnel) for fire safety purposes in parallel using parallel computers and distributed environments is also a challenging problem.

Author details

Ladislav Halada, Peter Weisenpacher and Jan Glasa
Institute of Informatics, Slovak Academy of Sciences, Slovakia

Acknowledgement

This work was partially supported by Slovak Scientific Research Agencies VEGA (project VEGA 2/0216/10) and ASFEU (project OPVV CRISIS ITMS 26240220060). Authors would like to thank to plk. Ing. Jaroslav Flachbart, PhD., the director of Fire Protection College of the Ministry of Interior of Slovak Republic in Zilina for his kind help during the full-scale automobile fire experiments conducted in cooperation with the research team from the University of Zilina led by Prof. Pavel Polednak, PhD.

8. References

[1] Pettersson, O.; Magnuson, S. E. & Thor, J. (1976). Fire Engineering Design of Structures, Swedish Institute of Steel Construction, Publication 50

[2] Babrauskas V. & Wiliamson, R. B. (1979). Post-flashover Compartment Fires: Application of a theoretical model, *Fire and materials*, Vol. 3, pp. 1-7, ISSN: 1099-1018

[3] Quintiere, J. (1984). A Perspective on Compartment Fire Growth. *Combustion Science and Technology*, Vol. 39, pp. 11–54, ISSN: 0010-2202

[4] Forney, G. P. & Moss, W. F. (1994). Analyzing and Exploiting Numerical Characteristics of Zone Fire Models, *Fire Science and Technology*, Vol. 14, pp. 49–60, ISSN: 0285-9521

[5] Rehm, R. G. & Baum, H. R. (1978). The Equations of Motion for Thermally Driven, Buoyant Flows, *Journal of Research of the NBS*, Vol. 83, pp. 297–308, ISSN: 1740-0449

[6] Oran, E. S. & Boris, J. P. (1987). *Numerical Simulation of Reactive Flow*, Elsevier Science Publishing Company, New York, ISBN: 0-521-58175-3

[7] CFX User's Guide, available at www.ansys.com

[8] Ewer, J., Galea, E., Patel, M., Jia, F., Grandison, A. & Wang, Z., SMARTFIRE - The Fire Field Modelling Environment (2010). The Fifth European Conference on Computational Fluid Dynamics, ECCOMAS CFD 2010 (Pereira, J. C. F., Sequeira, A, Pereira, J. M. C., eds.), Lisbon, Portugal, June 2010, ISBN 978-989-96778-1-4

[9] McGrattan, K., Klein, B., Hostikka, S. & Floyd, J. (2009). Fire Dynamics Simulator (Version 5): User's Guide, NIST Special Publication 1019-5, NIST, Washington

[10] Butcher, E. G., Langdon-Thomas, G. J. & Betford, G. K. (1968). Fire and Car Park Buildings. *Fire Note*, No. 10, HMSO, London, 24 p.

[11] Geiwan, R.G. (1973). Fire Experience and Fire Tests in Automobile Parking Structures. *Fire Journal*, Vol. 67, No. 4, pp. 50-54

[12] Bennetts, I. D., Proe, D. J., Lewins, R. & Thomas, I. R. (1985). Open-Deck Car Park Fire Tests, BHP Melb. Res. Lab. Rep. No. MRL/PS69/85/001, 24p.

[13] Mangs, J. & Keski-Rahkonen, O. (1994). Characterization of the Fire Behaviour of a Burning Passenger Car, Part. I: Car Fire Experiments, *Fire Safety Journal*, Vol. 23, No. 1, pp. 17-35, ISSN: 0379-7112

[14] Shipp, M. & Spearpoint, M. (1995). Measurements of the Severity of Fires Involving Private Motor Vehicles. *Fire and Materials*, Vol. 19, No. 3., pp. 143-151, ISSN: 1099-1018

[15] Kruppa, J., Joyeux, D. & Zhao, B. (1997). Evaluation of the Fire Resistance of a Car Park Structure Based on Experimental Evidences, *Proceedings of the 2nd Int. Conf. on Fire Research and Engineering* (August 1997), Gaithersburg, MD, pp. 416-426

[16] Steinert, C. (1999). Feuerubersprung und Abbrandverhalten von Personenkraftwagen Tl. 1. Stuttgart: Fraunhofer IRB Verlag, 82 p. + app. 110 p.

[17] Steinert, C. (2000). Feuerubersprung und Abbrandverhalten von Personenkraftwagen Tl. 1. Stuttgart: Fraunhofer IRB Verlag, 95 p. + app. 172 p.

[18] Steinert, C. (2000). Experimentalle Untersuchungen zum Abbrand- und Feuerubersprungs-verhalten von Personenkraftwagen. *Vfbd-Zeitschrift*, Vol. 49, No. 4, pp. 163-172, ISSN: 0042-1804

[19] Zhao, B. (2001). Numerical Modelling of Structural Behaviour of Open Car Parks under Natural Fire. Proceedings of the 9th Int. Conf. Interflam, Vol. 1 (September 2001), Edinburgh, Scotland, pp. 383-393, ISBN: 0-9532312-7-5

[20] Schleich, J. B., Cajot, L. G., Pierre, M., Brassauer, M., Franssen, J. M., Kruppa, J., Joyeux, D., Twilt, L., Van Oerle, J. & Aurtenetxe, G. (1999). *Development of Design Rules for Steel Structures Subjected to Natural Fires in Closed Car Parks*. Luxembourg, European Commission

[21] Kitano, T., Sugawa, O., Masuda, H., Ave. T. & Uesugi, H. (2000). Large Scale Fire Tests of 4-Story Type Car Park. Part. 1. The behaviour of structural frame exposed to the fire at the deepest part of the first floor. *Proceedings of the 4th Asia-Oceania Symposium on Fire Science and Technology* (Yamada, T., ed.), May 2000, Tokyo, Japan, pp. 527-538

[22] Okamoto, K., Watenabe, N., Hagimoto, Y., Chigira, T., Masano, R., Miura, H., Ochiai, S., Satoh, H., Tamura, Y., Hayano, K., Maeda, Y. & Suzuki, J. (2009). Burning Behaviour of Sedan Passenger Cars. *Fire Safety Journal*, Vol. 44, pp. 301-310, ISSN: 0379-7112

[23] Halada, L., Weisenpacher, P. & Glasa, J. (2010). Possible Use of Computer Fire Simulation for Automobile Fire Safety Purposes, *Proceedings of the 4th Int. Conf. on Fire Safety and Rescue Services*, Zilina, Slovakia, pp. 68-77, ISBN: 978-80-554-02028-6

[24] Weisenpacher, P., Glasa, J. & Halada, L. (2010). Computer Simulation of Fires in Automobile Engine Compartment, *Proceedings of the 4th Int. Conf. on Fire Safety and Rescue Services*, Zilina, Slovakia, pp. 78-87, ISBN: 978-80-554-02028-6

[25] Polednak, P. (2010). Experimental Verification of Automobile Fires (in Slovak). *Proceedings of the 4th Int. Conf. on Fire Safety and Rescue Services*, Zilina, Slovakia, pp. 248-255, ISBN: 978-80-554-02028-6

[26] Galla, S. (2010). Fire Intervention of Fire and Rescue Brigade (in Slovak), *Proceedings of the Workshop on Fire Risk and Damage Event Liquidation*, Bratislava, Slovakia

[27] Digges, K. H., Gann, R. G., Grayson, S. J., Hirschler, M. M., Lyon, R. E., Purser, D. A., Quintiere, J. G., Stephenson, R. R. & Tewarson, A. (2008). Human Survivability in Motor Vehicle Fires. *Fire and Materials*, Vol. 32, Iss. 4, pp. 249-258, ISSN: 1099-1018

[28] Ponce, I. & Polednak, P. (2006). Automobile Fires (in Slovak). *Proceedings of the 2nd Int. Conf. on Fire Safety and Rescue Services*, Zilina, Slovakia, pp. 269-274, ISBN: 80-8070-539-9

The New Use of Diffusion Theories for the Design of Heat Setting Process in Fabric Drying

Ralph Wai Lam Ip and Elvis Iok Cheong Wan

Additional information is available at the end of the chapter

1. Introduction

Hot air impingement is one of the most widely used methods for material drying. It is also the most traditional drying approach used in industrial process for various kinds of material, such as wood, paper, food, medicine and construction materials. Many research studies have been carried out to see how it is effectively used to process different types of material, and how it is implemented into the design of heat setting machines, such as spray dryers, conveyor dryers, tunnel dryers, fluidized bed dryers and drum dryers.

To study hot air impingement in textile and clothing industries, modeling of porous type fabric drying process would be the key study area. The heat and mass transfer principles are used as tools to assist with the investigation of the hot air impingement mechanism. The mechanism is usually treated as a mass transfer process of the moisture content from the porous material to the impinging air. The transfer of moisture content from the fabric material to the hot air stream is due to a heat transfer process under an in-equilibrium condition. The change of water phases is traditionally described by linear heat transfer equations. As a matter of fact, the driving force in the internal structure of porous materials is not a simple direct proportional relationship between energy exchange and the phase change of the interacting substances, i.e. air and water. Therefore non-linear analytical models based on the physical properties of fabrics will be proposed in this study to provide better simulation results. In the models, the parameters for modeling will be empirically determined and used to describe the drying phenomenon down to microscopic levels. The descriptions will involve the physical and mechanical properties of the drying materials such as mass density, flow viscosity, thermal conductivity, diffusion properties, cohesive properties and flow kinetics. Ip and Wan (2011) have suggested the strategies of using analytical techniques to determine the modeling parameters, and these methods will be investigated in greater depth in this research.

2. Three periods of a fabric drying cycle

Fabric is usually dried up for the purposes of storing or setting. Using thermal energy to dry up and perform setting has been the most traditional and effective method. In this study, heated air is used as a processing agent. Its physical properties will be changed in the gaining of moisture and the loss of thermal energy. The moisture in the fabric will change to vapor after gaining energy from air to create a mass transfer process. The reduction of moisture and increase of fabric temperature is a complicated heat/mass transfer process. Merely using linear conductive and convective heat transfer equations to model the process seems to be inadequate. Diffusion theories are therefore suggested to present the details of the drying process.

"Preheating", "Constant drying" and "Falling drying" are the three periods of a fabric drying cycle as shown in Fig. 1. In the preheating period, most thermal energy is absorbed by water on the fabric surface because air is a poor thermal conductor. The mass transfer rate of water is not high in this period. When more thermal energy is absorbed, water on the fabric surface will change to vapor by evaporation at a rapid mass transfer rate. The water loss rate will keep constant depending upon the air temperature, velocity and atmospheric pressure. As the mass transfer rate of water is constant in this period, it is labeled as constant drying period. While the moisture content in the fabric is going down from the initial θ_i to critical moisture content θ_k, the moisture on the fabric surface starts to separate and form dry/wet regions. Diffusion will appear at the dry/wet regions to form the falling drying period. Diffusion is a slow mass transfer process in comparison with evaporation happened at the second period, and the water transfer rate is correspondingly decreased to form a non-linear drying result until reaching the final moisture content at θ_o.

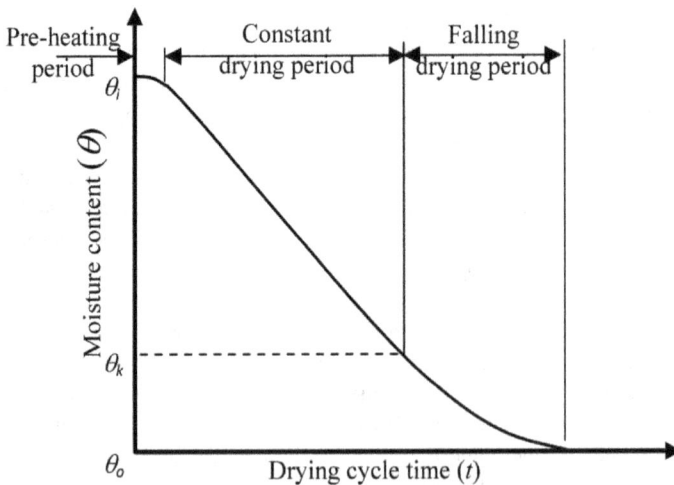

Figure 1. A typical fabric drying curve shows the three drying periods

Kowalski et al. (2007) has employed partial differentiation and numerical analysis tools to present thermo-mechanical properties of a drying process for porous materials. However, the modeling process is computational intensive and time consuming. It seems impractical to be used in the industrial drying process because a quick response is always needed to manage numerous varying conditions. The concept of using analytical approaches to model drying processes for fruits and sugar has been proved (Khazaei et al., 2008). However, its performance heavily relies upon the sample data and the reliability of the assumptions. The scopes of the research are therefore having rooms to improve the inadequacy. Objectives of the research are to explore and develop robust analytical models that can effectively simulate the characteristics of hot air impingement process for porous fabrics of different textile properties.

2.1. Research objectives

The research objectives in the Chapter are:

- to investigate how to present the drying characteristics of porous type fabric using non-linear analytical models,
- to evaluate the performance of the models in the simulation of drying process, and
- to comment their accuracies in the modeling of different fabric types under various air setting conditions.

In this research, four non-linear analytical models will be studied. The modeling parameters for the models will be empirically determined. The performance of the model will be examined through a careful comparison of testing results. A drying test will be set up to assist the determination of the modeling parameters, and evaluate the performance of the developed models.

2.2. Equations for moisture flow through control volume

Using the approach of control volume to the describe characteristics of moisture flow in porous materials would be close to the phenomenon of fabric drying (Kowalski, 2003). The total mass of the constituents within the control volume will remain unchanged when the porous fabric volume shrinks or otherwise. A set of mass balance equations for each individual constituent, i.e. water, vapor and air in a drying process is given as:

$$\rho^s \, \dot{X}^l = -div \; W^l + \psi^l \tag{1}$$

$$\rho^s \, \dot{X}^v = -div \; W^v + \psi^p \tag{2}$$

$$\rho^s \, \dot{X}^a = -div \; W^a \tag{3}$$

where ρ is partial mass density, $\dot{X}$ is mass content change rate, ψ is phase transition rate, s is solid fabric, l is liquid phase of water, v is vapor phase of water and a is air. The mass balance Equations (1) – (3) provides a platform for further investigation of moisture content change in a fabric drying process.

2.3. Boundary conditions for the period of constant drying

The water evaporated inside fabric material is much less than that on the boundary surface in the constant drying period. Assuming that the mass flux W of water vapor and air are negligible in this period, the relationship can be given as:

$$W^v = W^a = 0 \tag{4}$$

Then, the mass balance Equations (1) – (3) can be rewritten for the calculation of moisture content in the period of constant drying and give:

$$\rho^s \dot{X}^l = -div\ W^l + \psi^l \tag{5}$$

$$\rho^s \dot{X}^v = \psi^v \ll \rho^s \dot{X}^l \tag{6}$$

$$\rho^s \dot{X}^a = 0 \tag{7}$$

Equation (5) shows the phase transition of water inside the fabric. Equations (6) and (7) show the phase transition of vapor and air respectively inside the fabric.

2.4. Boundary conditions for the period of falling drying

When the fabric moisture content falls to the critical point at θ_k, meniscoidal water droplets will recede and the drying rate is slowing down until completely dry. The characteristics of fabric drying at this period could be divided into the hygroscopic and non-hygroscopic states, thus, the heat/mass transfer in this period is getting complex. At the initial stage of falling drying, the fabric is fully saturated and water flows in the form of liquid fluxes mainly due to capillary action. Air pockets gradually form at the second stage to replace some of the moisture to form small air bubbles inside the fabric pores. With further drying, the moisture decreases and the size of air bubbles considerably increases that could reduce the rate of heat transfer. As a result, the heat/mass transfer rate is correspondingly reduced since the thermal resistance of air is much higher than water. The drying cycle stops at the tertiary stage when moisture in the hygroscopic regions is totally removed, and a uniform non-hygroscopic property fabric is formed. The mass balance Equations (1) – (3) for the falling drying is rewritten to give:

$$\rho^s \dot{X}^l = \psi^\alpha \tag{8}$$

$$\rho^s \dot{X}^v = -div\ W^v + \psi^\alpha \tag{9}$$

$$\rho^s \dot{X}^a = -div\ W^a \tag{10}$$

where α in Equations (8) and (9) are all the constituents in the fabric.

2.5. Calculation of mass fluxes

The mass flux of water in Equation (5) at the constant drying period is given as:

$$W^l = -\wedge^l \left[\left(\frac{\partial \mu^l}{\partial T^l} \right) T^l + \left(\frac{\partial \mu^l}{\partial \theta^l} \right) \theta^l + \left(\frac{\partial \mu^l}{\partial X^l} \right) X^l - g \right] \tag{11}$$

$$C^l = \left(\frac{\partial \mu^l}{\partial \theta^l} \right) \tag{12}$$

In Equation (11), $\wedge$ is the coefficient of water diffusivity, μ is chemical potential of water, θ is relative moisture content, T is absolute temperature and g is gravitational acceleration. The equation shows the relationship between water mass flux W^l and the gradients of temperature, volume fraction and mass fraction. The water movement in the fabric is largely due to capillary forces and gravitational force at the constant drying period. C^l is the moisture coefficient related to the moisture cohesive force in the fabric. The mass flux of vapor in Equation (9) for the falling drying period is given as:

$$W^v = -\wedge^v \left[\left(\frac{\partial \mu^v}{\partial T^v} \right) T^v + \sum \left(\frac{\partial \mu^v}{\partial \theta^v} \right) \theta^v \right] \tag{13}$$

$$C^v = \left(\frac{\partial \mu^v}{\partial \theta^v} \right) \tag{14}$$

The generation of moisture is due to phase transition of water into vapor, in which, the efflux of vapor is significant. The coefficient C^v for water vapor could be experimentally determined.

3. Fabric drying tests

A series of experiments were conducted to measure the drying characteristics of a group of fabric samples. Cotton is the major studying material in the tests as it is used most widely in clothing industry. The objectives of the experimental tests are to examine the drying characteristics of the fabrics under different boundary conditions, such as fabric texture, density, thickness, air temperature and impingement velocity.

3.1. Drying test set up

Six cotton fabric samples were examined. The samples were labeled from A to F, and their properties are listed in Table 1.

The set up as shown in Fig. 2 is an air heater providing hot air stream for each drying test. The temperature and speed of the impinging air are adjustable to provide different boundary conditions for the study. Disc-shaped fabric samples of 100 cm² in area are mounted on a polystyrene backing plate with wire gauze facing the impinging hot air.

Fabric sample	Fabric texture	Yarn structure	Density (g/m³)	Thickness (mm)
A	Plain knitted	20 s/2	224.4	0.6594
B	Plain knitted	32 s/1	147.7	0.4363
C	Plain knitted	20 s/2	271.0	0.7769
D	Plain weaved	-	182.0	0.5638
E	Plain knitted	20 s/1	193.0	0.5025
F	Plain knitted	32 s/2	200.0	0.6188

Table 1. The properties of the fabric samples for drying tests

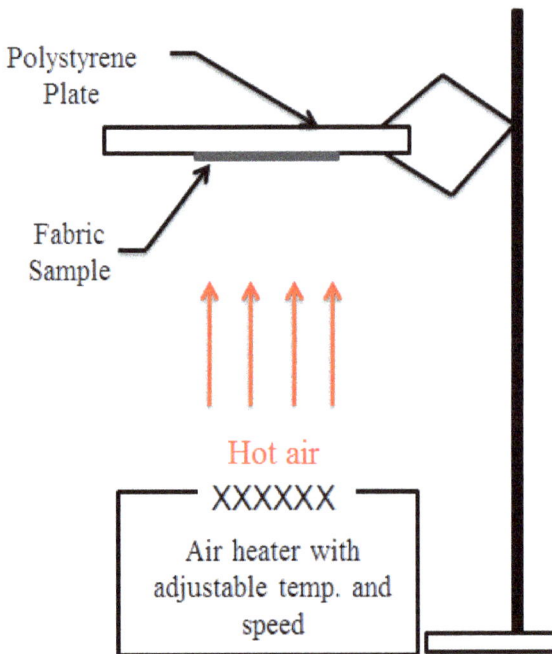

Figure 2. Schematic diagram of the drying test setup

In the tests, each fabric sample was being dried under eight conditions as listed in Table 2. The fabric weight was measured by an electronic balance every 30 seconds of drying under different setting conditions. The repeated drying and weight measurement procedures were conducted until all the moisture in the fabric samples was removed. The same testing procedures were repeated for all the fabric samples as given in Table 1.

Setting condition	Air temperature (℃)	Impinging velocity (m/s)
1	80.0	1.48
2	81.5	1.45
3	86.5	1.43
4	54.0	1.10
5	55.5	1.15
6	54.0	1.02
7	57.0	1.41
8	58.0	1.46

Table 2. The air setting conditions of drying tests for the six fabric samples

3.2. Results and discussions of the drying tests

Fig. 3 shows testing results from the six fabric samples under air setting condition 1 as listed in Table 2. The normalized water contents instead of the absolute values recorded from the tests were used in order to compensate the variation of fabric weight among the six samples.

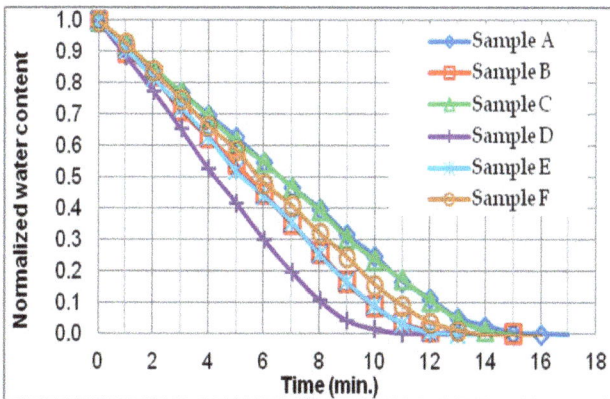

Figure 3. Drying curves for the six fabric samples under air setting condition 1

The testing results as illustrated in Fig. 3 for the tested fabrics have shown the relationships among water content, fabric density, texture, air temperature and impingement velocity. Their relationships are discussed in the following sections.

3.2.1. Drying rate versus fabric texture

Table 3 lists the results of four fabric samples tested under the air setting conditions of 1, 4, 6 and 8 as listed in Table 2. Among the tested samples, sample D is weaved fabric and the others are knitted fabrics. The drying rate of sample D is the highest among the others at the constant drying period.

Air setting condition	Sample B - Plain knitted		Sample D - Plain weaved	
	1/Drying time (min⁻¹)	Drying rate at constant drying period (g/min)	1/Drying time (min⁻¹)	Drying rate at constant drying period (g/min)
1	0.083	0.439	0.100	0.462
4	0.080	0.383	0.100	0.395
6	0.095	0.392	0.083	0.398
8	0.065	0.281	0.065	0.287
	Average drying rate = 0.374 g/min		Average drying rate = 0.386 g/min	
	Sample E - Plain knitted		Sample F - Plain knitted	
1	0.083	0.431	0.077	0.426
4	0.067	0.304	0.057	0.323
6	0.063	0.349	0.065	0.385
8	0.061	0.290	0.056	0.302
	Average drying rate = 0.344 g/min		Average drying rate = 0.359 g/min	

Table 3. Drying rate of the four tested samples with different fabric texture

3.2.2. Drying rate versus fabric density and thickness

Table 4 lists the testing results of two selected fabric samples A and C with same texture, yarn structure and different density and thickness under the air setting conditions of 1, 4, 6 and 8.

Air setting	Sample A - 224.4 g/m³, 0.6594 mm		Sample C - 271 g/m³, 0.7769 mm	
	1/Drying time (min⁻¹)	Drying rate at constant drying period (g/min)	1/Drying time (min⁻¹)	Drying rate at constant drying period (g/min)
1	0.071	0.416	0.067	0.443
4	0.063	0.407	0.063	0.391
6	0.065	0.378	0.053	0.396
8	0.047	0.281	0.044	0.289
	Average drying rate = 0.371 g/min		Average drying rate = 0.380 g/min	

Table 4. Drying rates of samples A and C for different air setting conditions

The results listed in Table 4 show a similar result at the constant drying for the fabrics with different density and thickness, but same texture and yarn structure.

3.2.3. Drying rate versus air temperature

Table 5 lists the results of the drying rate for all fabric samples tested under air setting conditions 1 and 8 with similar impingement velocity at 1.47 m/s and different temperature.

	Sample A		Sample B		Sample C	
Temp. (°C)	1/time (min^{-1})	Drying rate (g/min)	1/time (min^{-1})	Drying rate (g/min)	1/time (min^{-1})	Drying rate (g/min)
58	0.047	0.281	0.065	0.281	0.044	0.289
80	0.071	0.416	0.083	0.439	0.067	0.443
	Sample D		Sample E		Sample F	
Temp. (°C)	1/time (min^{-1})	Drying rate (g/min)	1/time (min^{-1})	Drying rate (g/min)	1/time (min^{-1})	Drying rate (g/min)
58	0.065	0.287	0.061	0.290	0.056	0.302
80	0.100	0.462	0.083	0.431	0.077	0.426

Table 5. Drying rate of fabrics under different air temperature

The drying rate at the constant drying period increases with the rise of air temperature for all fabric samples.

3.2.4. Summary of the experimental findings

The period of constant drying as illustrated in Figs. 1 and 3 has constituted a large portion of the drying cycle. The moisture reduction rate at the period could be used as an indicator to show the properties of the fabric, and conditions of the impinging air. The experimental findings in Tables 3 – 5 have shown the performance of the drying process against the boundary conditions including fabric texture, density, thickness, air temperature and impinging velocity. It has been observed that the increase of air temperature and velocity will speed up the drying rate. The fabric properties could also affect the drying rate but not as much as the air properties. These findings could be useful in the setting up of analytical models to simulate each period of a fabric drying cycle.

4. Development of non-linear analytical models to simulate the drying of porous type fabrics

As mentioned in Section 2, the drying rate of porous type fabrics has a non-linear relationship with time at the falling drying period. Some inaccurate results will be found if the traditional linear heat transfer equations are applied because the heat transfer coefficient changes with the change of the moisture contents at the falling drying period. To ensure an accurate modeling of the drying process, the heat transfer coefficient should be adjustable corresponding to the diffusion properties in the forming of dry/wet regions as mentioned in Section 2.4. A non-linear model is therefore used to describe the process characteristics (Haghi, 2006; Moropoulou, 2005). The knitted and weaved fabrics studied in this research are considered porous type materials because they contain unidirectional pores. The randomly distributed pores give an environment to establish a diffusion process when portions of water dry up. It is clear that the moisture diffusion rate has a close relationship to the size and number of fabric pores. The studied models given in Section 4.6 will address these essential modeling parameters. The other fabric parameters including texture, density and thickness that correlate to the drying rate will also be modeled by different modeling principles, and given in Sections 4.3 – 4.5.

4.1. Determination of the critical moisture content

The two periods of a fabric drying cycle as illustrated in Fig. 1 should be modeled separately. Traditional heat transfer equations could be used as modeling tools for the constant drying period, whilst, non-linear modeling equations should be considered when the moisture reduction rate varies with time in the falling drying period. The critical moisture content θ_k at the beginning of the falling drying period will be the separating point between the two periods. The finding of θ_k is given from the plotting of the normalized drying rate versus the moisture content in gram per gram of the dried fabric, see Fig. 4.

The moisture reduction rate $\dot{m}_n$ at the nth period of a time interval Δt can be expressed as:

$$\dot{m}_n = \frac{m_n - m_{n+1}}{\Delta t} \text{ for the first time interval,} \tag{15}$$

$$\dot{m}_n = \frac{(m_{n-1} - m_n) + (m_n - m_{n+1})}{2\Delta t} \text{ for a time interval } n > 0 \tag{16}$$

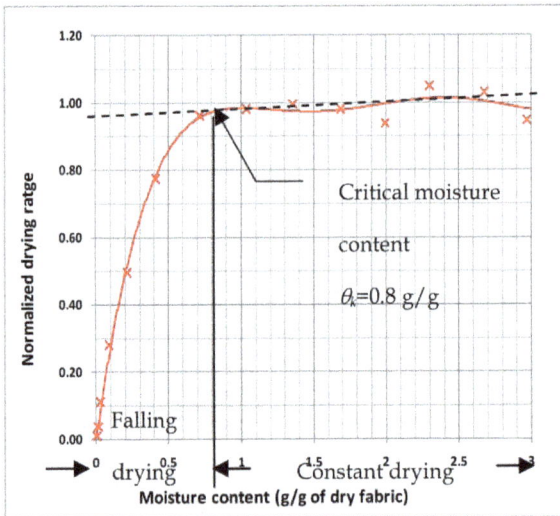

Figure 4. Plotting of normalized drying rate versus moisture content for fabric sample A

The critical moisture content θ_k can be identified from the curve as shown in Fig. 4 at the point of dramatic decrease of $\dot{m}_n$. The determined θ_k for fabric sample A under air setting condition 3 listed in Table 2 is 0.8 g/g. The testing results for other fabric samples under the same air setting condition are similar and listed in Table 6.

Fabric sample	Critical moisture content θ_k (g/g dry fabric weight)
A	0.8
B	0.7
C	0.8
D	0.8
E	0.7
F	0.7

Table 6. Critical moisture contents of the tested fabric samples

4.2. Using diffusion theories to model a fabric drying process

The boundary conditions for heat/mass transfer in the porous fabric have been discussed in Sections 2.2 to 2.5. However, they are not good enough to estimate the fabric moisture content during the drying process. It is necessary to have a further investigation to estimate the moisture content in individual period of drying. The authors have set-up a group of models based upon diffusion theories and Kowalski's (2003) boundary equations to simulate the moisture changing rate under various boundary conditions (Ip and Wan, 2011). The investigated drying models will be presented in differential forms to address the movement of moisture contents in fabric. The models are based on the principles of chemical diffusion mechanism to calculate the rate of moisture change (dM/dt) according to a set of modeling parameters empirically determined from drying cycles (Kowalski, 2000; Schlunder, 2004). Four non-linear analytical models, namely "Kinetics", "Diffusion", "Kinetics model based on the solutions of diffusion equations" and "Wet surface" have been developed, and the principles are given in the following sections.

4.3. First order kinetics model

Roberts and Tong (2003) have shown a successful result in the modeling of bread drying process using first order exponential equations. In their research, microwave was used as the drying agent, and the process has been assumed as isothermal. Unfortunately, the experiential setup is quite different from convective drying using impinging air in this study. It is therefore necessary to develop new modeling equations for porous type fabrics. Schlunder (2004) has stressed that the falling drying period should be considered as an isothermal process. First order exponential equations might be appropriate to describe the process, thus, the first model developed in this study is labeled as "First order kinetics model".

In the First order kinetics model, there is an assumption that the vaporization of water inside fabric can be described as a kinetic reaction motion of water molecules. The reaction rate is treated as the moisture reduction rate at the falling drying period. Thus, the water evaporation rate will correlate with the moisture content. The equation of the kinetic model is given as:

$$-\frac{dM}{dt} = kM^n \tag{17}$$

In Equation (17), M is the instant moisture content and $n = 1$ for the first order kinetics. If M_o is the initial moisture content at the beginning of the falling drying period, i.e. critical moisture content θ_k, the integration result of the differential form Equation (17) will be:

$$\frac{M}{M_o} = e^{-kt} \tag{18}$$

where k is the kinetic coefficient.

The testing results of moisture content as shown in Fig. 4 are further plotted in terms of drying cycle time t and given in Fig. 5. The red line in the figure represents the drying curve and the black line is the approximated drying rate at the constant drying period.

The kinetic coefficient k in Equation (18) at the falling drying period can be obtained by regenerating a new plotting from the results illustrated in Fig. 5. The ratio of M/M_o in the equation shows an exponential relationship with $-kt$. It can be converted into a linear relationship by applying logarithm for both sides of Equation (18). Fig. 6 shows a plotted graph of $\ln(M/Mo)$ versus the drying cycle time from the experimental records in Fig. 5.

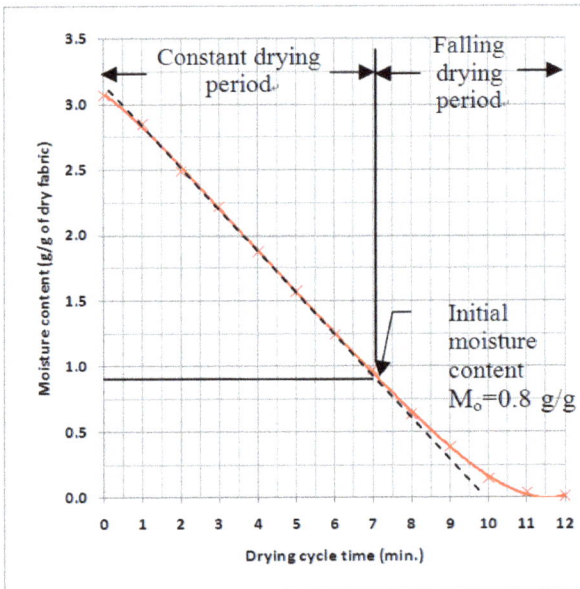

Figure 5. Experimental records of the drying of fabric sample A

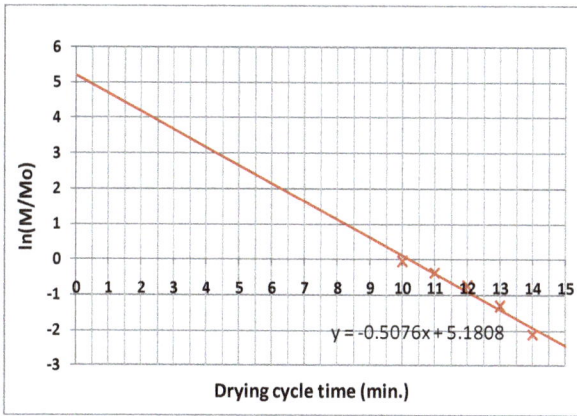

Figure 6. Determination of kinetic coefficient k for fabric sample A

A graphical method to determine the kinetic coefficient k is to measure the slope of the fitted line in Fig. 6. The slope is measured as -0.5076 that will be the tested fabric kinetic coefficient. An alternative method to determine k is to modify Equation (18) using Arrhenius relationship (Roberts and Srikiatden, 2005). It is a common method to correlate rate constant with reaction temperature T for kinetics reactions in chemistry. The Arrhenius form of equation in terms of k and T is:

$$k = Ae^{-E_a/RT} \tag{19}$$

where E_a is the activation energy, R is the universal gas constant at 8.314×10^{-3} kJ/mol K and A is a constant. Equation (19) gives the relationship of kinetic coefficient k in terms of air temperature T only, and does not include the impinging velocity V. However, V is also a key factor in the drying process and its effect could be empirically determined using linear regression methods. The regression equation for the calculation of k from the Arrhenius relationship in Equation (19) is given by taking natural algorithm of the equation.

$$\ln k = \frac{-E_a}{R}\frac{1}{T} + \ln A \tag{20}$$

A in the Arrhenius equation means reaction per time and is correlated to the impinging velocity in the studied drying models. Thus, the First order kinetics model in Arrhenius form can be written in terms of T and V to give:

$$\ln k = a + b\frac{1}{T} + c\ln V \tag{21}$$

If the Arrhenius relationship is applied to describe a fabric drying process, a plotting of $\ln k$ versus $1/T$ will give a straight line. The slope and intercept of the line are used to determine the correlation constants of E_a and b as given in Equations (20) and (21). The kinetic coefficient k of the fabric sample A calculated from Equation (19) under the air setting conditions listed in Table 2 are given in Table 7, and the corresponding values of $\ln k$, $1/T$ and $\ln V$ determined from Equation (21) are listed in Table 8.

Setting condition	AIR TEMPERATURE (K)	Impinging velocity (m/s)	k^*	k
1	353.0	1.48	0.5295	0.5198
2	354.5	1.45	0.5818	0.5230
3	359.5	1.43	0.5076	0.5336
4	327.0	1.10	0.5297	0.4634
5	328.5	1.15	0.5494	0.4667
6	327.0	1.02	0.5537	0.4634
7	330.0	1.41	0.3589	0.4699
8	331.0	1.46	0.3568	0.4722

Table 7. The calculated kinetic coefficients from experiments and Arrhenius equation for fabric sample A (k^* from experiential results, k from Arrhenius equation)

Setting condition	$\ln k$	$1/T$ (K⁻¹)	$\ln V$ (m/s)
1	-0.6358	0.0028	0.3920
2	-0.5416	0.0028	0.3716
3	-0.6781	0.0028	0.3577
4	-0.6354	0.0031	0.0953
5	-0.5989	0.0030	0.1398
6	-0.5911	0.0031	0.0198
7	-1.0247	0.0030	0.3436
8	-1.0306	0.0030	0.3784

Table 8. The calculated correlation constants for the tested fabric sample A

Fig. 7 illustrates the plotting of $\ln k$ versus $1/T$ from data in Table 8. Results from the plotting were used to calculate the activation energy E_a and A as given in Equation (20).

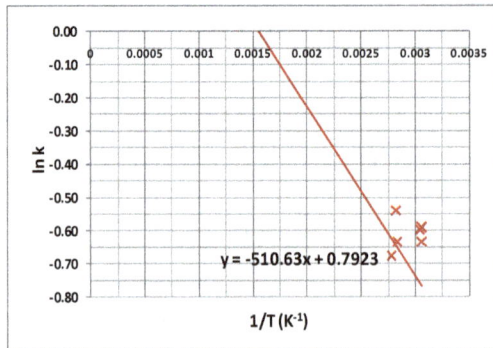

Figure 7. The plotting of $\ln k$ versus $1/T$ for fabric A under various air setting conditions

The calculated values for E_a and A are 4.2454 kJ/mole and 2.2085 respectively given from Fig. 7. The final form of the Arrhenius equation for fabric sample A will be given as:

$$\ln k = -510.63\frac{1}{T} + 0.7923 \tag{22}$$

Experimental results listed in Table 8 can be further used to determine the coefficients of a, b and c as given in Equation (21) by linear regression methods (Cohen, 2003). The regression results produced from MicroSoft Excel for fabric sample A are given in Fig. 8.

SUMMARY OUTPUT

Regression Statistics	
Multiple R	0.8912112
R Square	0.7942575
Adjusted R Square	0.7256766
Standard Error	0.1058416
Observations	9

ANOVA

	df	SS	MS	F
Regression	2	0.259478294	0.129739	11.58133
Residual	6	0.067214622	0.011202	
Total	8	0.326692916		

	Coefficients	Standard Error	t Stat	P-value
Intercept	5.5354099	1.333590724	4.150756	0.006006
1/T[k]	-1974.7449	429.2456499	-4.6005	0.00369
lnV	-1.5207901	0.359403041	-4.23143	0.005491

Figure 8. Regression table for fabric sample A determined from the First order kinetics model

Using results from the regression table, the model equation is given as:

$$\ln k = 5.535 - 1974.7\left(\frac{1}{T}\right) - 1.521\ln V, \text{ where } a = 5.535, b = -1974.7 \text{ and } c = -1.521 \tag{23}$$

Fabric sample	a	b	c
A	5.535	-1974.7	-1.521
B	5.740	-1993.9	-1.210
C	3.714	-1476.9	-0.911
D	4.430	-1584.6	-1.276
E	5.580	-2054.5	-0.494
F	6.087	-2223.7	-0.280

Table 9. Regression results of the tested fabric samples

Table 9 lists the regression results of all the fabric samples. A comparison of the differences of k determined from Arrhenius equation and regression model for fabric sample A is listed in Table 10.

Setting	AIR TEMP. (°C)	Velocity (m/s)	k1	k2	k1 Dev. (%)	k2 Dev. (%)
1	80.0	1.48	0.5198	0.5195	1.83	1.90
2	81.5	1.45	0.5230	0.5487	10.10	5.68
3	86.5	1.43	0.5336	0.6056	5.12	19.30
4	54.0	1.10	0.4634	0.5227	12.52	1.32
5	55.5	1.15	0.4667	0.5022	15.06	8.59
6	54.0	1.02	0.4634	0.5863	16.32	5.89
7	57.0	1.41	0.4699	0.3786	30.95	5.50
8	58.0	1.46	0.4722	0.3656	32.34	2.48
		Average	0.4890	0.5037	15.53	6.333

$k1$ is calculated from Arrhenius equation (Equation 19), $k2$ is calculated from regression model (Equation 21)

Table 10. Comparison of kinetic coefficient determined from Arrhenius equation and regression model for fabric sample A

The deviations of k for Arrhenius and regression models listed in Table 10 are calculated based on the results obtained from experiments listed in Table 7. The deviations calculated from regression model are much less than Arrhenius equation. The impinging velocity may have been considered in the regression model. Further study about the discrepancy of the modeling results between the regression model and the experimental records was performed.

The red curve illustrated in Fig. 9 is the modeled drying cycle obtained from the regression model for fabric sample A in the falling drying period with a kinetic coefficient k at 0.5037. Discrepancies have been found in comparison with the blue drying curve obtained from experiments. The average discrepancy and standard deviation of the comparison are 14.0139% and 7.8028% respectively.

In conclusion, the falling drying period in a fabric drying cycle can be modeled by the First order kinetics model using an exponential function. A coefficient k for the exponential function can be experimentally determined to model the drying characteristics under various air setting conditions. The coefficient can also be numerically determined from the regression model using air temperature and impinging velocity as the boundary conditions. It has a further relationship to the fabric density and thickness as illustrated in Fig. 10 other than temperature and velocity. Table 11 lists k determined from the regression model for all the fabric samples. It has been found that k decreases with the increase of fabric density and thickness. This relationship could be useful in the estimation of the drying cycle time for the fabrics with different thickness and density.

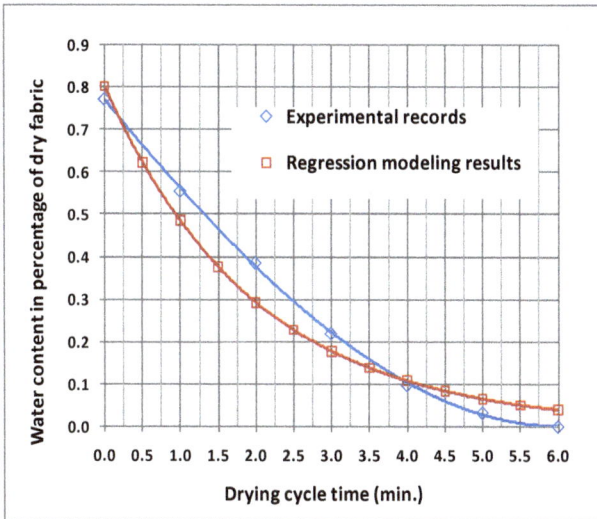

Figure 9. Comparison of experimental records and regression modeling results for fabric sample A

Fabric sample	Density (g/m³)	Thickness (mm)	k
A	224	0.6594	0.5037
B	148	0.4363	0.6328
C	271	0.7769	0.4136
D	182	0.5638	0.5607
E	193	0.5025	0.5469
F	200	0.6188	0.5869

Table 11. The fabric kinetic coefficient k determined from the regression model

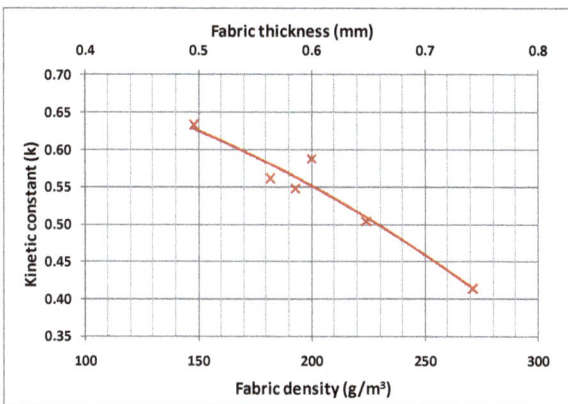

Figure 10. Relationship of k to the fabric density and thickness

4.4. Diffusion model

Most of the Diffusion models presenting the change of moisture content have been based upon Fick's law (Ramaswamy and Nieuwenhuijzen, 2002). The Fick's first law states the diffusion flux flowing from the regions of higher concentration to lower concentration obeying a magnitude proportional relationship to the concentration gradient. The one dimensional Fick's first law in differential form is given as:

$$J = -D\frac{\partial \phi}{\partial x} \tag{24}$$

where J is the diffusion flux in $m^{-2}s^{-1}$, D is the effective diffusion coefficient in $m^2\ s^{-1}$, $\emptyset$ is the concentration in m^{-3} and x is a linear distance in m. The Fick's second law given in Equation (25) shows the rate of concentration change. It is given by the derivative of Equation (24), with the assumption of D as a constant. The Fick's second law has been used commonly to simulate the drying process of agricultural products (Khazaei et al., 2008), such as seeds and grains.

$$\frac{\partial \phi}{\partial t} = D\frac{\partial^2 \phi}{\partial x^2} \tag{25}$$

If the Fick's second law is applied to model the process of drying porous fabric, the fabric will be considered as an infinite thin slab and dried from one direction. If heat transfer from the surrounding to the fabric is negligible, the integration result from Equation (25) will give the Diffusion model. The model equation given in Equation (26) is in terms of moisture content M, fabric thickness L, effective diffusion coefficient D and the drying cycle time t.

$$\frac{M}{M_0} = \frac{8}{\pi^2}\exp\left(-\frac{\pi^2 Dt}{4L^2}\right) \tag{26}$$

The Diffusion equation is similar to Equation (18) of the First order kinetics model. The only difference between the two model equations is the fabric thickness L included in the Diffusion model. As the same as in the First order kinetics model, the effective diffusion coefficient D in Equation (26) can be acquired from the plotting of ln (M/M_0) versus t as illustrated in Fig. 6. Equation (27) is obtained when a logarithm is applied to both sides of Equation (26).

$$\ln\frac{M}{M_0} = \left(-\frac{\pi^2 D}{4L^2}\right)t - 0.21 \tag{27}$$

The slope of the fitted straight line in Fig. 6 is -0.5076 for fabric sample A with a thickness of 0.6594 mm. D is then calculated by substituting the slope and L back to Equation (27), and the estimated value is 8.945×10^{-8}. The equation form of the Diffusion model and the First order kinetics is similar, D in the Diffusion model can be therefore calculated using the regression model. The Diffusion model in regression form is given in Equation (28).

$$\ln D = a + b\frac{1}{T} + c\ln V \qquad (28)$$

Using information in Table 8 to determine the constants of a, b and c. The determined regression model equation for fabric sample A is given as:

$$\ln D = 20.879 - 12070.75\frac{1}{T} - 2.173\ln V \qquad (29)$$

The regression results for all the fabric samples are listed in Table 12.

Fabric sample	a	b	c
A	20.88	-12070.75	2.173
B	10.59	-2006.16	1.215
C	11.43	-1495.8	0.9718
D	11.374	-1601.97	1.277
E	10.578	-1895.1	0.526
F	10.08	-2201.5	0.242

Table 12. Regression results of all the fabric samples modeled by Diffusion model

The effective diffusion coefficient D for fabric sample A calculated from various air setting conditions are listed in Table 13.

Setting	AIR TEMP. (°C)	Impinging velocity (m/s)	Effective diffusion coeff. D
1	80.0	1.48	7.03023×10^{-7}
2	81.5	1.45	8.49438×10^{-7}
3	86.5	1.43	1.40575×10^{-6}
4	54.0	1.10	8.83487×10^{-8}
5	55.5	1.15	9.49414×10^{-8}
6	54.0	1.02	1.04100×10^{-7}
7	57.0	1.41	7.20528×10^{-8}
8	58.0	1.46	7.46037×10^{-8}
		Average	4.24032×10^{-7}

Table 13. Effective diffusion coefficient D for fabric sample A under various conditions

Fig. 11 illustrates a comparison between the experiential records and the modeling results from regression model using D at 4.24032×10^{-7}.

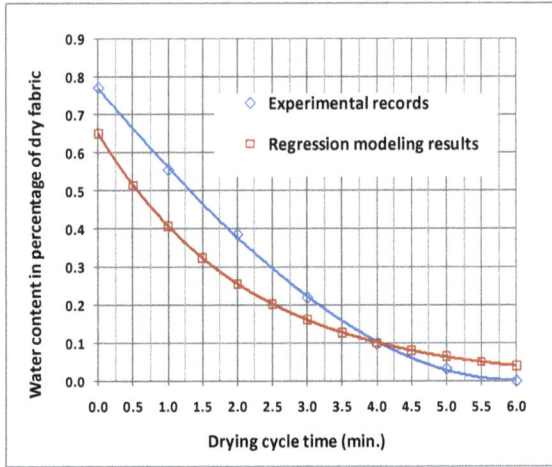

Figure 11. A comparison of regression results from Diffusion model and experimental records for fabric sample A

The Diffusion model has been applied to model each of the drying cycle for all fabric samples, the discrepancies between the modeling results and records from experiments are listed in Table 14.

Fabric sample	A	B	C	D	E	F
Discrepancy (%)	76.63	42.86	49.03	44.44	48.35	43.34

Table 14. Discrepancies of a comparison of Diffusion model and records of experiments

4.5. Kinetics model based on the solutions of diffusion equations

Fick's second law for diffusion applications is commonly used to simulate mass transfer process in convective drying. However, the exponential term in Equation (26) causes a restriction to the Diffusion model be applied in the falling drying period. A separate modeling process is needed to describe the constant drying period for completed modeling of a drying cycle. Efremov (1998, 2002) has proposed a mathematical solution to solve the Frick's law using integral error functions:

$$\frac{m}{m_o} = 1 - erf \frac{x}{2\sqrt{Dt}} \tag{30}$$

$$\frac{m}{m_o} = 1 - N_o \frac{t}{m_o} \tag{31}$$

where m is the moisture removal rate, m_0 is the initial moisture removal rate, N_0 is the drying rate at the constant drying period and x is the fabric thickness. Substituting the boundary conditions of $t = 0$ and $t = t_f$ for a drying cycle at the starting and ending points, a kinetics model equation developed from the diffusion model is given as:

$$\frac{m}{m_0} = 1 - N_0 \frac{t}{w_0} + \frac{N_0 \sigma \sqrt{\pi}}{2w_0}\left(1 - erf\frac{t_f - t}{\sigma}\right) \tag{32}$$

where σ is a characteristic drying time and expressed as:

$$\sigma = 2\frac{N_0 t_f - m_0}{N_0 \sqrt{\pi}} \tag{33}$$

The first and second terms in the right-hand-side of Equation (32) represent the characteristics in the constant drying period, and the third term represents the falling drying period. The new kinetics equation consists of two modeling sections to describe the linear and non-linear parts of a drying process. The drying rate at constant drying period N_0 and the drying cycle time t_f are needed to be predetermined when Equation (32) is applied. N_0 is the slope of the dotted line as shown in Fig. 5, they are given in Table 15.

Setting condition	Drying rate at constant drying period of the fabric samples (g/s)					
	A	B	C	D	E	F
1	0.0069	0.0073	0.0074	0.0077	0.0072	0.0071
2	0.0069	0.0079	0.0083	0.0080	0.0084	0.0071
3	0.0065	0.0067	0.0069	0.0069	0.0070	0.0065
4	0.0068	0.0064	0.0065	0.0066	0.0051	0.0054
5	0.0066	0.0065	0.0072	0.0066	0.0074	0.0074
6	0.0063	0.0065	0.0066	0.0066	0.0058	0.0064
7	0.0047	0.0046	0.0051	0.0047	0.0048	0.0051
8	0.0047	0.0047	0.0048	0.0048	0.0048	0.0050
Average	0.00618	0.00633	0.00660	0.00649	0.00631	0.00625

Table 15. The drying rate at constant drying period of the fabric samples under different air setting conditions

The drying rate at constant drying period N_0 listed in Table 15 for each fabric sample and their corresponding drying cycle time t_f are employed to assist the simulation of entire drying process. Equation (32) is the modeling tool to calculate the moisture removal rate m from t_0 to t_f. The red curve in Fig. 12 illustrates the modeling results determined from Equation (32). The values for N_0 and t_f are 0.00618 and 15 min. respectively. The modeling process has been repeated for the falling drying period using the final term of Equation (32), and the results are given in Fig. 13.

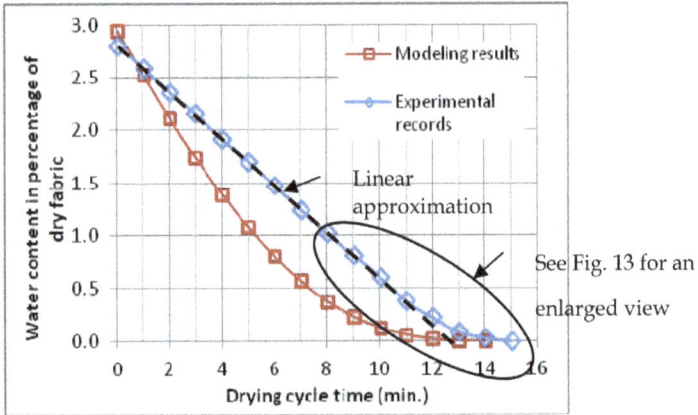

Figure 12. A comparison of modeling results from Equation (32) and testing records of a complete drying cycle for fabric sample A under air setting condition 3

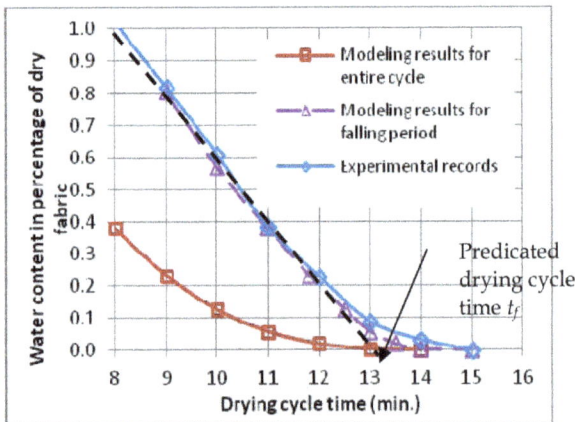

Figure 13. A comparison of modeling results from Equation (32) for the falling drying period and testing records of fabric sample A under air setting condition 3

Tables 16 and 17 list the discrepancies of the two modeling results from Figs. 12 and 13.

Fabric sample	A	B	C	D	E	F
Discrepancy (%)	41.17	30.25	40.53	30.21	36.63	38.06

Table 16. Discrepancies between the tests and modeling results for entire drying cycle

Fabric sample	A	B	C	D	E	F
Discrepancy (%)	8.70	7.92	10.12	11.69	8.87	7.15

Table 17. Discrepancies between the tests and modeling results at the falling drying period

4.6. Wet surface model

The fourth analytical model, Wet surface, was proposed by Schlunder (1988, 2004). He has proved that the remaining moisture in porous materials would be 20 to 30 % of the saturated moisture content at the initial stage when a drying process reaches the critical moisture stage. The material surface is unlikely to be fully wetted at this low moisture content condition. The Wet surface model is therefore designed to address the characteristics of the partially wet surface. The drying rate $\dot{M}_v$ is modeled directly in contrast with the calculation of the moisture content M required in the previous models. The Wet surface equation is given as:

$$\frac{\dot{M}_v}{\dot{M}_{vl}} = \frac{1}{1+\phi}$$

(34)

$$\phi = \frac{2r}{\pi\varepsilon}\sqrt{\frac{\pi}{4\varphi}}\left(\sqrt{\frac{\pi}{4\varphi}}-1\right)$$

(35)

$\dot{M}_v$ is the instant drying rate and $\dot{M}_{vl}$ is initial drying rate when the fabric surface is fully wetted, r is pore size, ε is viscous sub-layer thickness and φ is the fraction of a wet surface. It is assumed that the wet surface fraction φ is proportional to a ratio of moisture content θ to the critical moisture content θ_c in the falling drying period, thus the Equation (35) is rewritten as:

$$\phi = \frac{2r}{\pi\varepsilon}\sqrt{\frac{\pi}{4\frac{\theta}{\theta_c}}}\left(\sqrt{\frac{\pi}{4\frac{\theta}{\theta_c}}}-1\right)$$

(36)

The results of converting Equation (36) from the critical moisture content θ_c to the final moisture content give an expression as:

$$\theta - \theta_c + \frac{r}{2\varepsilon}\theta_c\ln\frac{\theta}{\theta_c} - \frac{2r}{\sqrt{\pi\varepsilon}}\sqrt{\theta_c\theta} + \frac{2r}{\sqrt{\pi\varepsilon}}\theta_c = \dot{M}_{vl}t$$

(37)

In Equation (37), the parameters for fabric drying modeling are the initial drying rate $\dot{M}_{vl}$, critical moisture content θ_c, pore size r and viscous sub-layer thickness ε. In fact, $\dot{M}_{vl}$ and θ_c can be experimentally determined as discussed in Section 4.1. The viscous sub-layer thickness ε is in terms of thermal conductivity of air c and heat transfer coefficient U, they are given as:

$$\varepsilon = c/U$$

(38)

$$U = \frac{\lambda N_o}{A\left(T_h - T_{SL}\right)}$$

(39)

where λ is the latent heat of vaporization, A is the fabric surface area, T_h and T_{SL} are temperature of the drying air and the temperature at saturated condition respectively. The values of λ, T_{SL} and c can be found from engineering handbooks. Table 18 lists the calculated ε for the fabric sample A from Equations (38) and (39).

Setting condition	N_o (g/s)	T_h (K)	T_{SL} (K)	λ (J/g)	u (W/m²K)	c (W/mK)	ε (mm)
1	0.0069	353.0	301	2310	30.79	0.028	0.91
2	0.0077	354.5	301	2304	33.29	0.028	0.84
3	0.0065	359.5	302	2291	25.70	0.028	1.09
4	0.0068	327.0	293	2372	47.35	0.028	0.59
5	0.0066	328.5	294	2372	45.32	0.028	0.62
6	0.0063	327.0	293	2372	43.89	0.028	0.64
7	0.0047	330.0	294	2366	31.04	0.028	0.90
8	0.0047	331.0	295	2366	30.75	0.028	0.91
						Average	0.8125

Table 18. The viscous sub-layer thickness ε for the fabric sample A

The final parameter to be determined for Equation (37) is the fabric pore size r. The pore size can be measured under microscope but it is not practical to determine through microscopic images. A better way to acquire the pore size can be done by graphical methods. Fig. 14 shows the plotting of the predicted drying rate from the Wet surface model versus the moisture reduction rate experimentally determined from Equation (15).

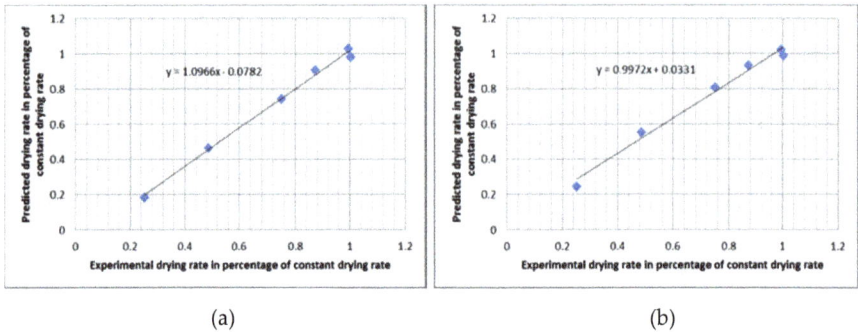

(a)   (b)

Figure 14. Relationship of predicted drying rate to experimentally determined drying rate for the assigned pore size r at (a) 0.5 mm and (b) 0.3478 mm for fabric sample A

The fitted line in Fig. 14(a) shows the results from an assigned pore size of 0.5 mm. Results from the plotting have shown that the slope of the fitted line is not a unity, and the y-intercept does not meet the origin. As a result, the calculated viscous sub-layer thickness ε from the pore size does not consist with the calculated value as listed in Table 18. Thus, a new assignment of 0.3478 mm was used to create another plotting. The new fitted line as illustrated in Fig. 14(b) is much closer to unity slope and zero y-intercept conditions. The new pore size could be used as the modeling parameter in the Wet surface model.

Table 19 lists the determined pore sizes for other fabric samples. The calculated pore sizes for each fabric sample are then substituted into Equation (37) to determine the drying rate $\dot{M}_v$. A comparison of the discrepancies between the modeled drying rate and recorded data from experimental tests for fabric sample A under the air setting condition 3 is given in Fig. 15.

Setting condition	Fabric sample					
	A	B	C	D	E	F
1	0.40	0.61	0.19	0.31	0.46	0.36
2	0.40	0.55	0.35	0.70	0.40	0.45
3	0.35	0.40	0.55	0.31	0.21	1.55
4	0.16	0.23	0.22	0.25	0.20	0.50
5	0.19	0.31	0.22	0.40	0.37	0.35
6	0.33	0.07	0.49	0.40	0.37	0.44
7	0.40	1.10	0.60	0.31	0.43	0.55
8	0.55	0.53	0.49	1.50	1.10	1.20
Average	0.3478	0.4750	0.3888	0.5225	0.4425	0.6750

Table 19. The determined pore size in mm for the tested fabric samples

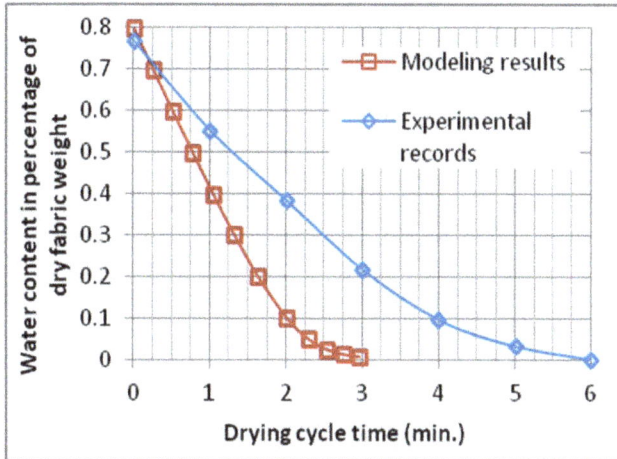

Figure 15. A comparison of the modeling results using Equation (37) and testing records for fabric sample A under test setting 3 for the falling drying period

The findings given in Fig. 15 have shown that the Wet surface model could not produce an accurate modeling result as the measured discrepancy is 77.53 % based upon the experimental records.

5. A performance evaluation of the studied models

The performance of the studied analytical models for the modeling of porous fabric drying process should be reviewed. The percentage of discrepancies from each modeling results in comparison with the experimental records are summarized in Table 20. It is made clear that the Kinetics model based on the solutions of diffusion equations has produced the best performance. The First order kinetics model has provided a better performance than the Diffusion model, and the Wet surface model has given the largest discrepancy from the statistical records listed in Table 20.

Model	Fabric sample						
	A	B	C	D	E	F	Average
First order kinetics	42.39	34.69	38.74	34.21	42.25	37.11	38.23
Diffusion	76.63	42.86	49.03	44.44	48.35	43.34	50.77
Kinetics based on the solutions of diffusion equations	8.70	7.92	10.12	11.69	8.87	7.15	9.08
Wet surface	77.53	37.47	93.50	56.00	69.87	64.02	66.40

Table 20. Summary of performance evaluation of the studied models

The findings have shown that the Kinetics model based on the solutions of diffusion equations could be the best one in the simulation of a porous fabric drying process among the others. The required condition for the model is to have a predictable drying cycle time t_f that could be obtained by a linear approximation of the experimentally determined drying curve as shown in Fig. 13. First order kinetics model has also produced a good performance in comparison with the Diffusion and Wet surface model. An empirically determined kinetic coefficient k is only needed for the modeling process, and the coefficient is highly correlated to temperature and impinging velocity of air. A less accurate modeling result observed from the Diffusion model is that a fabric thickness L is needed for the model equation. A significant change of the exponential index in Equation (26) would cause a large discrepancy in the modeling result if the thickness is inaccurately defined. It is understandable that the complexity of finding the pore size would cause unavoidable errors in the Wet surface model. To design experimental strategies determining the pore size to improve the model performance is some further work.

6. Conclusion

The principles of water mass movement due to phase change in the drying of porous fabrics have been studied. The boundary equations for mass transfer between water, vapor and air were used to support the establishing a new set of drying models using diffusion theories. Experiments were done to find information for the determination of the modeling parameters. The performance of the developed models has been evaluated. Among the four models, the Kinetics model based on the solutions of diffusion equations has produced the best performance. In the real life applications, they could act as a mathematical tool to assist a precise estimation of the moisture content in fabric drying or heat setting process under various processing conditions. Further work has been started to apply the developed drying models in the design of garment setting machines for clothing industry.

Author details

Ralph Wai Lam Ip and Elvis Iok Cheong Wan
Department of Mechanical Engineering, The University of Hong Kong, Pokfulam, Hong Kong SAR

7. References

Cohen, J., Cohen P., West, S.G., & Aiken, L.S. (2003). *Applied Multiple Regression/Correlation Analysis for the Behavioral Sciences.* (2nd ed.): Lawrence Erlbaum Associates, Hillsdale, NJ

Efremov, G.I. (1998). Kinetics of Convective Drying of Fibre Materials Based on Solution of a Diffusion Equation. *Fibre Chemistry*, Vol. 30, No. 6, pp. 417-422

Efremov, G.I. (2002). Drying Kinetics Derived from Diffusion Equation with Flux-Type Boundary Conditions. *Drying Technology*, Vol. 20, No. 1, pp. 55-66

Haghi, A.K. (2006). Transport Phenomena in Porous Media: A Review. *Theoretical Foundations of Chemical Engineering.* Vol. 40, No. 1, pp. 14-26

Ip, R.W.L. & Wan, I.C. (2011). New Use Heat Transfer Theories for the Design of Heat Setting Machines for Precise Post-Treatment of Dyed Fabrics. *Journal Defect and Diffusion Forum*, Vols. 312-315, pp. 748-751

Khazaei, J.; Chegini, G.R. & Bakhshiani, M. (2008). A Novel Alternative Method for Modeling the Effects of Air Temperature and Slice Thickness on Quality and Drying Kinetics of Tomato Slices: Superposition Technique. *Drying Technology*, Vol. 26, No. 6, pp. 759-775

Kowalski, S.J. (2000). Toward a Thermodynamics and Mechanics of Drying Processes. *Chemical Engineering Science*, Vol. 55, No. 7, pp. 1289-1304.

Kowalski, S.J. (2003). *Thermomechanics of Drying Processes*, Springer, ISBN 3-540-00412-2, Berlin, Germany

Kowalski, S.J.; Musielak, G. & Banaszak, J. (2007). Experimental Validation of the Heat and Mass Transfer Model for Convective Drying. *Drying Technology*, Vol. 25, No. 1-3, pp. 107-121

Moropoulou, A. (2005). Drying Kinetics of Some Building Materials. *Brazilian Journal of Chemical Engineering*, Vol. 22, No. 2, pp. 203-208

Ramaswamy, H.S. and van Nieuwenhuijzen, N.H. (2002). Evaluation and Modeling of Two-Stage Osmo-Convective Drying of Apple Slices. *Drying Technology*, Vol. 20, No. 3, pp. 651-667

Roberts, J.S. and Srikiatden, J. (2005). Moisture Loss Kinetics of Apple During Convective Hot Air and Isothermal Drying. *International Journal of Food Properties*, Vol. 8, No. 3, pp. 493-512

Roberts, J.S. & Tong,C.H. (2003). Drying Kinetics of Hygroscopic Porous Materials under Isothermal Conditions and the Use of a First-Order Reaction Kinetic Model for Predicting Drying. *International Journal of Food Properties*. Vol. 6, No. 3, p. 355-367.

Schlunder, E.U. (1988). On the Mechanism of the Constant Drying Rate Period and Its Relevance to Diffusion Controlled Catalytic Gas-Phase Reactions. *Chemical Engineering Science*, Vol. 43, No. 10, pp. 2685-2688

Schlunder, E.U. (2004). Drying of Porous Material during the Constant and the Falling Rate Period: A Critical Review of Existing Hypotheses. *Drying Technology*, Vol. 22, No. 6, pp. 1517-1532

Chapter 10

Blast Wave Simulations with a Runge-Kutta Discontinuous Galerkin Method

Emre Alpman

Additional information is available at the end of the chapter

1. Introduction

An explosion generates high pressure and temperature gases which expand into the surrounding medium to generate a spherical shock wave called a blast wave [1]. Experimental methods to simulate blast waves require handling real explosives. Hence, they may be dangerous and very expensive. On the other hand, computational fluid dynamics (CFD) could be used to generalize and support experimental results in simulating blast waves. Sharma and Long [2] used Direct Simulation Monte Carlo (DSMC) method to simulate blast waves. DSMC is a very powerful method for detonations and flows with chemical reactions [3], [4], [5]. However, being originally designed for rarefied gas dynamics, it can be very expensive for continuum flows. Also, if one makes a local thermodynamic equilibrium assumption in DSMC, the method becomes equivalent to solving Euler equations [1]. In references [6] - [10] blast wave simulations were performed by solving Euler equations using a finite volume method where a second order accuracy was achieved by extrapolating density, velocity and pressure at the cell interfaces. The fluxes at cell faces were calculated using the AUSM+ method [11] due to its algorithmic simplicity and suitability for flows with discontinuities. Resulting semi-discrete equations were solved using a 2 stage modified Runge-Kutta method [12].

Discontinuous Galerkin (DG) methods, which have the features of both finite volume and finite element methods, have become popular for the solution of hyperbolic conservation laws like Euler equations [13]. DG method represents solution on elements by a collection of piecewise discontinuous functions hence sometimes considered as a high order accurate extension of finite volume method [13]. One class of DG methods is the Runge-Kutta DG method (RKDG) where spatial discretization is performed using polynomials which are discontinuous across element faces. Then the resulting system of ordinary differential equations is solved using a total variation diminishing (TVD) Runge-Kutta scheme [14]. The details of the RKDG scheme can be found in references [15]-[20]. An alternative to the

RKDG method is the Space-Time Discontinuous Galerkin Method (STDG) in which the space and time discretizations are not separated [21], [22]. This method is particularly suitable for numerical solutions on adaptive meshes.

In this study RKDG method was selected for numerical solutions because of its relative programming simplicity and its similarity to the semi-discrete approach adopted for the finite volume solutions described above. Simulations were performed for one-dimensional moving rectangular and spherical shock waves (blast waves). Planar shock waves were simulated by solving planar shock tube problem with moderate and extremely strong discontinuities [23]. These two cases were compared in order to display the difficulties numerical methods may experience when the discontinuities in the flowfield are strong, as in the case of blast waves due to explosions. The predictions were also compared to the exact solutions by Sod [24]. Spherical shock waves were simulated by solving a spherical shock tube problem and blast waves generated from explosion of 1 kg of TNT. Here explosive material was assumed to transform gas phase instantly after the detonation, thus explosive was modeled as a high pressure sphere. This way, blast wave problem also became a spherical shock tube problem where an isobaric sphere of a high pressure gas expanded toward surrounding low pressure air.

Section 2 contains the theory of RKDG method and Section 3 describes the methodology followed for numerical solutions. Numerical solutions are presented and discussed in Section 4 and finally the conclusions are drawn in Section 5.

2. Theory

One-dimensional Euler equations are written as:

$$\frac{\partial \mathbf{q}}{\partial t} + \frac{1}{r^n}\frac{\partial \left(r^n \mathbf{f}\right)}{\partial r} + \mathbf{h} = 0 \tag{1}$$

with

$$\mathbf{q} = \begin{bmatrix} \rho & \rho u & \rho E \end{bmatrix}^T \tag{2}$$

$$\mathbf{f} = \begin{bmatrix} \rho & \rho u^2 + p & \rho u H \end{bmatrix}^T \tag{3}$$

$$\mathbf{h} = \frac{n}{r}\begin{bmatrix} 0 & -p & 0 \end{bmatrix}^T \tag{4}$$

In the above equations ρ is fluid density, u is fluid velocity, p is static pressure, E and H are total energy and total enthalpy per unit mass.

$$E = \frac{p}{\rho(\gamma - 1)} + \frac{\mathbf{V} \cdot \mathbf{V}}{2} \tag{5}$$

$$H = E + \frac{p}{\rho} \qquad (6)$$

In equation (4) $n = 0$ for Cartesian coordinates, $n = 1$ for cylindrical coordinates and $n = 2$ for spherical coordinates.

The DG method is derived by multiplying equation (1) by a test function $\phi(r)$ and integrating over a control volume (or element) (K)[20]

$$\int_K \frac{\partial \mathbf{q}}{\partial t} \phi(r) dK + \int_K \frac{1}{r^n} \frac{\partial(r^n \mathbf{f})}{\partial r} \phi(r) dK + \int_K \mathbf{h}\phi(r) dK = 0 \qquad (7)$$

where $dK = r^n dr$ for one-dimensional flow. Using this relation for dK, the second integral in equation (7) can be rewritten as for one-dimensional flow as:

$$\int_K \frac{1}{r^n} \frac{\partial(r^n \mathbf{f})}{\partial r} \phi(r) dK = \int_{r_{i-1/2}}^{r_{i+1/2}} \frac{\partial(r^n \mathbf{f})}{\partial r} \phi(r) dr = \int_{r_{i-1/2}}^{r_{i+1/2}} \frac{\partial(r^n \mathbf{f}\phi(r))}{\partial r} dr - \int_{r_{i-1/2}}^{r_{i+1/2}} r^n \mathbf{f} \frac{d\phi(r)}{dr} dr \qquad (8)$$

Therefore, for one space dimension equation (7) takes the following form for an element i [20]:

$$\int_{r_{i-1/2}}^{r_{i+1/2}} \frac{\partial \mathbf{q}}{\partial t} \phi(r) r^n dr + \left[\mathbf{f} r^n \phi(r) \right]_{i+1/2} - \left[\mathbf{f} r^n \phi(r) \right]_{i-1/2} - \int_{r_{i-1/2}}^{r_{i+1/2}} \mathbf{f} \frac{d\phi(r)}{dr} r^n dr + \int_{r_{i-1/2}}^{r_{i+1/2}} \mathbf{h}\phi(r) r^n dr = 0 \quad (9)$$

In the DG method the numerical solution $\mathbf{q}$ is represented using a collection of polynomials on an element. Since polynomial continuity across element faces is not enforced the solution may be discontinuous at the element faces and which results in a multi-valued the flux vector $\mathbf{f}$ at these locations. This problem is overcome by using numerical flux $\hat{\mathbf{f}}$ at the faces, which must be a conservative and upwind flux [19], [20]. Note that equation (9) reduces to finite volume formulation for $\phi(r) = 1$ [19].

Normally in a DG method the solution $\mathbf{q}$ and the test function $\phi(r)$ belongs to the space of polynomials of degree smaller or equal to some k. This way a $(k+1)$st order accurate approximation to the solution can be obtained [19],[20] The approximate solution can then be written as:

$$\mathbf{q}(r,t) = \sum_{j=1}^{k} \mathbf{q}^j(t) \phi_j(r) \qquad (10)$$

where $q^i(t)$ are called degrees of freedom of $\mathbf{q}$ and $\phi_j(r)$ are the bases of the polynomial solution space. Using the approximation given in equation (10) the first integral in equation (9) is written as:

$$\int_{r_{i-1/2}}^{r_{i+1/2}} \frac{\partial \mathbf{q}}{\partial t} \phi(r) r^n dr = \sum_{j=1}^{k} \frac{d(\mathbf{q}^j(t))}{dt} \int_{r_{i-1/2}}^{r_{i+1/2}} \phi_j(r) \phi_l(r) r^n dr \qquad (11)$$

Substituting the right hand side of equation (11) into equation (9) one can get a system of ordinary differential equations for the degrees of freedom of **q**. The integral on the right hand side of equation (11) is an inner product of the bases of the polynomial solution space with respect to the weight function r^n and it yields a matrix called the *mass matrix* [20]. When orthogonal polynomials are chosen as bases then this matrix becomes diagonal and can easily be inverted [13]. For solution in Cartesian coordinates ($n = 0$), the weighting function becomes unity therefore Legendre polynomials are typically used as bases [19] - [22]. For this case ($n = 0$) the mass matrix, M, for an element i becomes

$$M_{jl} = \int_{r_{i-1/2}}^{r_{i+1/2}} \phi_j(r)\phi_l(r)dr = \frac{\Delta r_i}{2j+1}\delta_{jl} \tag{12}$$

where δ_{jl} is the Kronecker's delta and $\Delta r_i = r_{i+1/2} - r_{i-1/2}$ is the grid size. Here the bases are defined as [20]

$$\phi_j(r) = P_j\left(\frac{2(r-r_i)}{\Delta r_i}\right) \tag{13}$$

where $P_j(r)$ is the Legendre polynomial of degree j. Therefore, a set of ordinary differential equations for the degrees of freedom can be obtained for Cartesian coordinates as:

$$\frac{dq^j}{dt} = -\frac{2j+1}{\Delta r_i}\left[\left[\hat{f}\phi_j(r)\right]_{i+1/2} - \left[\hat{f}\phi_j(r)\right]_{i-1/2} - \int_{r_{i-1/2}}^{r_{i+1/2}} f\frac{d\phi_j(r)}{dr}dr + \int_{r_{i-1/2}}^{r_{i+1/2}} h\phi_j(r)dr\right] \tag{14}$$

Using Legendre polynomials as bases not only yields a diagonal mass matrix, but also the diagonal elements have a simple form as displayed in equation (13) Unfortunately, this simplicity Legendre polynomials provide for Cartesian coordinates does not extend to cylindrical or spherical coordinates because Legendre polynomials are not orthogonal with respect to the weight function r^n when n is not equal to zero. The mass matrix corresponding to the weight r^n and $k = 1$ is given in equation (15)

$$M = \begin{bmatrix} \left(\frac{r^{n+1}}{n+1}\right)\Big|_{r_{i-1/2}}^{r_{i+1/2}} & \frac{2}{\Delta r_i}\left(\frac{r^{n+2}}{n+2} - r_i\frac{r^{n+1}}{n+1}\right)\Big|_{r_{i-1/2}}^{r_{i+1/2}} \\ \frac{2}{\Delta r_i}\left(\frac{r^{n+2}}{n+2} - r_i\frac{r^{n+1}}{n+1}\right)\Big|_{r_{i-1/2}}^{r_{i+1/2}} & \frac{4}{\Delta r_i^2}\left(\frac{r^{n+3}}{n+3} - 2r_i\frac{r^{n+2}}{n+2} + r_i^2\frac{r^{n+1}}{n+1}\right)\Big|_{r_{i-1/2}}^{r_{i+1/2}} \end{bmatrix} \tag{15}$$

For $k = 2$ the corresponding matrix will be 3x3 and the additional elements will be

$$M_{13} = M_{31} = \frac{6}{\Delta r_i^2}\left(\frac{r^{n+3}}{n+3} - 2r_i\frac{r^{n+2}}{n+2} + \left(r_i^2 - \frac{\Delta r_i^2}{12}\right)\frac{r^{n+1}}{n+1}\right)\Big|_{r_{i-1/2}}^{r_{i+1/2}} \tag{16}$$

$$M_{23} = M_{32} = \frac{12}{\Delta r_i^3} \left(\frac{r^{n+4}}{n+4} - 3r_i \frac{r^{n+3}}{n+3} + \left(3r_i^2 - \frac{\Delta r_i^2}{12} \right) \frac{r^{n+2}}{n+2} - \left(12r_i^3 - \frac{r_i \Delta r_i^2}{12} \right) \frac{r^{n+1}}{n+1} \right) \Bigg|_{r_{i-1/2}}^{r_{i+1/2}} \tag{17}$$

$$M_{33} = \frac{36}{\Delta r_i^4} \left(\begin{array}{l} \dfrac{r^{n+5}}{n+5} - 4r_i \dfrac{r^{n+4}}{n+4} + \left(\dfrac{r_i^2}{6} - \dfrac{\Delta r_i^2}{6} \right) \dfrac{r^{n+3}}{n+3} - \left(\dfrac{r_i^3}{9} - \dfrac{r_i \Delta r_i^2}{3} \right) \dfrac{r^{n+2}}{n+2} + \\[2mm] + \left(r_i^4 - \dfrac{r_i^2 \Delta r_i^2}{6} + \dfrac{\Delta r_i^4}{144} \right) \dfrac{r^{n+1}}{n+1} \end{array} \right) \Bigg|_{r_{i-1/2}}^{r_{i+1/2}} \tag{18}$$

Even for $k = 1$ the mass matrix is complicated for both cylindrical and spherical cases. Furthermore, the elements of the matrix are functions of element locations. Therefore, the elements of the inverse matrix must be computed for each element.

As an alternative approach one could search for a set of orthogonal polynomials with respect to the weight r^n. Reference [25] contains a method to derive such a set of polynomials starting from known orthogonal polynomials. Starting from Legendre polynomials and following the procedure described in [25], polynomials orthogonal with respect to the weight r^n can be derived as

$$\phi_0(r) = 1 \tag{19}$$

$$\phi_1(r) = \frac{2}{\Delta r_i} \left(r - \frac{n+1}{n+2} \frac{r_{i+1/2}^{n+2} - r_{i-1/2}^{n+2}}{r_{i+1/2}^{n+1} - r_{i-1/2}^{n+1}} \right) \tag{20}$$

$$\phi_2(r) = C_1 + C_2 \frac{2(r-r_i)}{\Delta r_i} + \frac{1}{2} \left(\frac{12(r-r_i)^2}{\Delta r_i^2} - 1 \right) \tag{21}$$

where [25]

$$C_1 = \frac{I_{0,1} I_{1,2} - I_{0,2} I_{1,1}}{I_{0,0} I_{1,1} - I_{0,1} I_{1,0}} \tag{22}$$

$$C_2 = \frac{I_{0,2} I_{1,0} - I_{0,0} I_{1,2}}{I_{0,0} I_{1,1} - I_{0,1} I_{1,0}} \tag{23}$$

and [25]

$$I_{r,s} = \int_{r_{i-1/2}}^{r_{i+1/2}} P_s \left(\frac{2(r-r_i)}{\Delta r_i} \right) r^{n+r} dr \tag{24}$$

In equation (24) $P_s(r)$ is the Legendre polynomial of degree s. One can clearly see from equations (19), (20) and (21) that these set of polynomials, especially the second order one, are more complicated compared to Legendre polynomials. Their advantage due to their orthogonality is offset by the complexity they bring to the terms on right hand side of the equation (9) for spherical or cylindrical coordinates. Therefore, this approach does not produce a good alternative.

Another approach for a simple formulization of equation (9) can be considered by changing the definition of the approximate solution originally given in equation (10). The first integral in equation (9) can be rewritten as:

$$\int_{r_{i-1/2}}^{r_{i+1/2}} \frac{\partial \mathbf{q}}{\partial t} \phi(r) r^n \, dr = \int_{r_{i-1/2}}^{r_{i+1/2}} \frac{\partial \left(r^n \mathbf{q} \right)}{\partial t} \phi(r) \, dr \tag{25}$$

Furthermore, if the definition of approximate solution is changed to be:

$$r^n \mathbf{q}(r,t) = \sum_{j=1}^{k} \mathbf{q}^j(t) \phi_j(r) \tag{26}$$

Then the following set of ordinary differential equations for the degrees of freedom can be obtained

$$\frac{d\mathbf{q}^j}{dt} = -\frac{2j+1}{\Delta r_i} \left[\left[\hat{\mathbf{f}} \phi_j r^n \right]_{i+1/2} - \left[\hat{\mathbf{f}} \phi_j r^n \right]_{i-1/2} - \int_{r_{i-1/2}}^{r_{i+1/2}} \mathbf{f} \frac{d\phi_j(r)}{dr} r^n \, dr + \int_{r_{i-1/2}}^{r_{i+1/2}} \mathbf{h} \phi_j(r) r^n \, dr \right] \tag{27}$$

Equation (27) provides a simple and similar formulation for all coordinate systems. Therefore, it looks like a very good approach. However, during the solutions, this approach leaded to difficulties in applying limiters (see Section 3 about limiters) during the calculation of numerical fluxes at the element faces.

Considering the approaches described above an approximation in the formulations is proposed which will use equation (10) and Legendre polynomials to define the solution in the elements however, will yield a set of ordinary differential equations similar to equation (27). This approximation is done in the first integral of equation (9). Here, the term r^n inside the integral is replaced by its average inside the element, which is nothing but the location of the center of that element r_i. This integral is approximated as:

$$\int_{r_{i-1/2}}^{r_{i+1/2}} \frac{\partial \mathbf{q}}{\partial t} \phi(r) r^n \, dr \approx r_i^n \int_{r_{i-1/2}}^{r_{i+1/2}} \frac{\partial \mathbf{q}}{\partial t} \phi(r) \, dr \tag{28}$$

Then using equation (10) to describe the solution in an element, the following equations are obtained for the degrees of freedom:

$$\frac{d\mathbf{q}^{j}}{dt} = -\frac{2j+1}{r_i^n \Delta r_i}\left[\left[\hat{\mathbf{f}}\phi_j r^n\right]_{i+1/2} - \left[\hat{\mathbf{f}}\phi_j r^n\right]_{i-1/2} - \int\limits_{r_{i-1/2}}^{r_{i+1/2}}\mathbf{f}\frac{d\phi_j(r)}{dr}r^n\,dr + \int\limits_{r_{i-1/2}}^{r_{i+1/2}}\mathbf{h}\phi_j(r)r^n\,dr\right] \quad (29)$$

The approximation in equation (29) not only provides a simple formulation for cylindrical and spherical coordinates, but also turns in to the exact formulation (equation (14)) for Cartesian coordinates.

3. Methodology

Numerical solutions were performed using an in-house computer code written in C language. The code solves equation (29) in Cartesian, cylindrical, and spherical coordinates and then constructs the numerical solution using equation (10). Numerical solution of equation (29) requires a numerical upwind flux function. The selected flux function affects the quality of the solution. Performance of RKDG method with different numerical fluxes can be found in [26]. Numerous different methods were used in literature for flow problems involving shock waves. Among these methods AUSM (Advection Upstream Splitting Method) -family of methods provide algorithmic simplicity by using a scalar diffusion term [27]. Hence, unlike many other upwind type methods, AUSM-family methods do not require the knowledge of the eigenstructure of the flow problem and this helps generate very efficient implementations. In this study the AUSM+ method [11] was used to calculate the numerical fluxes mainly because this method has previously been used by the author for moving shock wave simulations with satisfactory results [6] - [10]. Performance of this method was also tested against performance of van Leer [28], HLLE [29] and Roe's [30] upwind methods. Numerical solutions were obtained for $k = 1$ and 2 which correspond to second and third order spatial accuracies, respectively.

The integrals in equation (29) were obtained using a quadrature method. According to [19], for a given value of k, the applied quadrature rule must be exact for polynomials of degree $2k$ for the interior of the elements and three-point Gauss-Legendre rule [31] was used in that reference for $k = 2$. This method however, required calculation of flow variables at the interior locations of an element. The three-point Simpson's rule [31], on the other hand, used the flow variables at element faces, which were already computed for flux calculations, and element center. Thus, it provided programming simplicity. Therefore, this rule was also used in the solutions although it was exact up to the cubic polynomials only. Numerical solutions obtained using these two integration techniques were compared with each other.

A numerical method used for a flow with discontinuities should yield entropy satisfying weak solutions [32]. The so-called monotone methods yield such solutions, but they are only first order accurate, they perform poorly near discontinuities due to high amounts of artificial viscosity present in them [32]. High order accurate methods can be used to decrease artificial viscosity; however they yield dispersion errors which cause spurious oscillations in the vicinity of flow discontinuities. Therefore, high resolution methods, which provide high accuracy in smooth regions of the flow and apply some kind of stabilization near discontinuities, are required to get an accurate description of the flow [32]. Typically,

DG method is monotone for $k = 0$ only and solutions for $k > 0$ generate spurious oscillations in the vicinity of flow discontinuities [33]. Therefore, some kind of stabilization strategy is needed. Cockburn et.al. [13], [15], [20] adopted a generalized slope limiter with a TVB corrected minmod limiter [16], Calle et. al. used an artificial diffusion term in their stabilized discontinuous Galerkin method [34], van der Vegt and van der Ven [21], [22] used a stabilization operator, and Qui and Shu [35] used WENO type limiters to avoid spurious oscillations in their studies. Among these limiting approaches, slope limiting approach, which is also used in finite volume schemes, is computationally the simplest. Therefore, a slope limiting procedure was used in this study. In the slope limiting approach, gradients in each element (or cell) were limited by using a limiter function. There are many limiter functions available in the literature [36]. Details of limiting procedure and some comparisons can be found in [37].

It is known that minmod limiter is the most dissipative limiter [30], [37] and the TVB correction mentioned in [16], [19], [20] is problem dependent. On the other hand, Superbee limiter [30] is the least dissipative and is suitable for flows with very strong shocks like a blast wave [23]. In this study minmod and Superbee limiters along with the Sweby limiter [37], which is between Superbee and minmod limiters, were used for numerical solutions. During the limiting procedure, the ratios the successive gradients of $(j-1)^{st}$ degree of freedom of an element to the jth degree of freedom of the same element were calculated, and minimum of these ratios were supplied as input to limiting function. Then the jth degree of freedom was multiplied by the output of the limiter. This approach is similar to the one followed in [38].

After the above steps were followed, the resulting system of ordinary differential equations for the degrees of freedom (equation(29)) could be solved using a Runge-Kutta technique. It was shown in Ref. [14] if the Runge-Kutta method used does not satisfy the TVD property, spurious oscillations may appear in the solution of nonlinear problems, even when a TVD space discretization is used. Therefore, equation (29) was solved using a 2nd order accurate TVD Runge-Kutta scheme described in [14] .

4. Results and discussion

This section includes numerical simulations of planar and spherical shock waves using the RKDG method described above. Since the planar shock tube problem had exact solution; effects of limiters, flux functions and quadrature rule on solutions were analyzed and presented for that case.

4.1. Planar shock tube problem

Since blast wave simulation is basically a moving shock wave problem, numerical solution procedure was first tested for Sod's shock tube problem [24]. In this problem a stationary high pressure fluid was separated from a stationary low pressure fluid by a barrier. At $t = 0$ the barrier was removed. This leaded to a shock wave and a contact discontinuity move towards the low pressure region and an expansion fan move towards the high pressure region. In this study two shock tube problems with different initial conditions were solved. Predictions were compared with the exact solutions [24].

4.1.1. Shock tube with moderate discontinuities

In this case the initial conditions were given as follows:

$\rho = 1$ kg/m^3, $u = 0$, $p = 1$x10^5 Pa, $r \leq 1$ m

$\rho = 0.125$ kg/m^3, $u = 0$, $p = 1$x10^4 Pa, $r > 1$ m

These initial conditions were the original initial conditions studied by Sod [24] and this problem is a very popular test case for numerical methods for gas dynamics. The problem was solved in Cartesian coordinates and size of the solution domain was 5m. Fluid in the problem was calorically perfect air. Figure 1 shows density distribution at $t = 0.002$ seconds obtained with 1001 grid points. This corresponded to a mesh spacing of 0.5 cm. Predictions included RKDG method, finite volume (FV) [6], [7] solutions with minmod and superbee limiters, and first order accurate solutions. All the second order accurate methods yielded similar results except in the vicinity of flow discontinuities and the head expansion wave. On the other hand, dissipative nature of first order accurate solution could be easily seen in the vicinity of the contact surface. In order to better see the differences between the predictions of different methods, density distribution in the vicinity of the head expansion wave, contact surface and shock wave were displayed in Figure 2, Figure 3 and Figure 4. A close examination of these figures clearly indicates that RKDG method is less dissipative than the finite volume method using the same numerical flux function. Among the solutions RKDG method with superbee limiter is the least dissipative one. Spurious oscillations are evident for this case in the vicinity of the head expansion wave (Figure 2). But due its least dissipative nature, this method was able to resolve the contact discontinuity (Figure 3) and the shock wave (Figure 4) much better than the other methods did.

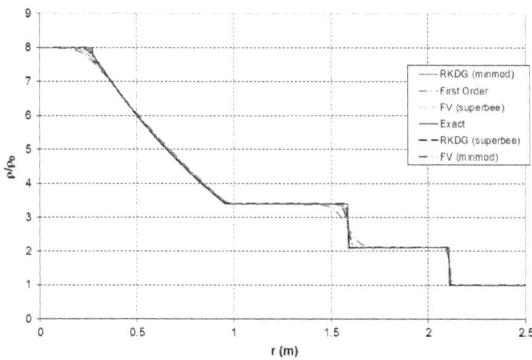

Figure 1. Density distribution at $t = 0.002$s. ($\rho_0 = 0.125$ kg/m^3)

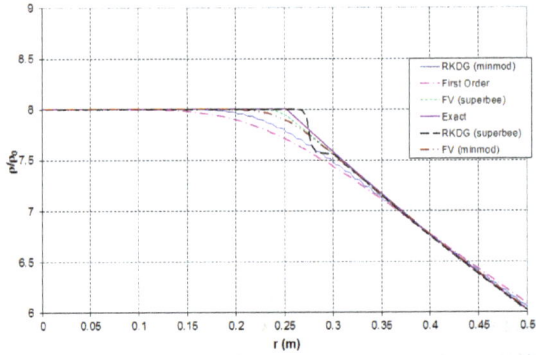

Figure 2. Density distribution at $t = 0.002$s in the vicinity of the head expansion wave ($\rho_0 = 0.125$ kg/m³)

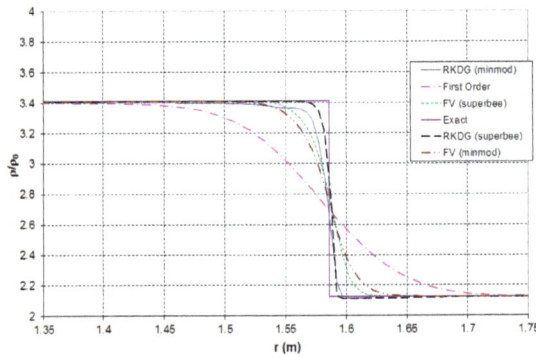

Figure 3. Density distribution at $t = 0.002$s in the vicinity of the contact surface ($\rho_0 = 0.125$ kg/m³)

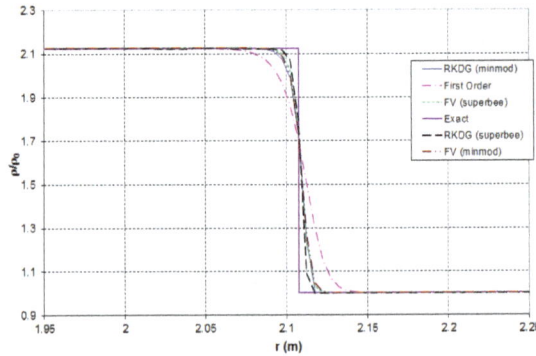

Figure 4. Density distribution at $t = 0.002$s in the vicinity of the shock wave ($\rho_0 = 0.125$ kg/m³)

4.1.2. Shock tube with extremely strong discontinuities

In this case the initial conditions were given as follows:

$\rho = 1$ kg/m^3, $u = 0$, $p = 1 \times 10^5$ Pa, $r \le 2$m

$\rho = 0.0001$ kg/m^3, $u = 0$, $p = 10$ Pa, $r > 2$m

These were the same conditions studied in [23] as an extremely strong discontinuity case. This case was selected because it more closely resembled a blast wave simulation problem and it constituted an extremely difficult case for numerical methods because of large gradients in the flowfield. The problem was solved in Cartesian coordinates, size of the solution domain was 7m and fluid was calorically perfect air. Numerical solutions were obtained using a mesh with $N = 701$ grid points. This corresponds to a mesh spacing of 1 cm. Figure 5 shows density distribution at $t = 1.2$ ms in the vicinity of the contact surface and shock wave. Predictions were obtained using FV and RKDG method with $k = 1$ using Superbee (sb), minmod (mm) and Sweby (sw) limiters. This way both methods would have the same spatial accuracy. Exact solution was also displayed for comparison. This figure clearly shows the difficulty the methods had experienced while resolving the strong discontinuities. Note the logarithmic scale used for the vertical axis. Predictions obtained using minmod limiter was nearly the same for FV and RKDG methods, in which, results showed a much smeared contact surface and over-predicted the shock location. FV method with Superbee limiter performed relatively well although the numerical solutions are on the overall poor. In order to improve the results solutions were performed on a finer mesh which was obtained by halving the mesh spacing using $N = 1401$ grid points. Density distributions for this case were displayed in Figure 6. Greatest improvement was observed for FV method with the Superbee limiter. Other solutions were also improved slightly but not as much. Overall, the RKDG method predictions were poor compared to FV method predictions obtained using the same limiter despite the fact that RKDG method with Superbee limiter had yielded the best predictions for the case with moderate discontinuities. It was also interesting to note that RKDG solutions took nearly 1.5 times more CPU time than FV solutions although this was not a big issue for one-dimensional flow simulations.

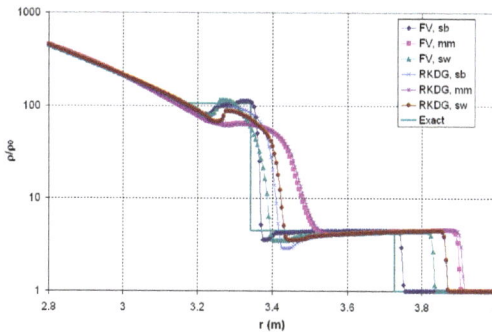

Figure 5. Density distribution at $t = 1.2$ ms in the vicinity of the contact surface and shock wave. ($\rho_0 = 0.0001$ kg/m^3, $N = 701$)

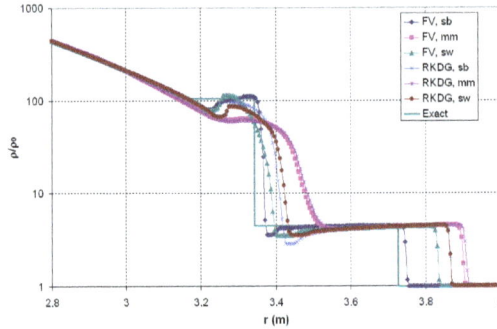

Figure 6. Density distribution at $t = 1.2$ ms in the vicinity of the contact surface and shock wave. ($\rho_0 = 0.0001$ kg/m³, $N = 1401$)

Pressure distributions at $t = 1.2$ ms obtained using $N = 1401$ grid points were shown in Figure 7. In this figure all numerical solutions experienced a spurious oscillation right after the expansion region. This location and the location of the shock wave were the places where discrepancies between numerical solutions were most severe. They were pretty much in agreement everywhere else in the solution domain. However, in Figure 6 there were also considerable differences between numerical solutions in the vicinity of the contact surface location which was not observed in Figure 7. Therefore, it was concluded that prediction of shock location could be improved if the contact surface could be resolved better.

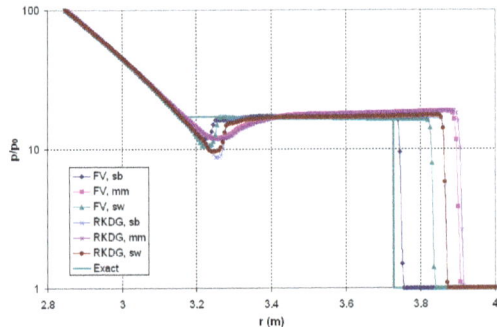

Figure 7. Pressure distribution at $t = 1.2$ ms. ($p_0 = 10$ N/m²)

Comparing Figure 6 and Figure 7 one can see that spurious oscillations were observed in the vicinity of the contact surface rather than the shock wave. These oscillations could be suppressed by using a more dissipative limiter however, this was shown to negatively affect the overall solution quality; solutions obtained using minmod limiter were the worst compared to others. Knowing that only density experiences a discontinuity across the contact surface while pressure and velocity remain continuous, an alternative limiting strategy was tested to see if it could improve the predictions. In this study limiting was performed on the primitive flow variables; density, velocity and pressure. Therefore, in this alternative strategy minmod limiter was applied for density only, while less dissipative Superbee and Sweby limiters were applied for pressure and velocity. Density distributions obtained using this limiting strategy was displayed in Figure 8. In the figure, legend first limiter was for density and second limiter was for pressure and velocity. The alternative limiting strategy, especially using minmod and Sweby limiters, improved RKDG predictions greatly. This strategy also suppressed oscillations observed in the previous FV solutions however; it also negatively affected the prediction of shock location. So far, the best predictions were obtained with the Superbee limiter for the FV method and with the minmod-Sweby combination for the RKDG method. These predictions were compared with exact solution in Figure 9. It is clear that RKDG method with this alternative limiting successfully suppressed oscillations after the expansion region and in the vicinity of the shock wave. Also it made a slightly better shock prediction compared to FV method. Considering this improvement, RKDG solutions will employ this alternative limiting with minmod and Sweby limiters from this point forward.

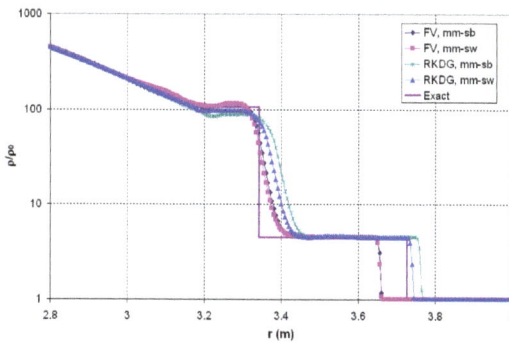

Figure 8. Density distribution at t = 1.2 ms in the vicinity of the contact surface and shock wave. (ρ_0 = 0.0001 kg/m³, N = 1401)

In order to see the effect of mesh spacing, solutions displayed in Figure 9 were repeated on the previously used coarse mesh and results were displayed in Figure 10. It is clear that RKDG solution was less sensitive to mesh spacing compared to FV method.

Spatial accuracy of RKDG methods can be easily increased by using a high order polynomial in the element. This avoids using a larger stencil as in the case of FV or finite difference method [16]. For the problem considered density distribution obtained using $k = 1$ and $k = 2$ were displayed in Figure 11. Results obtained with these two polynomials are nearly the same except $k = 2$ case placed the shock slightly ahead of the $k = 1$ case. The reason using high order polynomial did not bring much benefit was that the limiters applied degraded the high order accuracy in the vicinity of the discontinuities. This might have been overcome by using a limiter which preserves the accuracy like the TVB corrected minmod limiter of reference [19] however that correction was problem dependent.

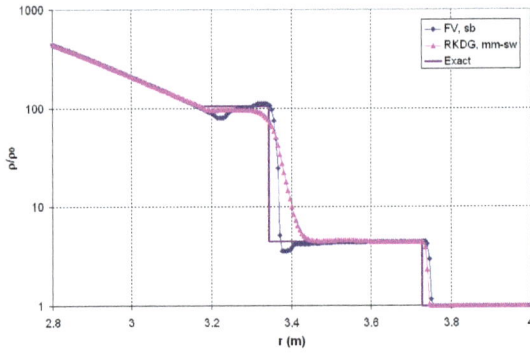

Figure 9. Density distribution at $t = 1.2$ ms in the vicinity of the contact surface and shock wave. ($\rho_0 = 0.0001$ kg/m^3, $N = 1401$)

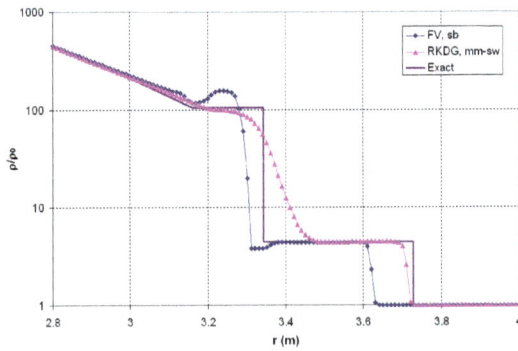

Figure 10. Density distribution at $t = 1.2$ ms in the vicinity of the contact surface and shock wave. ($\rho_0 = 0.0001$ kg/m^3, $N = 701$)

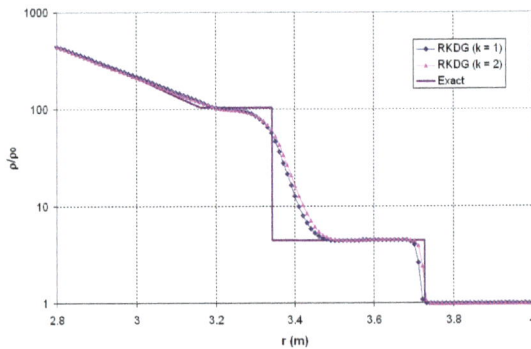

Figure 11. Density distribution at $t = 1.2$ ms obtained using $k = 1$ and 2. ($\rho_0 = 0.0001$ kg/m^3, $N = 701$)

4.1.3. Effect of the quadrature rule

Equation (29) contained integrals which were evaluated using a quadrature rule. The predictions displayed previously were obtained using three-point Simpson's rule of integration which is third-order accurate [31]. In order to see the effect of the quadrature rule on predictions, RKDG solution with $k = 2$ was also performed using three-point Gauss-Legendre rule (5th order accurate [31]) and compared to the one with Simpson rule in Figure 12. According to this figure, numerical solution was nearly insensitive to the quadrature rule employed. Since three-point Simpson rule used flow variables at the faces (already calculated for flux calculations) and the center of the element it did not require calculation of flow variables at the interior locations of as did three-point Gauss-Legendre rule. Therefore, it was mainly preferred and employed in the solutions.

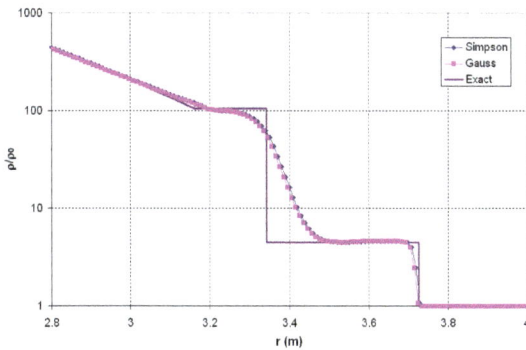

Figure 12. Density distribution at $t = 1.2$ ms obtained using different quadrature rules. ($\varrho0 = 0.0001$ kg/m3, N = 701)

4.1.4. Effect of numerical flux function

It is also known that the numerical flux function used in the computations affects the accuracy of the DG solutions. In order to see this effect, performance of the AUSM+ method [11] was compared to those of van Leer (VL) [28], HLLE [29] and Roe's [30] methods. Density predictions obtained using these flux functions and their comparisons with the exact solution were displayed in Figure 13. According to this figure, performances of these methods were close to each other except the HLLE method which clearly underpredicted the shock location. Among the flux functions tested, AUSM+ showed the best performance as expected.

4.2. Blast wave simulations

In this section blast waves generated by an explosion were analyzed. Here explosive material was assumed to transform gas phase instantly after the detonation, thus explosive was modeled as a high pressure sphere. This way, blast wave problem became a spherical shock tube problem where an isobaric sphere of a high pressure gas expanded toward surrounding low pressure air. Numerical solutions were obtained using RKDG method with $k = 1$, AUSM+ flux function and the alternative limiting technique which gave the best results for the planar shock tube problem analyzed in the previous section. A grid was used with a mesh spacing of 4 mm in a 10 m solution domain.

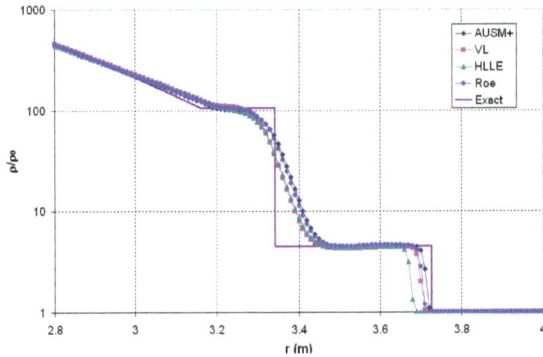

Figure 13. Density distribution at $t = 1.2$ ms obtained using different numerical flux functions. ($\varrho0 = 0.0001$ kg/m3, N = 701)

Numerical solutions were first compared with the approximate analytic solutions of blast waves by Friedman [39]. The problem considered contained a sphere of air compressed to 22 times the ambient air pressure. This sphere had a non-dimensional radius of one [40]. Temperature was taken to be uniform in the entire solution domain and air was assumed to be calorically perfect. Figure 14 contained loci of resulting primary shock, contact surface and secondary shock. Predictions were also compared with numerical solutions by Brode [41]. In this figure horizontal axis represented distance from the center normalized by the initial sphere radius, r_0 and vertical axis represented time normalized by r_0 and ambient speed of sound, a_0.

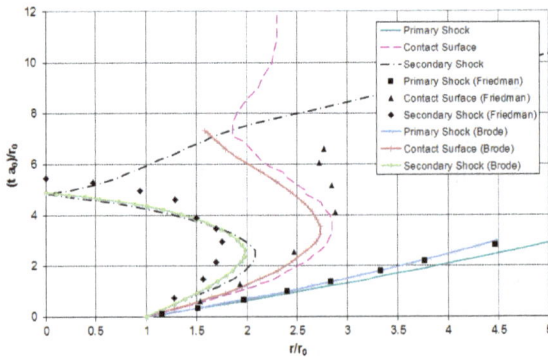

Figure 14. Loci of primary shock, contact surface and secondary shock.

According to this figure RKDG solutions showed only qualitative agreement with Friedman's approximate analytic result; however agreement with Brode's results were much better especially during the implosion of the secondary shock and its reflection from the center. Outward motion of the secondary shock after its reflection and its interaction with the contact surface were also observed in Figure 14.

Next, blast waves generated by explosion of 1 kg of TNT were analyzed and simulated. The explosive was modeled as an isobaric high pressure sphere in which the density was 1600 kg/m^3 and pressure was 8.8447 GPa which was obtained using the blast energy of TNT [42] and the JWL (Jones-Wilkins-Lee) equation of state [43]. Outside this sphere ambient air density and pressure were 1.225 kg/m^3 and 101320 Pa, respectively. Simulations involved two different fluids; detonation products and ambient air for which different state equations were used. For detonation products JWL equation of state was used mainly due to its popularity [44]. For surrounding air two different cases were followed. First case assumed ambient air to be calorically perfect. However, after an explosion like this one, ambient air temperature may easily become very high so that calorically perfect gas assumption might cease to be valid due to gas dissociations [45]. Therefore, as the second case, high temperature effects were included by assuming local chemical equilibrium meaning that the chemical reactions occur instantaneously [45]. Equilibrium relations given in reference [46] were used for air to calculate ratio of specific heat capacities and temperature in terms of pressure and density.

One of the important variables in blast wave simulation is the over-pressure, which is the rise of pressure above ambient pressure downstream of the primary shock wave. Figure 15 showed comparison of over-pressure predictions obtained for calorically perfect and equilibrium air cases with the data obtained from [42] which states that these results are curve fits to the data used in the weapons effect calculation program CONWEP [47]. Both cases over-predicted over-pressure with calorically perfect case being slightly better. The main discrepancy between both cases occurred between $r = 1$ and 3 m. Except this region both curves were nearly parallel.

Another important variable in blast wave simulation is the speed of the primary shock, which can be used to calculate shock arrival times. Primary shock speeds computed using RKDG method were compared with data from reference [42] in Figure 16. Here numerical solutions gave higher shock speeds close to the explosive but agreement with data from reference [42] became much better after $r = 2$ m, especially for calorically perfect air solution.

Overall solutions performed by treating air as a calorically perfect gas were in better agreement with data from reference [42]. But it is known that high temperatures encountered during detonation of high explosives like TNT makes calorically perfect assumption invalid. Since this assumption ignores any chemical reaction that might occur, it may yield unrealistically high temperatures after a shock wave. Such problems are usually encountered for hypersonic flows over reentry vehicles (see reference [48], chapter 16). In order to check this for our problem, temperature behind the primary shock wave predicted by the two cases of air were displayed in Figure 17. It was clear from this figure that temperatures predicted using the two assumptions were very close except in the region between $r = 1$ and 3 m, where major discrepancies between these numerical solutions took place. Since temperature predicted using calorically perfect and equilibrium air assumptions were very close, the potential negative consequences of using the former, was not experienced in this problem. Nevertheless, the predicted temperatures at lower r values were still high enough to make γ non-constant. In order to see the effect of temperature on γ, its values calculated using the equilibrium air assumption was compared with the constant value of 1.4 for calorically perfect gas assumption. This comparison was displayed in Figure 18 where a considerable drop in γ can be easily observed when r is less than 3m. This explains the relatively less pressure and temperature drop yielded by the equilibrium air assumption for the same amount of outward expansion. (See Figure 15 and Figure 17).

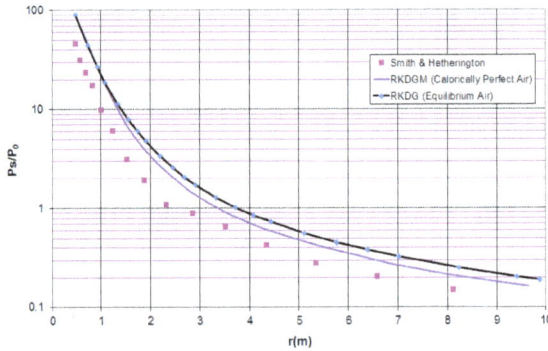

Figure 15. Variation of over-pressure with distance measured from the center.

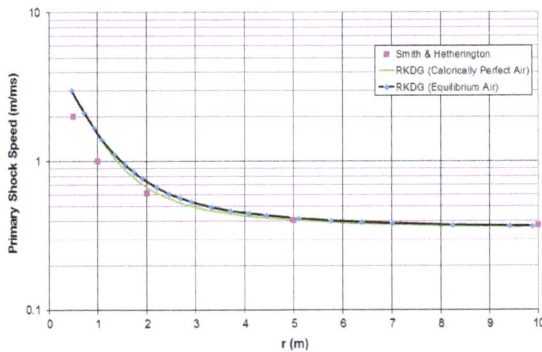

Figure 16. Variation of primary shock speed with distance from the center.

RKDG predictions for blast waves represented above were obtained using the alternative limiting approach because it yielded better results for planar shock wave problem where the strength of the discontinuities remain constant. However, in a spherical shock problem the strength of the discontinuities decrease as high pressure gases planar shock tube problem with moderate discontinuity strengths showed that the performances of different limiting strategies became closer to each other as the discontinuity strengths decreased. In order to see the effect of limiting procedure on blast wave simulations density distributions at t=0.1 and 1.6 ms were plotted in Figure 19 and Figure 20 for the blast wave generated by the explosion of 1 kg of TNT. Density was plotted in order to see the contact surface. Here the ambient air was assumed to be calorically perfect air. Predictions include RKDG solutions with minmod (mm), Sweby (sw), superbee (sb) and alternative (mm for density, sw for pressure and velocity) limiters. Finite volume solution with superbee limiter was also included in the figures. According to Figure 19 RKDG solutions were very close to each other especially for secondary shock wave. The major differences were in the vicinity of the contact surface where solutions with mm and mm-sw limiters yielded more smeared contact surfaces as expected. At the same time finite volume (FV) solution showed a considerable discrepancy with RKDG solutions. Nevertheless, the strength of the discontinuity at the contact surface was small compared to that of the planar shock tube problem studied above. Therefore, the discrepancies at the contact surface location did not affect the shock predictions considerably. This was also supported in Figure 20 which showed density distributions well after the secondary shock wave reflected from the origin and crossed the contact surface. Again RKDG predictions are very close to each other and disagreement with the FV solution is evident.

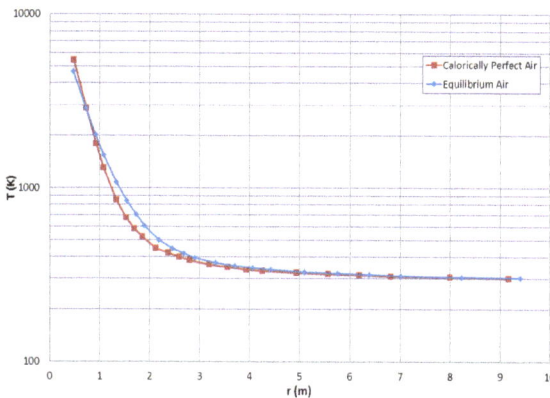

Figure 17. Variation of temperature behind primary shock with distance from the center

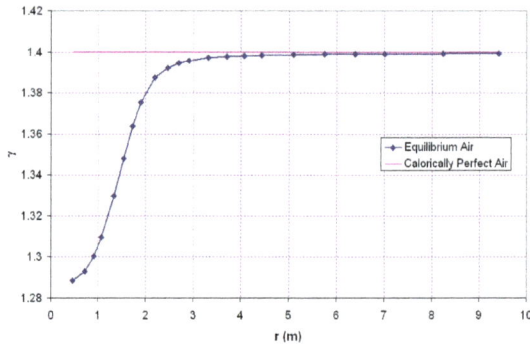

Figure 18. Variation of ratio of specific heat capacities behind primary shock with distance from the center.

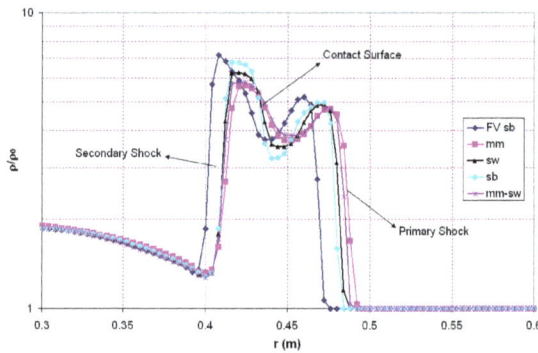

Figure 19. Density Distribution at t=0.1 ms

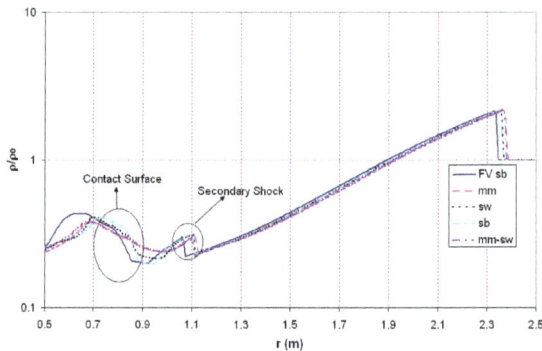

Figure 20. Density Distribution at t=1.6 ms

The blast wave solutions presented above were obtained by solving equation (29) which was derived by making the approximation described in equation (28). In order to see the effect of this approximation on the predictions, solutions of this approximate formulation were compared to the solutions obtained using exact formulation, i.e., by inverting the mass matrix given in equation (15). The loci of primary shock, contact surface and secondary shock wave predicted using these two approaches were displayed in Figure 21. Here solid lines corresponded to solutions of equation (29) and dashed lines corresponded to solutions of exact formulation. Examination of Figure 21 revealed that the approximation made in equation (29) had very little effect on the primary shock wave. Here shock speeds were nearly identical and this meant that the shock strengths were also predicted nearly equal. There were, however, visible discrepancies in the secondary shock predictions. These were evident especially after this shock reflected from the origin. Nevertheless, after the predicted secondary shocks interacted with the contact surface, their trajectories became nearly parallel. This also meant very nearly identical shock speeds and strengths. The most serious difference between predictions was observed in contact surface predictions. However, this had very little or no effect on the primary shock strength and speed which are the two most important parameters in blast wave simulations.

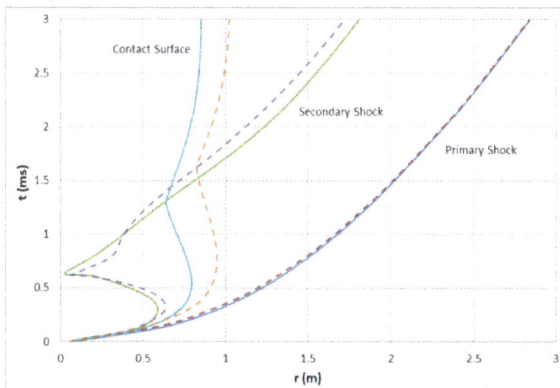

Figure 21. Loci of primary shock, contact surface and secondary shock wave. (Solid lines; approximate formulation (equation (29), dashed lines; exact formulation)

5. Conclusions

An implementation of RKDG method was performed to simulate moving planar and spherical shock waves (blast waves). The Legendre polynomials which were mainly used as basis polynomials for RKDG solutions in Cartesian coordinates did not yield a diagonal mass matrix when the formulation was extended to cylindrical and spherical coordinates. Attempts on inverting this non-diagonal mass matrix or finding a suitable set of orthogonal polynomials for cylindrical and spherical coordinates failed to yield a simplistic formulation which not only had the simplicity of the formulation in Cartesian coordinates, but also could be easily generalized for all coordinate systems. Therefore, an approximation to the integral containing the unsteady term was introduced which provided a formulation which possessed these properties. The resulting formulation was approximate for cylindrical and spherical coordinates; nevertheless, it was exact for Cartesian coordinates. In order to solve the resulting system of equations, an in-house computational fluid dynamics code which was able to solve Euler equations using a finite volume technique was further modified to include an RKDG method. The code was first tested with a Sod's shock tube problem with the so-called moderate extremely strong discontinuities [23]. Predictions were compared with the predictions obtained using a finite volume technique and to the exact solutions [24]. It was observed that the limiter used in the solutions clearly affected the overall quality of the predictions. The RKDG method was not able to produce satisfactory results with neither of the limiter functions used and was out-performed by FV method using the same limiter. After observing that spurious oscillations occur around the contact surface rather than the shock wave and alternative limiting strategy which used more dissipative minmod limiter [37] for density and less dissipative Superbee or Sweby limiter [37] for pressure and velocity was tested. This strategy clearly improved the solutions of RKDG method and RKDG with minmod-Sweby combination gave the best results. It was interesting to see that this alternative limiting strategy did not produce a similar improvement for the FV method. Overall RKDG method with this so-called alternative limiting outperformed FV method for the analyzed shock tube problem with very strong discontinuities and it was shown to be less sensitive to grid coarsening compared to FV method.

The formulation derived contained integrals which were evaluated using a quadrature method. Here the performance of three-point Simpson's rule was compared to that of three-point Gauss-Legendre rule and solutions were found to be in very close agreement with each other. Therefore, the simpler Simpson's rule was mainly employed. The resulting method was tested for different flux functions and comparison of the results showed that the AUSM+ [11] method performed relatively better compared to the HLLE [29], van Leer [28] and Roe's [30] methods.

Blast waves were simulated by modeling the explosion problem as a spherical shock tube problem. For blast waved generated by explosion of 1kg of TNT, the numerical solutions were performed with and without including the high temperature effects for the surrounding air. Although high temperatures were encountered in the solutions, using calorically perfect gas assumption for air did not produce negative effects, and even gave better agreement with blast wave data taken from Ref. [42]. This may be due to the fact that equilibrium relations given in [46] were obtained for hypersonic flow at the upper levels of the atmosphere where pressure and density is low.

Finally, the effect of the approximation made in the formulation on the predictions was analyzed by comparing the blast wave solutions obtained using exact and approximate formulations. There was very little influence on the primary shock wave, although some discrepancies were observed for the secondary shock and the contact surface. Nevertheless, the approximate formulation provided a simple formulation for all coordinates and yielded satisfactory results.

Author details

Emre Alpman
Department of Mechanical Engineering, Marmara University, Istanbul, Turkey

6. References

[1] Dewey, J. M. (2001) Expanding Spherical Shocks (Blast Waves). Handbook of Shock Waves. 2: 441 – 481.

[2] Sharma, A., Long, L. N. (2004) Numerical Simulation of the Blast Impact Problem using the Direct Simulation Monte Carlo (DSMC) Method. Journal of Computational Physics. 200: 211 – 237.

[3] Long, L. N, Anderson, J. B. (2000) The simulation of Detonations using a Monte Carlo Method. Rarefied Gas Dynamics Conference, Sydney, Australia.

[4] Anderson, J. B., Long, L. N. (2003) Direct Monte Carlo Simulation of Chemical Reaction Systems: Prediction of Ultrafast Detonations. Journal of Chemical Physics. 118: 3102 – 3110.

[5] Anderson, J. B., Long, L. N. (2002) Direct Simulation of Pathological Detonations. 18th International Symposium on Rarefied Gas Dynamics, Vancouver, Canada.

[6] Alpman E. (2009) Blast Wave Simulations using Euler Equations and Adaptive Grids. 5th Ankara International Aerospace Conference, AIAC-2009-093, Ankara, Turkey.

[7] Alpman E. (2009) Simulation of Blast Waves using Euler Equations. 17th National Thermal Science and Technology Conference, Sivas, Turkey. (in Turkish)

[8] Chen, C.C., Alpman, E., Linzell, D.G., and Long, L.N. (2008) Effectiveness of Advanced Coating Systems for Mitigating Blast Effects on Steel Components. Structures Under Shock and Impact X, Book Series: WIT Transactions on the Build Environment. 98: 85 - 94.

[9] Chen, C.C., Linzell, D. G., Long, L. N., Alpman, E. (2007) Computational studies of polyurea coated steel plate under blast loads. 9th US National Congress on Computational Mechanics, San Francisco, CA.

[10] Alpman, E., Long, L, N., Chen, C. –C., Linzell, D. G. (2007) Prediction of Blast Loads on a Deformable Steel Plate Using Euler Equations. 18th AIAA Computational Fluid Dynamics Conference, Miami, FL.

[11] Liou, M. -S. (1996) A sequel to AUSM: AUSM+. Journal of Computational Physics. 129:364 – 382.

[12] Hoffmann, K. A., Chiang, S. T. (2004) Computational Fluid Dynamics for Engineers, Volume I. Engineering Education System™.

[13] Cockburn, B. (2001) Devising Discontinuous Galerkin Methods for Non-Linear Hyperbolic Conservation Laws. Journal of Computational and Applied Mathematics. 128:187 – 204.

[14] Gottlieb, S., Shu, C. W. (1998) Total Variation Diminishing Runge-Kutta Schemes. Mathematics of Computation. 67:73 – 85.

[15] Cockburn, B., Shu, C. W. (1991) The Runge-Kutta Local Projection P^1-Discontinuous Galerkin Method for Scalar Conservation Laws. M^2 AN. 23:337.

[16] Cockburn, B., Shu, C. W. (1989) TVB Runge-Kutta Local Projection Discontinuous Galerkin Finite Element Method for Scalar Conservation Laws II: General Framework. Mathematics of Computation. 52:411.

[17] Cockburn, B., Lin, S. Y., Shu, C. W. (1989) TVB Runge-Kutta Local Projection Discontinuous Galerkin Finite Element Method for Scalar Conservation Laws III: One-dimensional Systems. Journal of Computational Physics, 84:90.

[18] Cockburn, B., Hou, S., Shu, C. W. (1990) TVB Runge-Kutta Local Projection Discontinuous Galerkin Finite Element Method for Scalar Conservation Laws IV; The Multidimensional Case. Mathematics of Computation. 54:545.

[19] Cockburn, B., Shu, C. W. (1998) The Runge-Kutta Discontinuous Galerkin Methods for Conservation Laws V: Multidimensional Systems. Journal of Computational Physics. 141:199 – 224.

[20] Cockburn, B., Shu, C. W. (2001) The Runge-Kutta Discontinuous Galerkin Methods for Convection-Dominated Problems. Review Article, Journal of Scientific Computing. 16:173 – 261.

[21] van der Vegt, J. J. W., van der Ven, H. (2002) Space-Time Discontinuous Galerkin Finite Element Method with Dynamic Grid Motion for Inviscid Compressible Flows: I. General Formulation. Journal of Computational Physics. 182: 546 – 585.

[22] van der Vegt, J. J. W., van der Ven, H. (2002) Space-Time Discontinuos Galerkin Finite Element Method with Dynamic Grid Motion for Inviscid Compressible Flows: II. Efficient Flux Quadrature. Computer Methods in Applied Mechanics and Engineering. 191: 4747 – 4780.

[23] Tai, C. H., Chiang, D. C., Su, Y. P. (1997) Explicit Time Marching Methods for The Time-Dependent Euler Computations. Journal of Computational Physics. 130: 191 – 202.

[24] Sod, G. A. (1978) A survey of Several Finite Difference Methods for Systems of Nonlinear Hyperbolic Conservation Laws. Journal of Computational Physics. 27: 1 – 31.

[25] Price, T. E. (1979) Orthogonal Polynomials for Nonclassical Weight Functions. SIAM Journal on Numerical Analysis. 16: 999 – 1006.

[26] Qiu, J., Khoo, B. C., Shu, C. W. (2006) A Numerical Study for the Performance of the Runge-Kutta Discontinuous Galerkin Method based on Different Numerical Fluxes. Journal of Computational Physics. 212: 540 – 565.

[27] Liou, M. -S. (2000) Mass Flux Schemes and Connection to Shock Instability. Journal of Computational Physics. 160: 623 – 648.

[28] van Leer, B. (1982) Flux-Vector Splitting for the Euler Equations. Lecture Notes in Physics, Springer - Verlag, 170: 507 – 512.

[29] Einfeldt, B. (1988) On Godunov Type Methods for Gas-Dynamics. SIAM Journal of Numerical Analysis. 25, 294.

[30] Roe, P. L. (1986) Characteristic-based Schemes for Euler Equations. Annual Review of Fluid Mechanics. 18:337 – 365.

[31] Chapra, S. C., Canale, R. P. (2006) Numerical Methods for Engineers. 5th Edition, Mc Graw-Hill, p. 626.

[32] LeVeque, R. J. (1992) Numerical Methods for Conservation Laws. 2nd Edition, BirkHauser.

[33] Flaherty, J. E., Krivodonova, L., Remacle, J. F., Shephard, M. S. (2002) Aspects of Discontinuous Galerkin Methods for Hyperbolic Conservation Laws. Finite Elements in Analysis and Design. 38: 889 – 908.

[34] Calle, J. L. D. , Devloo, P. R. B., Gomes, S. M. (2005) Stabilized Discontinuous Galerkin Methods for Hyperbolic Equations. Computer Methods in Applied Mechanics and Engineering. 194:1861 – 1874.

[35] Qiu, J., Shu, C. W. (2005) Runge-Kutta Discontinuous Galerkin Method using WENO limiters. SIAM Journal of Scientific Computing. 26: 907 – 929.

[36] http://en.wikipedia.org/wiki/Flux_limiter

[37] Sweby P. K., (1984) High Resolution Schemes using Flux Limiters for Hyperbolic Conservation Laws. SIAM Journal on Numerical Analysis. 21: 995 – 1011.

[38] Biswas, R., Devine, K. D., Flaherty, J. E. (1994) Parallel, Adaptive, Finite Element Methods for Conservation Laws. Appl. Numer. Math. 14: 255, 1994.

[39] Friedman, M.P. (1961) A simplified Analysis of Spherical and Cylindrical Blast Waves. Journal of Fluid Mechanics. 11: 1.

[40] Sachdev, P. L., (2004) Shock Waves and Explosions. Monographs and Surveys in Pure and Applied Mathematics. Chapman & Hall/CRC, pp. 208 – 224.

[41] Brode H. L., (1957) Theoretical Solutions of Spherical Shock Tube Blasts. Rand Corporation Report. RM – 1974.

[42] Smith, P. D., Hetherington J. D. (1994) Blast and Ballistic Loading of Structures, Butterworth-Heinemann. pp. 24 – 62.

[43] Dobratz, B. M., Crawford, P. C., (1985) LLNL Explosives Handbook – Properties of Chemical Explosives and Explosive Simulants. Lawrence Livermore National Laboratory.

[44] Kubota, S., Nagayama, K., Saburi, T., Ogata, Y. (2007) State Relations for a Two-Phase Mixture of Reacting explosives and Applications. Combustion and Flame. 74 – 84.

[45] Hoffman, K., A., Chiang, S. T. (2000) Computational Fluid Dynamics, Volume II, Engineering Education System™.

[46] Tannehill, J. C., Mugge P. H. (1974) Improved Curve Fits for the Thermodynamic Properties Equilibrium Air Suitable for Numerical Computation using Time Dependent and Shock-Capturing Methods. NASA CR-2470, 1974.

[47] Hyde, D. W. (1991) CONWEP: Conventional Weapons Effects Program. US Army Waterways Experimental Station.

[48] Anderson, J. D. (2004) Modern Compressible Flow with Historical Perspective, 3rd Edition, McGraw Hill, p. 587.

pMDI Sprays:
Theory, Experiment and Numerical Simulation

Ricardo F. Oliveira, Senhorinha Teixeira, José C. Teixeira,
Luís F. Silva and Henedina Antunes

Additional information is available at the end of the chapter

1. Introduction

The inhalation therapy is a cornerstone in asthma treatment. A disease characterized by episodes of wheezing, breathing difficulties, chest tightness and coughing. Essentially it is a chronic inflammatory disorder associated with airway hyper responsiveness [1]. This disease already affects over 300 million people worldwide, growing at a rate of 50% per decade, and causes the death of 220 thousand per year [1].

Therefore it is of utmost importance to improve the delivery effectiveness of the devices used in the inhalation therapy, which is the most used way of treatment, thanks to its greater efficiency, faster action, lower toxicity and minor drug waste [2].

Anti-inflammatory and bronchodilator drugs are used with the objective of reducing the inflammation of the pulmonary tissue, which cause the diameter reduction of the bronchus [1,3].

In the inhalation therapy procedure, the following devices are mostly used: nebulizers, Dry Powder Inhalers (DPI), and pressurized Metered-Dose Inhalers (pMDI), which can include a spacer attached (add-on device).

The nebulizer presents a simpler usage procedure but most of models available need to be attached to the power plug. They also present a greater waste of drug and a slower duration for each session of treatment and a lower efficiency when compared with other devices. The DPI is the most efficient device but also the most expensive, requiring a minimum inspiratory flux by the patient in order to work properly which is not expected in young children and elderly. Regarding the pMDI, it has a low cost and it is the most used by the medical community and its efficiency is acceptable. Its major drawbacks are the high spray velocities, which creates the so called "cold-Freon" sensation on the back of the throat, and the necessity of inspiratory coordination with the priming action, resulting that the younger and/or older patients may not be able to use the device properly. This problem is highlighted by the fact that the implementation of classes to teach the correct usage technique provide increased efficiency [4–6].

With the objective of solving some of the pMDI problems, the spacers were created and they are presented into three categories: simple tube, valved holding chamber and reverse flow. The most common type is the valved holding chamber, inside which the drug spray is released. This device leads to the reduction of the high velocity impact of the spray in the throat and the need for inspiratory coordination, allowing the patient to breath normally from the other side of the chamber. The one-way valve makes the normal respiratory cycle possible by not allowing the expiration flux to go back inside the chamber [5–8].

1.1. History of inhalation therapy

According to Crompton, inhalation therapy can be traced as far back as 4000 years ago in India, and many *'hundreds of ingenious devices and hopeful medications'* (according to Sanders) have been discovered so far. Inhalation therapy is so important nowadays that is hard to imagine the treatment of asthma without a proper inhalation therapy procedure, inhalation device, and a necessary drug [9,10].

Briefly reviewing the antecedents of contemporary inhalation therapy and the therapies and medicines used for the management of asthma over the last 50 years, Crompton presents the 19th century hand-held glass bulb nebuliser up to the introduction of the first pMDI, which came into common clinical practice in 1956 [9]. Sanders, detailing the inhalation therapies since ancient times, refers several inhaler devices, some of which are the C. Bennet's inhaler (dated from 1654), the J. Mudge's inhaler (from 1778, and adapted from a pewter tankard), the first pressurised inhaler (presented in 1858, in Paris, France, by Sales-Girons), the Improved Nelson inhaler (proposed in 1865 by S. Maw & Sons, London) which is still being manufactured up to this day, and the Siegle's steam spray inhaler (a German invention from early 1860's that marked the beginning of the nebuliser therapy) [10].

Many contributions have been given so far, and many medications are available nowadays as inhaled treatments, as well as many inhaler devices (Rotahaler® and Accuhaler® are just two examples).

1.2. pMDI devices

The pMDI, shown in Figure 1, is the most commonly used aerosol delivery device in the world for the treatment of asthma. In 2000, an estimated production of 800 million units have been reported [5,11]. It has been the backbone of inhalation therapy for asthma for approximately 50 years [6]. The pMDI contains 100 - 400 doses in a small and (very) portable device that can be easily concealed in a pocket. The best feature of a pMDI device is that is always ready to be used, making it a magnificent example of an engineering solution.

Figure 1. Three commercial pMDI actuators with its drug canisters attached

1.2.1. Components and their role

The pMDI device essentially incorporates a disposable canister, where the drug formulation is stored, which can be replaced for a new one at any time, mounted in an actuator with a mouthpiece zone and usually a dust cap is included [6,12]. The four basic components that can be found in all pMDIs are: the formulation canister; the metering valve; the actuator; and the container [13]. On the following list are described the components of a hydrofluoroalkane (HFA) pMDI device, published in [5]:

- Canister with O-Ring
- Formulation
  - Drug substance
  - Propellant(s)
  - Surfactant(s)
  - Lubricant(s)
  - Co-Solvent(s)
- Ferrule Gasket
- Complete Valve
  - Mounting cup
  - Valve core assembly
    - Diaphragm
    - Stem
    - Spring
    - Seal
  - Metering tank
  - Retaining cup
- Adaptor with dust cap

The spray pattern and Mass Median Diameter (MMD) are influenced by the ambient temperature, the design of the actuator nozzle and valve stem, also by the vapour pressure of the propellant in the formulation [5].

The formulation is composed by the drug, the propellants gases and it often contains surfactant and other excipients. In the formulation, the propellant is the component which creates vapour pressure inside the canister, allowing the drug to exit the pMDI and form a spray plume every time it is pressed. When the solid drug particles exit the pMDI they are encapsulated in a droplet made of propellant, which evaporates as it travels through air. Some of the components in the formulation of a common used drug in asthma treatment, Ventolin®, are listed in Table 1.

	Ventolin® HFA	Ventolin® CFC
Active ingredient	Albuterol sulphate	Albuterol sulfate
Excipient	None	Oleic acid
Propellants	HFA-134a	CFC-11/12
Formulation type	Suspension	Suspension

Table 1. Ventolin®'s formulation components present is two propellant types, HFA and chlorofluorocarbon (CFC). Adapted from [12]

The canisters for a pMDI are typically made of aluminium and they are designed to be light, compact and strong to hold the high internal vapour pressure of 3 to 5 bar made by the propellant in the formulation [6,12]. Its regular capacity ranges 15-30 mL [13].

The metering valve of a pMDI is crimped into the container. There are several designs of valves but they all work under de same principle: while the canister is in the inverted position, by gravity action, the valve fills with the formulation through a channel archiving the desired measured dose; at the priming moment, this channel closes and another one opens in the opposite side of the metering valve, allowing the formulation to rapidly expand into the expansion chamber. The actuator pit and the valve stem constitute the expansion chamber where the propellant begins to boil [12,13]. This process is often referred as the primary atomisation, where a flash evaporation of the propellant takes place [13,14]. After it exits the expansion chamber through the actuator nozzle it rapidly expands in the shape of a solid-cone plume [12,13]. In this way, the metering valve of a pMDI is a critical component in the effectiveness of the delivery system, due to the fact that its main functions are [4–6,11–13]:

- deliver, in an accurately and reproducibly way, a measured volume (20-100 µL), containing between 20 - 5000 µg of the dispersed drug;
- form a propellant-tight seal for high pressure in the canister.

The actuator is normally a single plastic piece produced by injection moulding that consists of a mouthpiece, body and nozzle [6,11–13,15]. When the pMDI reached the market in 1956, it was composed of an elongated mouthpiece (around 8 cm); nowadays, it was reduced (from 2 to 3 cm) to a more compact size to improve the portability [12].

- the mouthpiece is the interface part to the patient mouth;
- the body provides support for the canister;

- the nozzle has a very important role in controlling the atomisation process, to guarantee a spray plume formation. The nozzle diameter interferes directly with the particles size distribution.

1.2.2. Formulation: CFC versus HFA

On the 16th September of 1987, the Montreal Protocol was signed, with the objective to drastically reduce the CFC emission to atmosphere which was contributing to the destruction of the ozone layer. Being the CFC propellant the main component in the pMDI formulation, it was clear that some big changes were required to the pharmaceutical industry. The chosen propellants for this transition were the HFA. Although these greenhouse gases contribute to the global warming, their effect at this scale is considered negligible [6]. During this transition not every single drug type available with CFC propellant was transferred to the HFA formulation, reducing the types of drugs available to be prescribed [3].

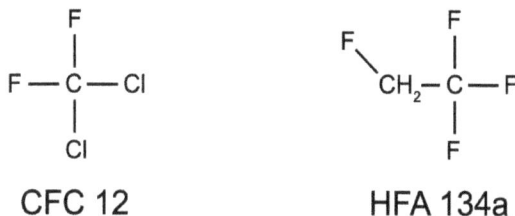

Figure 2. Chemical structures of the two most common types of propellants (CFC-12 and HFA-134a) used in pMDI

The CFC gas meets the criteria to be used as pMDI propellant, such as, the constant vapour pressure during the product's life ensuring consistent dose; needs to be nontoxic, non-flammable; correct boiling point and densities and also needs to be compatible with the drug solution. These reasons made CFC the propellant of choice, it is formulated using the CFC-12 (Figure 2) as the major component, added with CFC-11 and/or CFC-114 to regulate the vapour pressure of the formulation [12]. Its substitute, the HFA-134a (Figure 2) or HFA-227, resulted in a reformulation of all the pMDI components due to the serious changes in the spray plume shape, particle size distribution and amount of drug dose delivered [7,11,16,17]. These HFA propellants forced the redesign of the actuator and valve components, essentially the gaskets [7]. The HFA-134a has similar thermodynamic properties to the CFC-12, but there is not a substitute HFA component to the CFC-11 or CFC-114. So there is the need to introduce excipients with lower volatility in the mixture to modify the vapour pressure, contributing to the development of new methods for the pressure-filling process [12]. The transition progressed in two different ways. Some companies decided to develop bioequivalent products using the HFA propellant, while others decided to take the opportunity to make a new product which is clinically more efficient in the delivery to the patient lungs [6,12].

Table 2 describes some of the most important physicochemical properties of the two main pMDI propellants, CFC-12 and HFA-134a. The HFA propellant products require a higher vapour pressure to reach the liquid state in comparison to the CFC, resulting in a higher speed spray [7,17]. This forced the redesign of the metering valves so they could deliver the same amount of formulation volume *per* puff, leading to changes in the particle size distribution, because the HFA formulation produces smaller particles [5,7].

Property	CFC 12	HFA 134a
Boiling point (ºC)	-30	-26
Vapor pressure (kPa)	566	572
Density (g/cm^3)	1.33	1.23
Viscosity (mPa.s)	0.20	0.21
Surface tension (mN/m^2)	9	8
Dielectric constant	2.1	9.5
Water solubility (ppm)	120	2200

Table 2. Some physicochemical properties for two main pMDI propellants (CFC-12 and HFA-134a). Adapted from [11]

1.2.3. Mechanism and technique

The pMDI device contains the drug formulation inside the canister at a high pressure, which at moment of priming flows rapidly into the metering chamber and to the expansion chamber of the actuator, and then throughout the nozzle. At this point, the drug particles became airborne forming the spray plume, travelling through the air. The sudden pressure change makes the liquid propellant of the formulation to rapidly evaporate, going from a high diameter droplet into a smaller drug particle and/or droplet.

The pressing, also called priming, of the top of the metal canister against the plastic actuator, makes the spray plume discharge, also known as puff, into the air. This priming needs to follow some procedure steps to result in an efficient delivery to the patient. These steps are normally the following [18]:

1. Shake inhaler
2. Exhale
3. Breathe in deeply and slowly
4. Press on inhaler (co-ordinating actuation with inhalation)
5. Hold breath
6. Breath out slowly
7. One actuation *per* inhalation
8. Wait before second actuation

By studying the technique of the pMDI used by the patients, it was found that the majority of the patients have difficulty to execute all the steps correctly. The study included children (n= 26), adolescent (n= 23), adult (n= 209) and elderly (n= 66) patients. Among them, the children are the ones with more difficulty in the step 4, and the elderly have more difficulties executing the steps 7 and 8. The adolescents and adults are the patients with better abilities to use the pMDI device correctly [18].

2. Spray characterisation

In this section, all the parameters and variables needed to describe the spray will be presented and discussed, which will deal with its characterization, spray plume formation and position, droplet size distributions and mass flux.

2.1. Spray characteristics

The production of aerosol in a pMDI is a complicated process, which is influenced by several factors. The vapour pressure is the most important factor in the pMDI mechanics, and it influences the distribution and shape of the plume. Higher vapour pressure inside the canister results in a faster velocity plume and with faster evaporation rate [11]. Clark developed a mathematical equation based on empirical results from CFC formulation, although it performs well for HFA propellants too. Equation 1 allows the calculation of the MMD produced by the pMDI [11,13,14].

$$D_{0.5} = \frac{8.02}{q_{ec}^{0.56}\left[\left(p_{ec} - p_\infty\right)/p_\infty\right]^{0.46}} \tag{1}$$

where the MMD is represented as $D_{0.5}$, the q_{ec} is the quality of the flow from the Rosin-Rammler distribution, p_{ec} is the pressure in the expansion chamber and the p_∞ represents the ambient pressure.

The interaction of the formulation with the other pMDI components defines the final spray plume form and characteristics. An relationship can be used to predict the percentage of fine-particle fraction in a HFA-134a spray, see Eq. 2, by using the dimensions of its components, as described in [12].

$$fine-particle\ fraction\ (\%) = 2.1 \times 10^{-5} A^{-1.5} V^{-0.25} C_{134}{}^{3} \tag{2}$$

Where, A represents the actuator nozzle diameter (in mm), V is metered volume size (in µL) and C_{134} is the HFA content (in percent) [12].

2.2. Velocity

Within the pMDI spray plume, the droplet velocity varies accordingly to the position in it. In the middle of the plume cone the velocity is maximum, while in the outer zones it gets its lower values (see Fig. 3).

The droplet velocity also decelerates along the axial distance from the nozzle, due to momentum exchange with the air. As reported by Dunbar, using phase-Doppler particle analysis measurements of a HFA-134a spray plume during an actuation, values were taken at different distances from the nozzle of the actuator (see Fig. 4) [13].

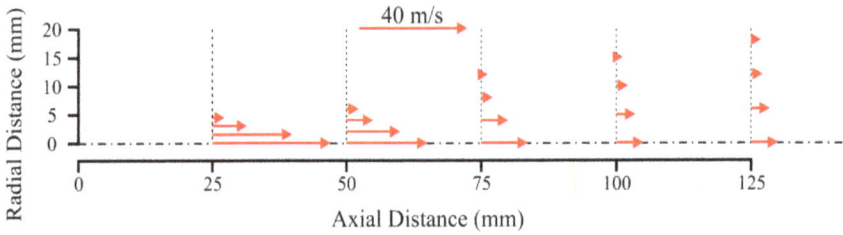

Figure 3. Droplet velocity vectors represented in half of a mid-section plane of pMDI plume using a formulation only with HFA-134a, measured at the last moment of the actuation. Adapted from [13]

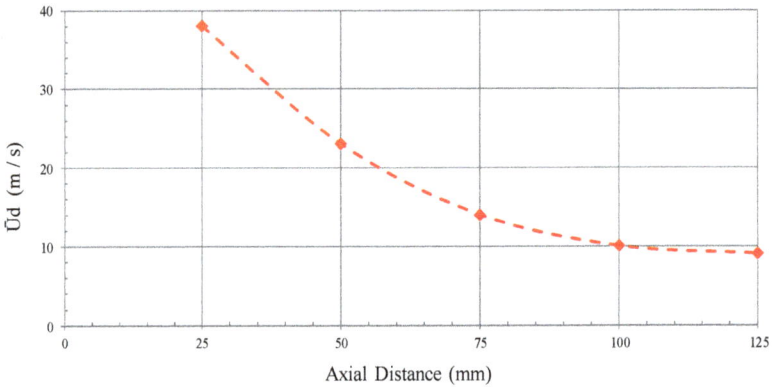

Figure 4. Droplet mean velocity along the axial distance from the nozzle of a pMDI using a formulation only with HFA-134a. Adapted from [13]

According to the measurements made by Dunbar, the author concluded that a HFA propellant formulation produces a spray with higher velocities than a CFC formulation. This fact is attributed to the higher vapour pressure used in the HFA formulation. The plume behaves like a spray up to the distance of 75 mm from the nozzle and as an aerosol afterwards that distance, where the droplet motion is being influenced by the gas [13].

2.3. Droplet size distributions

As the spray is formed downstream of the nozzle of the plastic actuator, it has suffered the influence of the drug formulation, valve and actuator design dimensions. These factors create a resulting plume of spray with a characteristic particle size distribution. The spray formation process causes an interesting variation of the MMD of the dose, having a value of 35 μm at the nozzle which reduces to 14 μm at just 10 cm from it, due to evaporation process of the propellant [19].

Particle size distributions can be represented in two forms, as a Probability Density Function (PDF) or as a Cumulative Distribution Function (CDF). Normally, an independent variable (x) and two more adjustable parameters, representing the particle size and the distribution of particle sizes are used [20]. There are several types of mathematical distributions to describe spray particles/droplets, being the Log-Normal, Rosin-Rammler and Nukiyama-Tanasawa, the most common.

It is a common idea that pharmaceutical aerosols can be accurately described by the Log-Normal function fitting the measured data to the cumulative mass distribution (see Eq. 4). The Log-Normal PDF (see Eq. 3) was developed from the normal distribution [20].

$$f(x;\mu,\sigma) = \frac{1}{x\sigma\sqrt{2\pi}}e^{-(\ln x-\mu)^2/2\sigma^2}, \quad x > 0 \tag{3}$$

$$F(x;\mu,\sigma) = \frac{1}{2}erfc\left[-\frac{\ln x-\mu}{\sigma\sqrt{2}}\right] \tag{4}$$

where σ_g is the geometric standard deviation (which shall be $\neq 0$), the μ represents the mean diameter and *erfc* is the complementary error function.

Using data from coal powder, an empirical function was developed to describe the distribution, this function is called Rosin-Rammler, also known as Weibull distribution (see Eq. 5). It has been widely applied in the atomisation field, using an independent variable (x) and other two to describe the distribution, λ represents the mean diameter and k is the distribution spread. The CDF is simply given by Eq. 6, making it easy for graphical representation [20,21].

$$f(x;\lambda,k) = \begin{cases} \frac{k}{\lambda}(x/\lambda)^{k-1}e^{-(x/\lambda)^k} & ,x \geq 0 \\ 0 & ,x < 0 \end{cases} \tag{5}$$

$$F(x;\lambda,k) = 1 - e^{-(x/\lambda)^k} \tag{6}$$

In 1939, Nukiyama and Tanasawa developed a function for characterisation of twin-fluid atomizers (see Eq. 7). It makes use of the gamma distribution function, $\Gamma(x)$, and two other parameters: b that represents the size and δ that is the distribution parameter. In the CDF form it is described by Eq. 8 [20,21].

$$f(x;\delta,b) = \frac{\delta b^{(6/\delta)}}{\Gamma(6/\delta)}x^5e^{-bx^\delta} \tag{7}$$

$$F(x;\delta,b) = 1 - \frac{\Gamma(6/\delta,bx^\delta)}{\Gamma(6/\delta)} \tag{8}$$

Dunbar and Hickey studied the best distribution and fitting method for empirical data obtained from a pMDI spray using an Andersen eight-stage cascade impactor. They concluded that the best fitting method is through the nonlinear least squares of the PDF and that the Log-Normal and Nukiyama-Tanasawa PDFs produced better fitting results to the data than the Rosin-Rammler function [20].

The authors measured a pMDI HFA-134a formulation of salbutamol using the laser diffraction analysis technique (throughout a particle sizer Malvern 2600), using an independent model for data processing. The data were fitted to the three distribution models described above (see Eqs. 4, 6 and 8) using the least squares method. A comparison is shown in Figure 5.

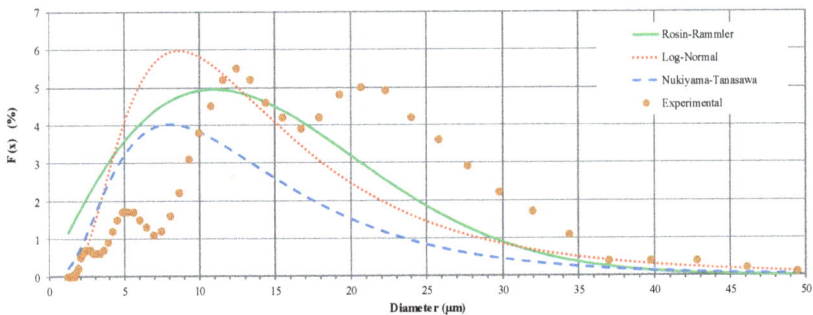

Figure 5. Graphical representation of the pMDI HFA-134a salbutamol experimental data and its fitting for the Rosin-Rammler, Log-Normal and Nukiyama-Tanasawa distributions. Measurements obtained at 100 mm from the laser beam

2.4. Mass flux

Mass (or volume) fluxes are an important characteristic of any spray. In particular for drug delivery applications, such knowledge is of utmost importance to the understanding of droplet/particle transport through the respiratory system and any delivery devices.

Space averaged measuring systems (such as laser diffraction based) only provide a crude estimative of droplet/particle concentration. The user has to specify the sample length illuminated by the laser beam and the concentration (estimated for the duration of the measurement throughout the entire sample) is calculated as the ratio of the light intensity collected by the sample over the total light intensity existing in the absence of any sample. One must remember that the light array detector includes a central detector where the undiffracted light is focused.

Because the technique only measures particle size, the procedure does not yield data for mass flux. Furthermore, the information is a time-space averaged quantity which is of limited value, particularly for transient phenomena such as that occurring during the operation of inhalator devices.

Phase Doppler anemometry may overcome such limitations, because it is a point measuring procedure and, crucially, includes data for velocity. In addition, because the sampling occurs over a period of time, one may capture time variations in properties such as the droplet mass flux.

The mass flux results from the integration of the total droplet volume (mass) over a certain area for a period of time. Various sources of errors make this calculation difficult. Crucial to this procedure is the correct determination of the effective area. The probe volume results from the intersection of two Gaussian laser beams (of diameter d_e) which has an ellipsoidal shape. The actual probe volume is the intersection of this ellipsoid with the slit length (L) in the direction of the receiving optics which is oriented at an angle φ from the forward direction. A simple approximation for the cross area is the projected area ($\approx d_e L$) of the volume described above. However one should take into account the cross section area of the volume perpendicular to the particle velocity. This only coincides with the former in a one dimensional flow if the main axis is oriented with this velocity. Therefore one must take into account the actual velocity vector of each particle which is problematic because most of the phase Doppler anemometry configurations are 2D. Zhang and Ziada refer a method for overcoming this limitation [22].

In addition, the droplets used for the calculation of the mass flux may not cross the probe volume at the same rate throughout its area. Therefore, the droplets used for this calculation are allocated to a certain position in the probe volume by taking into account the Doppler burst length. The calculation of the number of droplets in a certain time is based on the number of droplets during the elapsed time of the measurement. It is argued that this is not accurate because in that time more particles have been through the probe volume than those actually accounted for and this ratio depends upon the validation rate of the experiment. The *attempted/validated samples* ratio should correct this estimate. Furthermore, other sources of uncertainty are caused by the fact that not all the particles may trigger the system. This may result in a bias against the contribution of small particles.

Because the particle counting is based on average values recorded in size bins, other sources of uncertainty may come from high turbulence and insufficient number of particles per size bin. This may be a serious drawback because bins for large particles are most likely to have small number of droplets and, on the other hand these droplets are those which most contribute to the total mass/volume fluxes. Concerns regarding this estimative have been raised by other authors [23].

Figure 6 shows an example of the limitations and uncertainties in the calculation of a mass flux. The data reports the variation of the local liquid mass fluxes of a twin fluid atomizer for various gas flow rates (the liquid flow rate is constant at 1 L/min), which shows a considerable scatter for the various Air to Liquid Ratio (ALR).

Nevertheless, the integration of the liquid fluxes over the diameter should yield a constant value for all the cases. As shown in Fig. 7, the results (normalized) may differ by a factor of 5. It should be stated that some factors may contribute to such differences: very high turbulence levels, dense sprays (particularly at high ALR's).

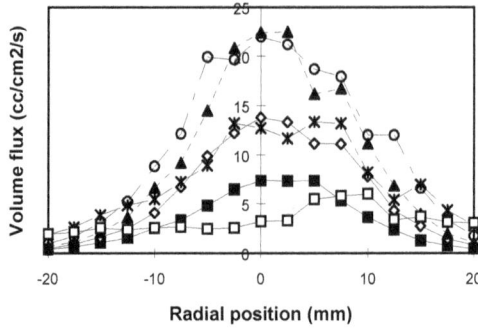

Figure 6. Mass flux as a function of ALR

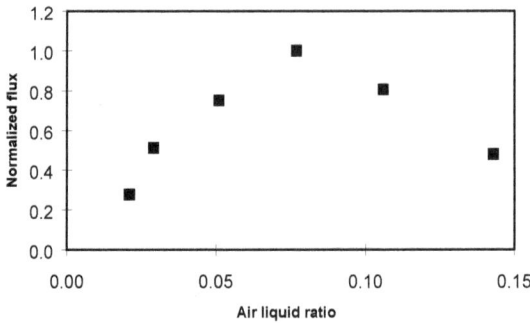

Figure 7. Integration of the mass fluxes at constant liquid flow rate

In conclusion, this information must always be treated with caution and may only provide an indication of the relative value of the droplet mass fluxes.

3. Experimental techniques

The drug delivery through the oral system may be put in a simple manner as the transport of a dispersed phase through the upper and lower respiratory tracts. In addition it may include additional systems such as a spacer. Therefore it is widely accepted that particle size plays an important role in determining where the aerosol particles are deposited. The respiratory system acts as a filter that progressively filters particles of smaller size as the inhaled air flow reaches the lower tracts of the lungs. Particles with an aerodynamic diameter above 6 μm will deposit in the oropharyngeal region while those below in the range 2-4 μm will reach the alveoli being the most effective as drug delivery agents. In short, particle size is of the greatest relevance for accessing the effectiveness of the drug delivery systems.

In a brief sentence, droplet sizing is one of the most sensitive parameter to measure in a wide range of dispersed multiphase flows. The problem is compounded if the flow is confined by walls.

Over the last decades, a variety of techniques have been developed in order to overcome some basic limitations: the required accuracy for measuring small objects; the statistical requirement of very large samples; the fact that droplets may be fast moving in the flow stream; the spray may be dense; possible sampling interference.

Over the years, various literature reviews have been published. As an example, one should refer the work of Black et al, Hong et al and Mitchell and Nagel [24–26].

3.1. Laser diffraction techniques

Coherent light (wavelength λ) diffracted by a particle can be used to measure its size. Two kinds of diffraction must be considered. In Figure 8 (a) light diffracted by an angle θ interferes with the undiffracted light, forming an interference pattern on the screen. This is near field or Fresnel diffraction. However, it can be observed that particles with the same size but positioned at different places along the light axis (diffracting by the same angle) will give a different radial position on the detector, which can cause confusion.

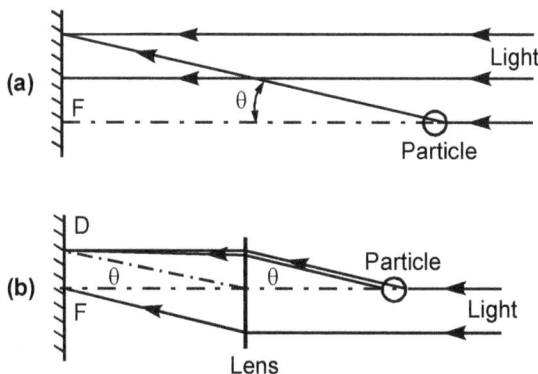

Figure 8. Fresnel (a) and Fraunhofer (b) diffraction of light

Alternatively, if the detector plane is moved away from the particle (ideally at an infinite distance) this problem will not occur. This is known as the far field or Fraunhofer diffraction. In practice this can be achieved by placing a lens in the light path so that the undiffracted light will be focused onto a central spot F and light diffracted by an angle θ is focused at the point D. The distance DF is related to the diffraction angle θ by the expression $DF \approx f\theta$, where f is the lens focal length – Figure 8 (b). In the Fraunhofer approximation that only describes diffraction of light at the contour of the particle, it is assumed that all particles are much larger than λ and that only near forward scattering is considered (i.e. θ is small).

In the method proposed by Shifrin the light intensity pattern on the focal plane is measured at different radial positions by means of a photo multiplier, moving this unit along the focal plane [27]. The minimum particle diameter measurable is proportional to the light wavelength by a factor of $10/\pi$. This method is more suitable for measuring the diffracted light at small angles.

The method developed by Swithenbank et al relates the measured light energy at the lens focal plane to the particle size [28]. This technique is commercially available in an instrument manufactured and distributed by Malvern Instruments Ltd. (England). The technology employed is based on the Fraunhofer diffraction of a parallel beam of monochromatic light by a moving particle. The diffraction pattern produced consists of a series of alternate light and dark concentric rings, the spacing of which is dependent on the drop diameter. This results in a series of overlapping diffraction rings, each of which is associated with a characteristic particle size range, depending on the focal length of the Fourier transform lens. This lens focuses the diffraction pattern onto a photo detector that measures the light energy distribution. Typically, the photo detector comprises 31 semi-circular rings surrounding a central circle. Each ring is therefore most sensitive to a specific size of droplets. The output of the photo detector is multiplexed through an A/D converter, which is connected to a computer. The computer provides an instant display of the measured distribution of drop sizes. Thus, a major advantage of the instrument is the speed at which data can be both accumulated and analysed. Another important attribute is that the diffraction pattern is independent of the position of the drop in the light beam. This allows measurement of size distributions to be made with drops moving at any speed. Lefebvre presents a very complete description of the most important characteristics of the instrument. Figure 9 shows such instrumentation [29].

Figure 9. View of the Malvern 2600 HSD

The relationship between the size distribution $[D]$ and the light intensity distribution $[I]$ in each ring is of the form:

$$[D].[A]=[I] \tag{9}$$

in which $[A]$ is a square matrix whose coefficients depend upon the optical configuration of the instrument and of the sensitivity of each one of the photodetectors. By selecting an appropriate focusing lens, the sizing range can be adjusted. Typically the dynamic range is of approximately 100:1. For instance, by selecting a 63 mm lens the measuring range will fit in between 0.5 and 100 μm. Because of the nature of the coefficients, equation 9 is solved by fitting a probability distribution (see section 2) to the data. In addition, the occurrence of multiple diffraction due to the spray denseness, may result in a bias towards smaller diameters (outer rings).

Figure 10 shows such influence upon the droplet distribution. It is accepted that only if the fraction of light collected by the sample exceeds 50% of the incident, there is cause for concern. This is rarely the case for the aerosols such as those produced by pMDIs.

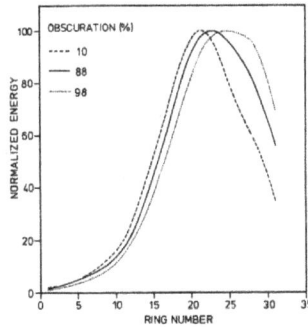

Figure 10. Influence of light obscuration

3.2. Phase doppler anemometry

Phase Doppler Anemometry (PDA) is an extension of the Laser Doppler Anemometry (LDA) principle, capable of measuring simultaneously the size and velocity of individual droplets. The velocity measurement principle is the same used for LDA, making use of one detector for each velocity component to be measured. However, to perform size measurements two detectors are needed, in principle, as described below.

Figure 11 represents light scattered from a spherical particle, collected by two detectors. Since the detectors are at different positions, the optical path lengths for reflection from the two incident beams are not equal. Consequently, when a particle crosses the probe volume both detectors receive a Doppler burst of the same frequency, but with a phase difference.

The phase difference between the two Doppler bursts depends on the size, or its surface curvature, of the particle. If Δt is the time lag separating the wave fronts reaching the two detectors, the corresponding phase difference is $\Phi_{12} = 2\pi f \Delta t$. For a spherical particle, the phase difference increase with its size.

The relationship between size and phase of a particular optical arrangement is described by the Mie scattering theory. However, a geometrical optics approximation can provide a good basis for analysis.

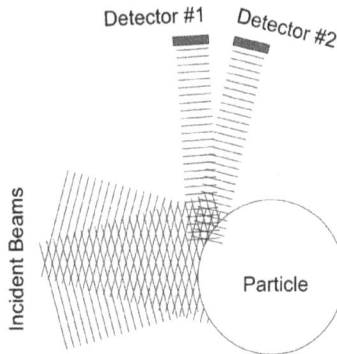

Figure 11. Phase difference in scattered light from a spherical particle

The phase of the Doppler burst received at detector i is expressed as:

$$\Phi_i = \pi \frac{n_1}{\lambda} D \cdot \beta_i \tag{10}$$

where n_1 is the refractive index of the scattering medium and β_i is a geometrical factor, which depends on the scattering mode and the collection arrangement.

As shown in Figure 12, when an incident light strikes a particle, several modes of scattering may occur. The light may be reflected or suffer 1st, 2nd or nth order of diffraction. The phase difference on the detector will depend upon the dominant scattering mode as well as the collection arrangement.

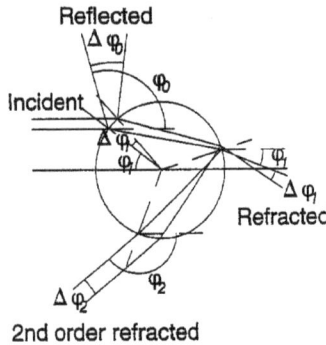

Figure 12. Light scattering modes from a spherical particle

The angle of intersection between the two beams, θ, is determined by the beam separation, St, and the lens focal length. This angle determines the fringe separation. The collection angle, ϕ, and the elevation angle, ψ, define the direction towards the photo-detectors from the measuring volume.

For the reflection mode the geometrical factor is given by:

$$\beta_i = \sqrt{2}.\left[\sqrt{1 + \sin\frac{\theta}{2}.\sin\phi_i.\sin\psi_i - \cos\frac{\theta}{2}.\cos\phi_i} - \sqrt{1 - \sin\frac{\theta}{2}.\sin\phi_i.\sin\psi_i - \cos\frac{\theta}{2}.\cos\phi_i}\right] \quad (11)$$

This expression shows that, for a particular optical arrangement, β_i is a constant factor, which gives a linear relationship between phase and size. Furthermore, the refractive index of the particle, n_2, does not appear, making this scattering mode useful in situations where the value of the refractive index is not known.

On the other hand for 1st order refraction it takes the form:

$$\beta_i = 2.\left[\sqrt{1 + n_{rel}^2 - \sqrt{2}.n_{rel}.\sqrt{f_{i+}}} - \sqrt{1 + n_{rel}^2 - \sqrt{2}.n_{rel}.\sqrt{f_{i-}}}\right] \quad (12)$$

where:

$$n_{rel} = \frac{n_2}{n_1}$$

n_2 = particle refractive index

$$f_{i\pm} = 1 \pm \sin\frac{\theta}{2}.\sin\phi_i.\sin\psi_i + \cos\frac{\theta}{2}.\cos\phi_i$$

For the other modes of scattering similar relationships may be derived [30]. As expressed in the equations for β_i above, changing any of the angles θ, ϕ or ψ, will affect the geometrical factor and, consequently, the sensitivity and range of the PDA.

In practice, there are some restrictions in the selection of the geometrical optical parameters. For instance, the selection of scattering angle, ϕ, is quite restricted, either to ensure a specific scattering mode or a sufficient signal-to-noise ratio, or from practical considerations of the measurement situation (optical access and working distance). The required working distance also affects the possible range of θ and ψ.

The scattering angle, ϕ, and the elevation angle, ψ, are those with the most important effect upon the parameter β_i. Once the elevation angle is selected (usually $0°$), the scattering angle is crucial in defining the optical set-up. As explained before, although light scattering is fully described by the Mie theory, the PDA instrumentation relies upon the simpler approach based on the geometric optics principles. The question here is to decide about the reliability of this approach. Regarding this, the scattering angle plays an important role.

In a measurement situation all scattering modes are present. Depending on the collection angle, some may be dominant relatively to the others or may be of equal importance. A phase Doppler system should be set up so that the relationship between particle diameter and phase difference is linear. In general, this is obtained by choosing an angle where one single mode dominates the scattered light received by the receiving optics, and where the signal–to-noise ratio is as high as possible. Tayali and Bates have shown that, depending on the refractive index of both the medium and the particles, the scattering angle should be such that only one scattering mode is clearly dominant [31]. In this way the geometric optics principles will approach the exact theory and processing errors will be smaller. This shows that one condition for a reliable PDA set-up is the knowledge of the particle and medium refractive index.

Finally, if two detectors are present, the phase difference is calculated by subtracting the phase between each of the detectors (see Eq. 10). The phase difference of the larger particles may fall beyond the 2π range. Thus, there is no way one can tell whether the diameter is D3 or D3'. This is called the 2π ambiguity.

Therefore, with two photo-detectors, there is a compromise between, on the one hand, high sensitivity and small working range and, on the other hand, a larger working range at the expense of the sensitivity.

This problem is overcome by using three detectors asymmetrically positioned, as shown in Figure 13, and measuring more than one phase difference. Therefore, the phase difference between the two closer will extend the working range and the phase difference from the detectors further apart will give the necessary resolution.

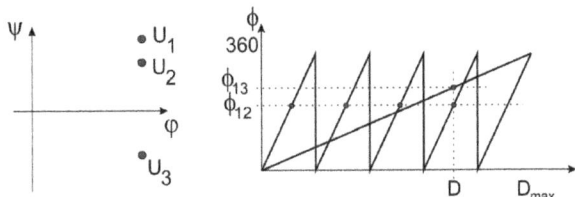

Figure 13. Removing the 2π ambiguity

This solution has another useful feature. The phase difference corresponding to each pair of detectors gives information about the curvature of the particle surface. This can be used to deal with non-spherical particles, either by measuring their sphericity ratio or by establishing a criterion for sphericity acceptance.

This technique was applied, by Liu et al, specifically to pMDI for measuring the droplets instantaneous velocity and mean diameter of the spray distribution [32]. A total of nine commercial pMDI products were tested, showing different droplet velocities for each one. It is also noticed the decrease in the spray droplets along the axial distance from the pMDI nozzle, as previously reported by Dunbar [13].

3.3. Uncertainties in PDA measurements

In applying a technique such as the PDA, one must be aware of the problems and potential limitations; errors may result of inappropriate configuration of the instrument. The major question concerns the fact that any sample will observe a limited slice window of the population; the question balances the various compromises necessary to carry out any experimentation.

In this topic one looks into the influence of PDA configuration and set up on the measured data. Particular attention is given to parameters that do affect the measurements and discussion is provided into these factors based on experimental data. For that purpose, the tests were carried out at constant gas/liquid flow rates in one nozzle. In addition, all the tests were carried out at the centre of the spray. Various parameters were slightly changed ant their effect upon the mean diameters, data rates and droplet fluxes are discussed. Insight is given into the best approach to obtain consistently reliable results.

Figure 14 (a) and (b) show the influence of the influence of the collection angle in the mean diameter and the mass flux, respectively. It must be stressed that in the case of a more complex flow (with walls for example) the difficulties should be greater and the results would show higher deviations. As a general conclusion, it can be concluded that considerable spread can be obtained unintentionally: a factor of two in diameter and a factor of six in the mass flux is possible. The various possible causes for such observations are discussed below.

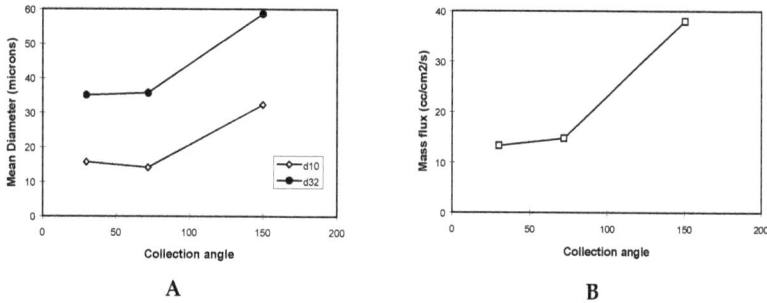

Figure 14. (a) Effect of collection angle on droplet size; (b) Effect of collection angle on mass flux

It has already been mentioned that the collection angle is by far the single most important factor in deciding an optical set up. However, in certain conditions, other angles could be more attractive and have been extensively used in the past. For example, if windows are necessary (Internal Combustion engines as an example) one single window would be very helpful in the rig construction. Therefore an angle of 150º (close to back scatter) is attractive. Other angles have also been employed, being the 30º one of the most common.

Results from a 10 μm water droplet show that reflection goes to zero at 72º (Brewster angle) and that 1st order refraction is over three orders of magnitude greater than any other non-zero mode. Thus, at this angle the geometric optics algorithm should not be affected by the presence of other scattering modes as they are of negligible intensity.

However, the detailed analysis of the droplet size PDFs also shows other interesting features, see Fig. 15. Firstly, it is obvious that the PDF at 150º is totally absent of small droplets. The problem is further compounded by the fact that, at that angle, the intensity of the scattered light is much lower than in the forward direction and therefore small droplets may not trigger the processors.

Although the mean diameters at 30 and 72º are close to each other, a close inspection in Figure 15 shows another story. Firstly, by altering the collection angle, the sizing range has dropped from 195 μm down to 134 μm which means that the distribution measured at 30º is truncated. If one removes the individual droplets above 134 μm in the 72º experiments, the mean diameter (d32 in particular) drops by approximately 10%! In Fig. 15 (c), the width of each bin in the distribution is made equal to that of the 72º collection angle. Therefore the counting in each bin can be directly compared. Although the 30º measurements should see more small droplets (close to forward collection where the signal intensity is higher which is supported by the higher data rates - 140 kHz as opposed to 84.5 kHz at 72º) the number of droplets at the lower end of the range is less than at 72º. This is clearly because at that angle the approximate solution is less accurate than at 72º. As expected, the data rates at 150º are much lower (33 kHz). However the mass flux shows that at this angle of collection, higher fluxes are observed (see Fig. 14 (b)). This is certainly due to the fact that the distributions contains a higher proportion of larger droplets and the flux (either on a mass or volume basis) are weighted by those large droplets.

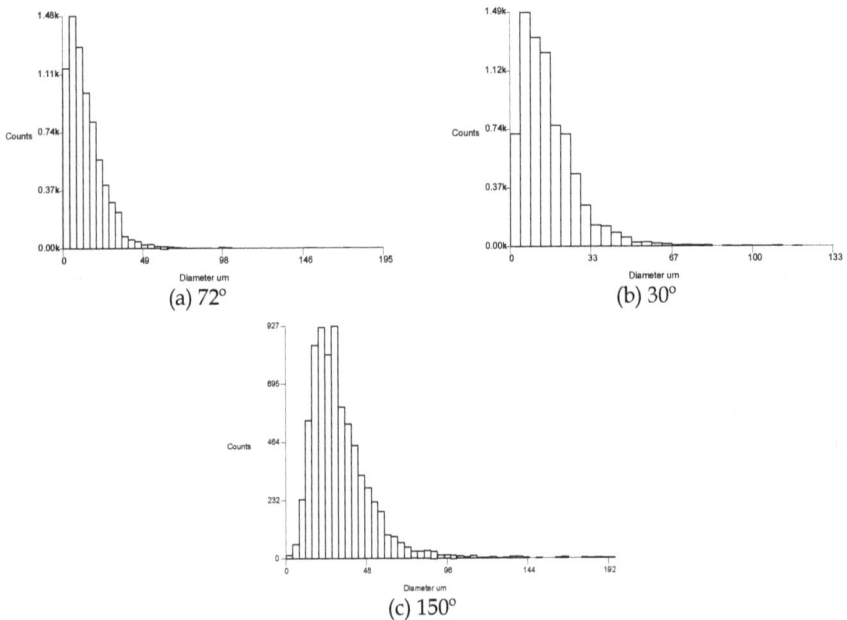

(a) 72°   (b) 30°

(c) 150°

Figure 15. Effect of collection angle upon the droplet probability density function

3.4. Photography and high-speed camera

Photography can be used for particle sizing through its application to small and fast moving particles requires extreme care in order to ensure an acceptable level of accuracy. Because of the particles motion, a very short exposure time is required to "freeze" the image. This requires both a high level of illumination intensity and a wide lens aperture. This means that the depth of field is small and only a fraction of the total sample is likely to be in focus at any moment.

For smaller particles, the turbulence induced motion in a direction perpendicular to the focusing plane compounds the problem because they are prone to be out of focus. One technique proposed to overcome this problem consisted in producing multiple focusing planes that overlap each other. Another major hurdle is due to the limited amount of data that can be experimentally retrieved. The measuring process is very time consuming rendering the effective amount of data of limited statistical value. Furthermore, although the sizing range is high, the lower end of the limit (of approximately 5 μm) makes the technique of limited interest in drug aerosols measurements.

This technique can also be extended to measure the particle velocity by increasing the exposure time in such a way that the path described by the particle during the lens aperture is correlated with the particle velocity. Additionally, a double flash (with a controlled time delay) can be deployed to superimpose two photographs in the frame. This technique has the advantage of a more accurate determination of particle size.

Nonetheless the development of other techniques that provide better resolution (such as the phase Doppler) has rendered this technique of limited use for the last decades.

However the advent of high speed digital video camera can successfully be used to gather meaningful insight into spray dynamics. The problems associated with sample illumination are ever present but this technique, which can record up to 10,000 frames per second, is very useful for understanding and capturing details of transient phenomena. In particular for aerosol delivery of drugs the very nature of the spray resulting from a fast operating canister suits the potential of such technique. When combined with a plane illumination from a light sheet the particles can be easily identified and tracked to measure their velocity. Also, the frame high resolution coupled with filtering algorithms for identifying the contours of particles, are useful for measuring their size. However the most important feature of this technique is its ability to capture the transient nature of the aerosol formation over the delivery time.

Using a high-speed camera (FASTCAM-APX RS 250KC), the authors recorded a puff event from a Ventolin® HFA-134a pMDI. Results are shown in Figure 16, the images presented have an interval of 0.02 seconds. The images were taken at a rate of 6000 frames *per* second, allowing us to confirm the duration of the spray (0.1 seconds) and to calculate the approximated angle (10 degrees).

3.5. Impacting techniques and collectors

Impacting collectors have been used to measure particle size for a wide range of sprays. In particular for inhaler derived aerosols, it has been the most widely used method. Mitchell and Nagel have extensively reviewed this type of particle size analysers, their strengths and limitations [33].

0.00s

0.06s

0.02s

0.08s

0.04s

0.10s

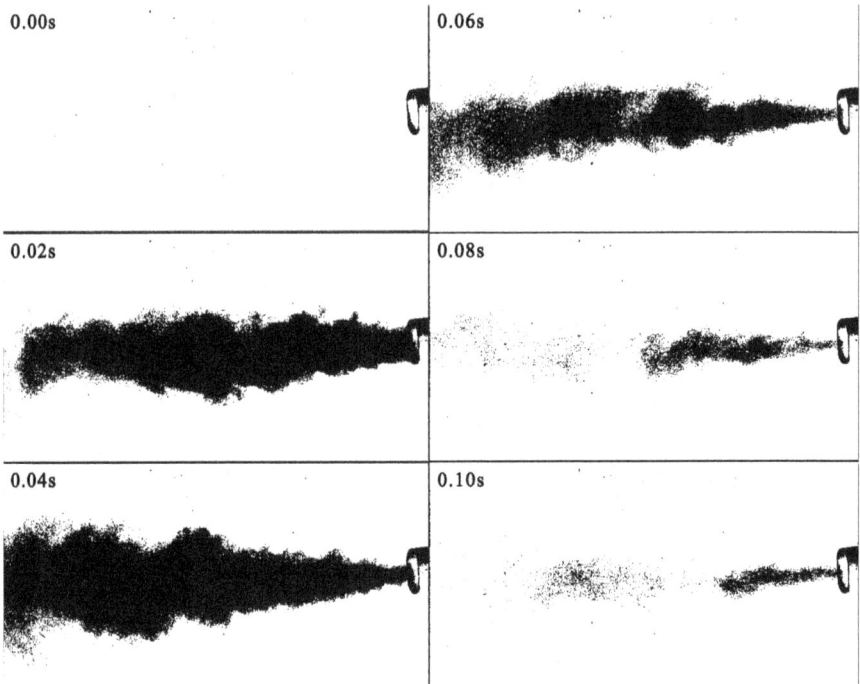

Figure 16. High speed images of a puff taken from a Salbutamol HFA-134a pMDI (Ventolin®). These images were treated with greyscale and inverted colours after application of a threshold filter for easier visualisation of the plume

The theoretical principles of an impactor are based on the solution of the two dimensional Navier Stokes equations of a fluid around a target oriented perpendicularly to the impinging flow. This flow field is subsequently coupled with Newton's equation of motion to model the trajectories of particles/droplets through various stages and geometries.

In its basic form, the impactor consists of a jet of a known diameter that exits from an orifice diameter (D) located at a certain distance (S) from the flat target that acts as a collector (see Fig. 17).

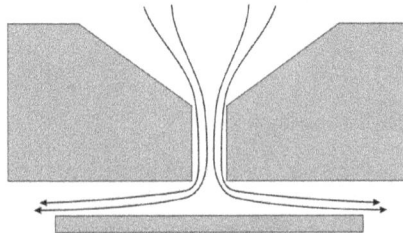

Figure 17. Basic geometry of an impactor

As the flow exits the orifice, its streamlines are diverged in the vicinity of the collection surface. The high inertia particles move through the streamlines and impact on the collection surface. The Stokes number (St) defines the critical particle size (aerodynamic diameter d_a) that will impact the collection surface through

$$St = \frac{\rho_a C_a d_a^2 U}{2\mu D} \tag{13}$$

where C_a is the Cunningham slip, ρ_a is the particle density, U is the gas velocity at the orifice exit and μ is the fluid viscosity.

The particle collection efficiency of an ideal impactor will increase in a step from 0 to 100% at a critical St. For each impactor stage, the corresponding cut-off diameter (d_{50}) can be calculated through equation 14.

$$\sqrt{C_{50}} d_{50} = \left[\frac{9\pi\mu n D^3}{4\rho_a C_a Q} \right]^{(1/2)} \sqrt{St_{50}} \tag{14}$$

where n accounts for the number of circular orifices in the nozzle plate, Q refers to the flow rate and subscript 50 identify conditions at 50% efficiency.

A cascade impactor consists of several stages with progressively decreasing cut sizes, assembled in series, so that an aerosol is separated into various fractions according to the number of stages assembled. The design of the cascade can be tailored to measure the size down to sub-micron particles. The two stage impactor is used to separate the large from the small particles. The former are deposited in the upper tract of the respiratory system while the later are most likely to reach the alveoli. For inhaler testing, it is desirable that at least 5 stages are assembled with cut off aerodynamic diameters within the range 0.5-5.0 μm. One of the most widely used is the Anderson type impactor with 8 stages. The testing of aerosols usually involves the operation at a constant flow rate (28 or 60 L/min) although this does not represent the conditions typical of a tidal respiratory cycle. The combination of a breathing machine (Harvard type) and a vacuum pump to draw air through the impactor at a constant rate enables the measurement of aerosol deposition during the breathing tidal cycle [34].

4. Numerical studies

Some preliminary studies have been implemented in a Computational Fluid Dynamics (CFD) software code creating a computational model for a pMDI spray. In this way, a simple geometry test case (named "testbox") was created and tested. These assumptions and simplifications are important in order to analyse spray flow without interference of the room walls.

The results given by the CFD analysis are normally a realistic approximation of a real-life system. The user has a wide choice regarding the level of detail of the results, since it is possible to simulate any fluid that cannot be reproduced experimentally due to economical or physical reasons. These advantages makes the CFD software to be a very powerful tool in the engineering research field [35,36].

The conservation equations for mass and momentum are solved in Fluent. Fluent is a finite volume based code, which uses the integral form of the conservation equations as its starting point. For this, a suitable grid for this simple geometry was generated. The solution domain is divided into a finite number of contiguous control volumes and the conservation equations are applied to each one. The governing equations are discretized on a curvilinear grid and Fluent uses a nonstaggered grid storage scheme to store the discrete values of dependent variables (velocities, pressure and scalars). A choice of interpolation schemes is then available.

By default, Fluent uses a Semi-Implicit Method for Pressure-Linked Equations (SIMPLE) algorithm type to solve in a sequential way the discretized equations for mass and momentum. The algebraic equations are then solved iteratively using a Line-Gauss-Seidel solver and Multigrid acceleration techniques.

In the present application (spray in air flow), a dispersed second phase is included. Fluent models the dispersed second phase using a Lagrangian approach. The dispersed second phase is assumed sufficiently dilute and so, particle-particle interactions and the effects of the particles volume fraction on the gas phase are negligible. The calculations for the dispersed phase include the Lagrangian trajectory solution with stochastic tracking to account for the effects of gas.

The coupling between the phases and its impact on both the dispersed phase trajectories and the continuous phase flow can be included.

4.1. Gas flow and spray simulation

The air flow inside the "testbox" is solved as transient, incompressible, Newtonian and viscous turbulent. The governing equations of conservation of mass and momentum are solved with appropriate modelling procedures to describe the effects of turbulence fluctuations.

The spray modelling includes the calculation of the dispersed phase trajectories with the turbulence effects. The dispersed phase is included in the model by defining the initial conditions which will be used to initiate the second phase calculations. Initial and boundary conditions also need some attention.

The main equations are now presented, as well as, a brief description of the boundary conditions and numerical solution.

4.1.1. Mathematical modelling

The fluid conservation equations are simplifications to the Navier-Stokes equations, which are not possible to solve by conventional numerical means, each of them implies several considerations according to the specifics of the problem.

Mass conservation implies that the mass entering a control volume equals the mass flowing out, creating a balance between input and the output flows for a certain volume. This concept is mathematically expressed by Eq. 15, assuming constant the fluid properties [35–37].

$$\frac{\partial \rho}{\partial t} + \rho \frac{\partial u_i}{\partial x_i} = 0 \tag{15}$$

where ϱ stands for density, t for time, x_i (i = 1, 2, 3) or (x, y, z) are the three-dimensional Cartesian coordinates and u_i or (u_x, u_i, u_z) are the Cartesian components of the velocity vector u.

The conservation equations used in Fluent for turbulent flows are obtained from those for laminar flows using a time averaging procedure usually known as Reynolds averaging.

For the momentum equations, the velocity at a point is considered as a sum of the mean and the fluctuating components:

$$u_i = \bar{u}_i + u_i' \tag{16}$$

Dropping the overbar on the mean velocity, $\bar{u}$, the ensemble-averaged momentum equations for predicting unsteady state turbulent flows are described by Eq. 17 [35-37]:

$$\rho \frac{\partial u_i}{\partial t} + \frac{\partial}{\partial x_j}\left(\rho u_i u_j\right) = \frac{\partial}{\partial x_j}\left[\mu\left(\frac{\partial u_i}{\partial x_j} + \frac{\partial u_j}{\partial x_i}\right) - \left(\frac{2}{3}\mu\frac{\partial u_l}{\partial x_l}\right)\right]$$
$$-\frac{\partial p}{\partial x_i} + \rho g_i + F_i + \frac{\partial}{\partial x_j}\left(\overline{\rho u_i' u_j'}\right) \tag{17}$$

Where μ represents the viscosity, p represents static pressure, g represents gravitational acceleration and F the source term for the momentum equation. The effect of turbulence is incorporated through the "Reynolds stresses" terms, $\overline{\rho u_i' u_j'}$. Fluent relates the Reynolds stresses to mean flow quantities via a turbulence model. The most commonly used model for turbulence calculations is the k-ε with applications in several fields of engineering. It is classified as a Reynolds-Averaged Navier–Stokes (RANS) based turbulence model, considering the eddy viscosity as linear, it is a two equation model. This model accounts for the generation of turbulent kinetic energy (k) and for the turbulent dissipation of energy (ε). The model formulation described is typically called the Standard k-ε, and it is calculated by Eq. 18 and 19 [37].

$$\frac{\partial}{\partial t}\left(\rho k\right) + \frac{\partial}{\partial x_i}\left(\rho k u_i\right) = \frac{\partial}{\partial x_j}\left[\left(\mu + \frac{\mu_t}{\sigma_k}\right)\frac{\partial k}{\partial x_j}\right] + G_k - \rho\varepsilon \tag{18}$$

$$\frac{\partial}{\partial t}\left(\rho\varepsilon\right) + \frac{\partial}{\partial x_i}\left(\rho\varepsilon u_i\right) = \frac{\partial}{\partial x_j}\left[\left(\mu + \frac{\mu_t}{\sigma_\varepsilon}\right)\frac{\partial\varepsilon}{\partial x_j}\right] + C_{1\varepsilon}\frac{\varepsilon}{k}G_k - C_{2\varepsilon}\rho\frac{\varepsilon^2}{k} \tag{19}$$

In these equations, G_k represents the generation of turbulent kinetic energy due to the mean velocity gradients, σ_k and σ_ε are the turbulent Prandtl numbers for the k and ε, respectively. $C_{1\varepsilon}$ and $C_{2\varepsilon}$ are model constants. Tubulent viscosity, μ_t, is modelled according to the Eq. 20, by combining the values of k and ε.

$$\mu_t = \rho C_\mu \frac{k^2}{\varepsilon} \tag{20}$$

This model uses, by default, the following values for the empirical constants:

$$C_{1\varepsilon} = 1.44, \quad C_{2\varepsilon} = 1.92, \quad C_\mu = 0.09, \quad \sigma_k = 1.0, \quad \sigma_\varepsilon = 1.3$$

Fluent allows for inlet, outlet, symmetry, cyclic and wall boundary conditions for treatment of the continuous phase. Two different inlet boundaries can be considered: inlet (velocity) boundary or pressure boundary. An inlet boundary is a boundary at which flow enters (or exits) at a known velocity, and turbulence parameters. At the pressure boundary, the fluid pressure is defined instead of velocity. At wall boundaries, the normal velocity component is zero.

The CFD software favours the use the Euler-Lagrange approach to track the particle trajectory through the fluid domain. The dispersed phase can exchange momentum and mass with the fluid phase. The particle or droplet trajectories are computed individually at specified intervals during the fluid phase calculation by integrating its equation of motion, which one is written in a Lagrangian reference frame. A force balance on the droplet immersed in a turbulent gas flow equates the particle inertia with the forces acting on the droplet, formulated (for the x-direction in Cartesian coordinates) in Eq. 21 [37].

$$\frac{du_p}{dt} = F_D\left(u - u_p\right) + g_x \frac{\rho_p - \rho}{\rho_p} + F_x \tag{21}$$

where the first term in the right side accounts for the drag force, the second one for the gravity effects and F_x includes additional forces which can be important only in certain circumstances. In the first term,

$$F_D = \frac{18\mu}{\rho_p d_p^2} \frac{C_D \, \text{Re}}{24} \tag{22}$$

where Re is the relative Reynolds number defined by:

$$\text{Re} = \frac{\rho d_p \left| u_p - u \right|}{\mu} \tag{23}$$

When the gas flow field is turbulent, Fluent can predict the trajectories of the second phase using only the mean fluid phase velocity in the trajectory equations or optionally it can include the instantaneous value of the fluctuating gas flow velocity in order to predict the dispersion of the droplets due to turbulence.

In the present application, the inclusion of the turbulence effect on the droplets has been considered important. By computing the droplets trajectories for a sufficient number of representative droplets, the stochastic nature of turbulence can be modelled. The stochastic tracking (random walk) model includes the effect of instantaneous turbulent velocity fluctuations on the particle trajectories.

The dispersed second phase is introduced in the computational domain at a certain position (with initial conditions) and its trajectory is then computed along the continuous gas flow field.

Fluent provides different ways for input of the initial conditions for the dispersed phase. These initial conditions provide the starting values for all the dependent dispersed phase variables above mentioned.

Each set of these initial conditions represents an 'injection' of 'droplet type' identified in Fluent by a label 'injection number'. Each injection or parcel means thousands of droplets assumed to have the same size, diameter, temperature, starting at the same spatial position with the same velocity components. The mass flow rate indicates the amount of liquid that will follow the same calculated droplet trajectory.

The user can define any number of different sets of initial conditions for dispersed phase droplets provided that sufficient memory has been allocated.

The treatment of the dispersed phase near a boundary of the computational domain is a very difficult task. In Fluent, different dispersed phase boundary conditions can be applied when a droplet reaches a physical boundary. These boundary conditions are defined by the user at each cell-type and they can be: 'reflect', 'trap', 'escape' and 'saltation'. The 'reflect' condition rebounds the particle/droplet off the boundary with a change in its momentum defined by a coefficient of restitution; the 'trap' condition terminates the trajectory calculation and if it is a case of an evaporating second phase, their entire mass is converted into vapour phase; with the 'escape' condition the trajectory is simply terminated; with 'saltation', the particle/droplet is placed in the gas field at a small distance from the wall and again a coefficient of restitution is used to modified the particle momentum.

4.1.2. Numerical solution

As mentioned before, Fluent uses a control volume based technique to solve the conservation equations for mass, momentum, and turbulence quantities. The domain is divided into discrete control volumes where the governing equations are integrated to obtain the algebraic equations for the unknowns (velocities, pressure and scalars). The integration of the differential equations in each control volume yields a finite-difference equation that conserves each quantity on a control-volume basis.

For the temporal discretization of the first term in the conservation equations, the implicit Euler scheme has been used.

Because Fluent defines the discrete control volumes using a non-staggered storage scheme (all variables are stored at the control volume cell center), interpolation schemes are needed to determine the face values of the unknowns from the stored values at the cell center.

The discretized equations are solved sequentially and the SIMPLE algorithm has always been used in the present application. This type of algorithm is based on using a relationship between velocity and pressure corrections in order to recast the continuity equation in terms of a pressure correction calculation. In this way, the calculated velocity and pressure fields satisfy the linearized momentum and continuity equations at any point.

Fluent does not solve each equation at all points simultaneously and so an iterative solution procedure is used with iterations continuing until the convergence criteria specified has been achieved.

The algebraic equation for each variable is solved using a Line Gauss-Seidel procedure (LGS) and the user can specify the direction in which the lines are solved (the direction of the flow or alternate directions) and the number of times the lines are solved in order to update a given variable within each global iteration cycle. To speed up the convergence achieved by the LGS procedure, Fluent uses a Multigrid acceleration technique by default to solve the pressure and enthalpy equations.

The new calculated values of a given variable obtained in each iteration by the approximate solution of the finite difference equations are then updated with the previews values of the variable using a under relaxation technique. The user can choose the best relaxation factors for each variable in order to achieve a better convergence.

From the above, it can be noted that the trajectory equation for each parcel droplet is solved by step-wise integration over discrete time steps. Small steps have to be used to integrate the equations of motion for the droplet and Fluent allows the user to control the integration time step size when the equations are analytically solved.

The second phase trajectory calculations are based on the local gas flow field conditions, but Fluent does not include the direct effect of the droplets on the generation or dissipation of turbulence in the continuous phase. It keeps track of mass and momentum gained or lost by each droplet injection, as its trajectory is calculated. These quantities can be incorporated in the subsequent continuous phase calculations and, in this way, to study the effect of the dispersed phase on the continuous gas flow field. This two-way coupling is essential to calculate an accurate solution for the two-phase flow.

The momentum transfer from the continuous phase to the dispersed phase is computed in Fluent by examining the change in momentum of a droplet as it passes through each control volume in the computational domain. This momentum change is computed by the Eq. 24 [37].

$$F = \sum \left(\frac{18\mu C_D \, \mathrm{Re}}{\rho_p d_p^2 24} \left(u_p - u \right) \right) \dot{m}_p \Delta t \tag{24}$$

The mass flow rate of each injection, which has no impact on the droplet trajectory calculation, is now used in the calculation of the second phase effect on the gas phase. It must be noted that when different trajectories are calculated for the same injection, in order to simulate the turbulence effects of the gas phase, the mass flow rate of that injection is divided equally by the number of stochastic tracks computed for that injection and so the exchange terms of momentum, mass and heat are calculated using the divided mass flow rate. This momentum exchange acts as a momentum sink in the gas momentum conservation equation (see Eq. 16).

The coupling between the two phases can be automatically simulated in Fluent by solving alternately the dispersed and continuous phase equations, until the solutions in both phases have stopped changing. Usually, the second phase calculations are made after a certain number of gas phase iterations.

Also, the exchange of momentum and mass terms calculated during the second phase simulation is not introduced directly in the gas phase conservation equations because Fluent allows for the under relaxation of the interphase exchange terms during the subsequent calculations. A small under relaxation factor seems to improve the convergence of the solution

4.2. Spray plume simulation

The present simulation aims to highlight the configuration parameters that better fits the real-life case of a pMDI spray, by using commercially available CFD software, such as Fluent version 14.0 from ANSYS®.

In this research work, the Ventolin® was used because it is one of the most common drugs used in developed countries to treat asthma in children and adults. It mainly consists of salbutamol, which is the most frequently prescribed Short-Acting β-Agonist (SABA) [9,38,39]. The used pMDI actuator was the one that is typically sold with the Ventolin®, both produced by the GlaxoSmithKline® company.

4.2.1. "Testbox" geometry and grid

The "testbox" consists of a simple parallelepiped form with the dimensions of 0.2 x 0.2 x 0.3 (m) representing a sample of a room environment, with a pMDI actuator and canister in the middle of it. The spray injection point, the exit of the actuator's nozzle, is located in the origin point (0,0,0), see Figure 18.

The geometry was drawn using an external design program and then loaded into the ANSYS® meshing software. The generated mesh for the numeric calculations was composed by tetrahedral and wedge elements, with sizes ranging from 0.1 mm to 20.0 mm, resulting in a 3D computational grid with 3060339 elements and 1022403 nodes. Several refinements near to wall zones of high proximity and curvature were included. The quality report showed that mesh had a good quality according to the skewness parameter, with an average values located at 0.21 (see Table 3).

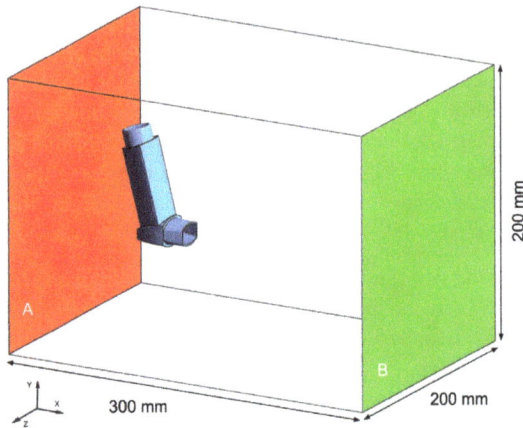

Figure 18. A "testbox" representation constituted, the red plane (A) is the boundary condition 'Velocity Inlet' and the green plane (B) is the boundary 'Outflow'. The other four outer walls of the domain have 'Symmetry' properties

	Skewness	Element Quality
Minimum	0.0000080602	0.065609
Maximum	0.9172	0.9999
Average	0.2129	0.6079
Standard Deviation	0.1292	0.2641

Table 3. Mesh quality criteria parameters: skewness and element quality

The boundary conditions were defined in two opposite faces, one as a 'Velocity Inlet' (see Fig. 2 A), forcing air to move inside the "testbox" at 0.01 m/s and the other as an 'Outflow' (see Fig. 2 B), enabling the freely motion of the air, as well as particles. For the remaining four faces, a 'Symmetry' boundary condition was assumed. The pMDI actuator and canister boundaries were considered 'Wall', trapping all the particles that collide with them.

4.2.2. Drug properties and spray characterisation

As mentioned previously, Ventolin® was used since it is a common SABA drug applied in the treatment of asthma with a pMDI. The Ventolin® most important characteristics for the simulation are listed in Table 4.

Characteristic	Value	Refs.
Propellant	HFA 134a	[11,40]
Salbutamol density (kg/m³)	1230	[11]
Actuation dose (μg)	100	[7,40,41]
Actuation time (s)	0.1	[15]

Table 4. Ventolin® drug properties

The particle size data were measured using the laser diffraction technique, using an independent model for data processing. Data were fitted to the cumulative function of Rosin-Rammler distribution, using the least squares method [20]. The CFD solver accepts this distribution type by inserting its corresponding parameters, see Table 5 [37].

Parameter	Value
Diameter distribution	Rosin-Rammler
Minimum diameter (μm)	1.22
Maximum diameter (μm)	49.50
Mean diameter (μm)	16.54
Spread parameter	1.86

Table 5. Particles diameter distribution parameters

4.2.3. Numerical parameters in Fluent

The spray parameters used to configure the solver were obtained from various references, though some caution is required, such as, the angle of the spray was considered to be 10 degrees, estimated by high speed image analysis. Assumptions were made, such as, the spray particles to be solid instead of the well-known liquid droplets. The reason for this consideration is simple: when the drug exits the metering chamber of the pMDI, it undergoes a flash evaporation. Because this is an instantaneous process, it is assumed that no heat transfer between the gas and liquid phases [5,13,14]. Using the drug amount of a puff (100 μg) and divide it by the duration of a puff (0.1 s), it is possible to calculate the spray flow rate, see Table 1. The main spray parameters used to configure the solver are summarized in Table 6, such as the spray angle, maximum velocity of the spray plume, shape of the plume, as well as, the dimension of the actuator's nozzle.

Parameter	Value	Refs.
Spray type	solid-cone	[12,37]
Angle ($^{\circ}$)	10	
Velocity (m/s)	100	[12,42]
Radius (m)	0.00025	[12,15,42]
Flow rate (kg/s)	1e^{-6}	

Table 6. Solver spray configuration parameters

The solution of the differential equations for mass and momentum was done in a sequential manner, using the SIMPLE algorithm [35–37,43]. The standard discretisation scheme was used for the pressure and the second order upwind scheme for the momentum, turbulent kinetic energy and turbulent dissipation rate equations. Convergence was reached in the simulation by using a criterion value of 1.0e-5 for the continuity (pressure), x, y and z velocity, and for k and ε turbulence parameters.

The time step used in the simulation was of 0.01 s, during 11 time steps with a maximum limit of 3000 iterations for each time step. Also the gravitational acceleration was assumed 9.81 m/s^2 in the y axis direction, although for these particles diameter range, this effect might be negligible.

Some parameters used to configure the Discrete Phase Model (DPM) model in the solver are listed in Table 7. Although different configuration parameters have been tested, the ones presented here seemed to be the most appropriate to be used in the simulation. Due to the fact that the injection occurs in a limited period in time (0.1 s), the unsteady particle tracking was used. The interaction with the continuous phase was also included to approximate the simulation to the real event, because the velocity difference between the two phases is high.

Parameter	Value
Interaction with Continuous Phase	On
Unsteady Particle Tracking	On
Inject Particles at	Particle Time Step
Particle Time Step Size (s)	0.001
Drag Law	Spherical
Two-way coupling turbulence	On

Table 7. Parameters for the DPM model

The drag law used (spherical) is the simplest and most used law, once it is the one that better fits the presented model amongst the four different options available in the solver. This drag law considers the particle as a sphere, which is an acceptable simplification for the drug particles that exit the pMDI nozzle [14,37].

The total number of particle streams injected during the simulation was approximately 164000.

4.2.4. Results

The CFD results obtained are represented in the form of images, showing the contours of velocity along a longitudinal mid-section plane and as streams of tracked particles along the fluid domain at the last injection moment.

Figure 19 shows the velocity magnitude of air in the domain at the time of 0.11 seconds, which coincides with the end of the injection of the spray. It is possible to observe the high velocity that air shows at the nozzle zone, accelerated by the discharge of particles, reaching a velocity of 10 m/s and lowering its velocity until reaches an equilibrium, the "testbox" air velocity.

In Figures 20 and 21, the representation of the particles as spheres scaled by its diameter provides a direct reading of the particles diameter and positioning at the end of the injection. The bigger particles are at the end of the plume and the smaller ones are next to the nozzle. Although the particle size distribution configured provides the expected result, there are several factors that might influence numerical shape of the spray plume, such as the solver injects always the same quantity of particle streams at each instant of the injection, which is known not to be the spray real behaviour.

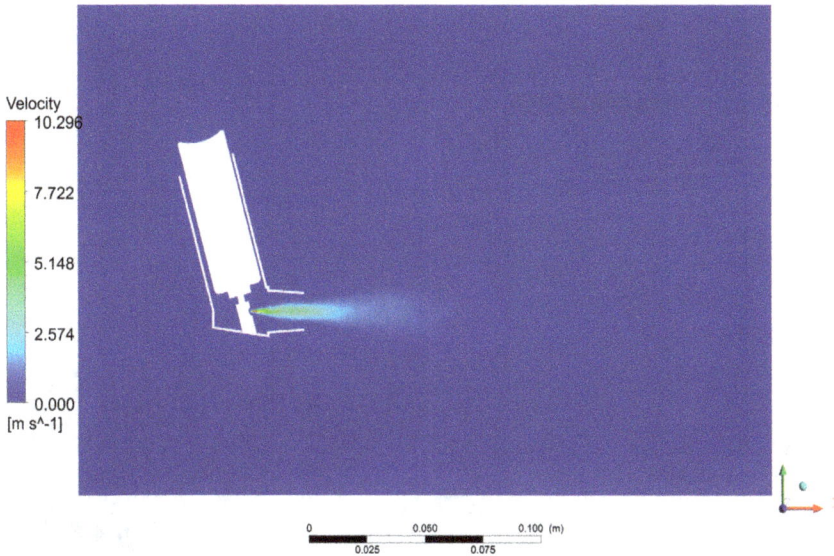

Figure 19. Velocity magnitude contours of the air in the domain, along the longitudinal midsection plane (*t*=0.11 s)

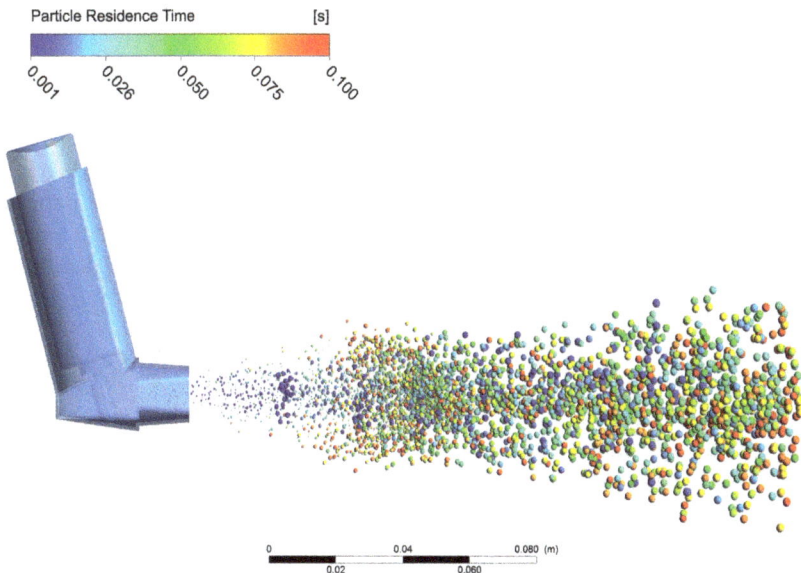

Figure 20. Representation of the particle streams at the end of the injection (*t*=0.11 s), image shows the particles coloured by residence time inside the domain. The particle streams are draw as spheres with proportional size scaled 50 times more than the real diameter

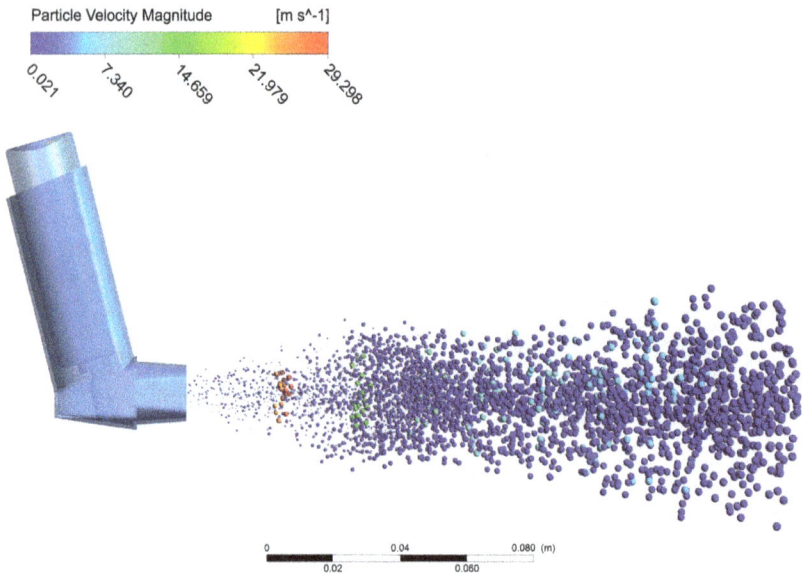

Figure 21. Representation of the particle streams at the end of the injection (*t*=0.11 s), image shows the particles coloured by its velocity magnitude. The particle streams are draw as spheres with proportional size scaled 50 times more than the real diameter

In Figure 21, it can be observed that particles velocity drastically reduces after the injection into the still air, which was expected, but might not be fully accurate due to the simplification of the drag law used. In Figure 20 the residence time of each particle stream in the domain is coloured, showing that all diameters take the maximum time of 0.1 seconds, the duration of the injection, which means that particle streams are inside the domain since the beginning of the injection.

5. Conclusions

The authors believe this chapter will provide a meaningful insight into the pMDI spray thematic, with a significant description that covers several pMDI spray topics, such as, the components, mechanisms, distributions, different measurement techniques and numerical simulations, regarding the gas flow and spray simulation as well as the spray plume simulation. This review clearly highlights that a wide variety of techniques can be applied to the study of the drug delivery systems through the respiratory system. The experimental and numerical techniques complement each other to enhance the understanding of drug transportation into the respiratory tract.

The computational techniques can be successfully used to study complex flows in a cost effective manner. The availability of high performance computing tools enables the introduction of complex physical models into even more complex geometries that are representative of *in vivo* systems.

Nonetheless, experimentation is still a cornerstone approach to validate models, to turn them into more reliable and accurate simulation tools that should closely characterize the physical events they represent. The most recent techniques can be used to provide very detailed data of flow patterns, such as particle size and motion, amongst other parameters. Different techniques have been referred and summarized in the study herein reported, regarding laser diffraction techniques, phase Doppler anemometry and the use of photography and high-speed cameras to determine particle sizing and moving, as well as impacting techniques, to measure spray particle sizes.

The CFD spray study proved to be useful in the study of add-on devices to the pMDI, such as holding chambers (or spacers) design. The velocity contours determined and the streams of tracked particles, obtained at a time corresponding to the end of the spray injection, suggests interesting aspects about the behaviour and performance of the spray and particles positioning and distribution, although other simulations need to be carried out in the future with the introduction of the spacer device in the "textbox" geometry. These aspects are currently under investigation by the authors and they represent a further contribution given to understand drug delivery through the oral system and to help the design of even more efficient delivery instruments.

Author details

Ricardo F. Oliveira, José C. Teixeira and Luís F. Silva
Mechanical Engineering Department, University of Minho, Portugal

Senhorinha Teixeira
Production and System Department, University of Minho, Portugal

Henedina Antunes
Pediatric Department, Braga Hospital, Portugal,
Life and Health Sciences Research Institute, School of Health Sciences, University of Minho, Portugal,
ICVS/3B's - PT Government Associate Laboratory, Braga/Guimarães, Portugal

6. References

[1] Global Initiative for Asthma, 2010, Global strategy for asthma management and prevention, GINA.

[2] Virchow J. C., Crompton G. K., Dal Negro R., Pedersen S., Magnan A., Seidenberg J., and Barnes P. J., 2008, "Importance of inhaler devices in the management of airway disease," Respiratory medicine, 102(1), pp. 10-19.

[3] Dolovich M. B., 2004, "In my opinion - Interview with the Expert," Pediatric asthma Allergy & Immunology, 17(4), pp. 292-300.

[4] Tashkin D. P., 1998, "New devices for asthma," Journal of Allergy and Clinical Immunology, 101(2), p. S409-S416.

[5] Dolovich M. B., and Fink J. B., 2001, "Aerosols and devices," Respiratory care clinics of North America, 7(2), pp. 131-73, v.

[6] Newman S. P., 2006, "Aerosols," Encyclopedia of Respiratory Medicine, G.J. Laurent, and S.D. Shapiro, eds., Elsevier, pp. 58-64.

[7] Terzano C., 2001, "Pressurized metered dose inhalers and add-on devices.," Pulmonary pharmacology & therapeutics, 14(5), pp. 351-66.

[8] Newman S. P., 2004, "Spacer devices for metered dose inhalers.," Clinical pharmacokinetics, 43(6), pp. 349-360.

[9] Crompton G., 2006, "A brief history of inhaled asthma therapy over the last fifty years.," Primary Care Respiratory Journal, 15(6), pp. 326-31.

[10] Sanders M., 2007, "Inhalation therapy: an historical review," Primary care respiratory journal, 16(2), pp. 71-81.

[11] Smyth H., 2003, "The influence of formulation variables and the performance of alternative propellant-driven metered dose inhalers," Advanced Drug Delivery Reviews, 55, pp. 807-828.

[12] Newman S. P., 2005, "Principles of metered-dose inhaler design," Respiratory Care, 50(9), pp. 1177-1190.

[13] Dunbar C. A., 1997, "Atomization mechanisms of the pressurized metered dose inhaler," Particulate Science and Technology, 15(3-4), pp. 253-271.

[14] Finlay W. H., 2001, The mechanics of inhaled pharmaceutical aerosols: an introduction, Academic Press.

[15] Smyth H., Hickey A. J., Brace G., Barbour T., Gallion J., and Grove J., 2006, "Spray pattern analysis for metered dose inhalers I: Orifice size, particle size, and droplet motion correlations," Drug development and industrial pharmacy, 32(9), pp. 1033-1041.

[16] McDonald K. J., and Martin G. P., 2000, "Transition to CFC-free metered dose inhalers — into the new millennium," International Journal of Pharmaceutics, 201(1), pp. 89-107.

[17] Tiwari D., Goldman D., Malick W. a, and Madan P. L., 1998, "Formulation and evaluation of albuterol metered dose inhalers containing tetrafluoroethane (P134a), a non-CFC propellant.," Pharmaceutical development and technology, 3(2), pp. 163-74.

[18] Pereira L. P., Clement Y., and Simeon D., 2001, "Educational intervention for correct pressurised metered dose inhaler technique in Trinidadian patients with asthma," Patient Education and Counseling, 42, pp. 91-97.

[19] Newman S. P., and Clarke S. W., 1983, "Therapeutic aerosols 1 - physical and practical considerations," Thorax, 38(12), pp. 881-886.

[20] Dunbar C. A., and Hickey A. J., 2000, "Evaluation of probability density functions to aporoximate particle size distributions of representative pharmaceutical aerosols," Journal of Aerosol Science, 31(7), pp. 813-831.

[21] Mugele R. A., and Evans H. D., 1951, "Droplet Size Distribution in Sprays," Industrial & Engineering Chemistry, 43(6), pp. 1317-1324.

[22] Zhang Z., and Ziada S., 2000, "PDA measurements of droplet size and mass flux in the three-dimensional atomisation region of water jet in air cross-flow," Experiments in Fluids, 28(1), pp. 29-35.

[23] Dodge L. G., and Schwalb J. A., 1989, "Fuel Spray Evolution: Comparison of Experiment and CFD Simulation of Nonevaporating Spray," Journal of Engineering for Gas Turbines and Power, 111(1), p. 15.

[24] Lee Black D., McQuay M. Q., and Bonin M. P., 1996, "Laser-based techniques for particle-size measurement: A review of sizing methods and their industrial applications," Progress in Energy and Combustion Science, 22(3), pp. 267-306.

[25] Hong M., Cartellier A., and Hopfinger E. J., 2004, "Characterization of phase detection optical probes for the measurement of the dispersed phase parameters in sprays," International Journal of Multiphase Flow, 30(6), pp. 615-648.

[26] Mitchell J. P., and Nagel M. W., 2004, "Particle Size Analysis of Aerosols from Medicinal Inhalers," KONA, 22, pp. 32-65.

[27] Bayvel L. P., 1980, "Application of the laser beam scattering technique method for multi-phase flow investigations," Proceedings of the conference on multiphase transport, fundamentals and reactor safety applications, Y.N. Veziroglu, ed., New York.

[28] Swithenbank J., Beer J. M., Taylor D. S., Abbot D., and McCreath G. C., 1976, "A laser diagnostic technique for the measurement of droplet and particle size distribution," American Institute of Aeronautics and Astronautics, 14th Aerospace Sciences Meeting, Washington, D.C.

[29] Lefebvre A. H., 1989, Atomization and Sprays, Hemisphere Publishing Corp., New York and Washington, D.C.

[30] Dantec, 1995, PDA user's guide, Dantec information.

[31] Tayali N. E., and Bates C. J., 1989, Scattered light intensities and phase difference derived using geometrical optics.

[32] Liu X., Doub W. H., and Guo C., 2012, "Evaluation of metered dose inhaler spray velocities using Phase Doppler Anemometry (PDA)," International journal of pharmaceutics, 423(2), pp. 235-239.

[33] Mitchell J. P., and Nagel M. W., 2003, "Cascade impactors for the size characterization of aerosols from medical inhalers: their uses and limitations.," Journal of Aerosol Medicine, 16(4), pp. 341-77.

[34] Foss S. A., and Keppel J. W., 1999, "In Vitro Testing of MDI Spacers: A Technique for Measuring Respirable Dose Output with Actuation In-Phase or Out-of-Phase with Inhalation," Respiratory Care, 44(12), pp. 1474-1485.

[35] Ferziger J. H., and Peric M., 2003, Computational methods for fluid dynamics, Springer.

[36] Versteeg H. K., and Malalasekera W., 1995, An introduction to computational fluid dynamics: the finite volume method, Longman, Harlow, England.

[37] ANSYS, 2009, ANSYS FLUENT Theory Guide, ANSYS Inc, Canonsburg, PA, USA.

[38] Jepson G., Butler T., Gregory D., and Jones K., 2000, "Prescribing patterns for asthma by general practitioners in six European countries," Respiratory medicine, 94(6), pp. 578-583.

[39] Zuidgeest M. G., Smit H. A., Bracke M., Wijga A. H., Brunekreef B., Hoekstra M. O., Gerritsen J., Kerkhof M., Jongste J. C., and Leufkens H. G., 2008, "Persistence of asthma medication use in preschool children," Respiratory Medicine, 102, pp. 1446-1451.

[40] Dubus J. C., Guillot C., and Badier M., 2003, "Electrostatic charge on spacer devices and salbutamol response in young children," International Journal of Pharmaceutics, 261(1-2), pp. 159-164.

[41] Verbanck S., Vervaet C., Schuermans D., and Vincken W., 2004, "Aerosol Profile Extracted from Spacers as a Determinant of Actual Dose," Pharmaceutical Research, 21(12), pp. 2213-2218.

[42] Clark A. R., 1996, "MDIs: physics of aerosol formation.," Journal of aerosol medicine : the official journal of the International Society for Aerosols in Medicine, 9 Suppl 1(s1), pp. S19-26.

[43] Abreu S., Silva L. F., Antunes H., and Teixeira S. F. C. F., 2008, "Multiphase flow inside the Volumatic spacer: a CFD approach," Proceedings of the 10th International Chemical and Biological Engineering Conference - CHEMPOR 2008, E.C. Ferreira, and M. Mota, eds., Braga, p. 6.

Chapter 12

A New Correlation for Prediction of Viscosities of Omani Fahud-Field Crude Oils

Nabeel Al-Rawahi, Gholamreza Vakili-Nezhaad,
Ibrahim Ashour, Amin Fatemi

Additional information is available at the end of the chapter

1. Introduction

Crude oil viscosity is an important physical property that controls and influences the flow of oil through porous media and pipes. The viscosity, in general, is defined as the internal resistance of the fluid to flow. Viscosity is an extremely important property from process and reservoir simulations to the basic design of a pipeline. Experimental liquid viscosities of pure hydrocarbons and their mixtures under pressure are important to simulating the behaviour of the fluid at reservoir conditions. Also, experimental measurements over a wide range of temperature and pressure are needed to test the effectiveness of semi-theoretical and empirical viscosity models [1]. Oil viscosity is a strong function of many thermodynamic and physical properties such as pressure, temperature, solution gas-oil ratio, bubble point pressure, gas gravity and oil gravity. Usually oil viscosity is determined by laboratory measurements at reservoir temperature. Viscosity is usually reported in standard PVT analyses. Oil viscosity correlations all belong to three categories: dead oil, saturated oil and undersaturated oil viscosity correlation. Numerous correlations have been proposed to calculate the oil viscosity. There have been a number of empirical correlations developed for medium and light crude oils [2]. However their applicability is limited to specific oils due to the complex formulation of the crude oils. These correlations are categorized into two types. The first type which refers to black oil type correlations predict viscosities from available field-measured variables including reservoir temperature, oil API gravity, solution gas- oil ratio, saturation pressure and pressure [3-7]. The second type which refers to compositional models is derived mostly from the principle of corresponding states and its extensions. In these correlations beside previous properties, other properties such as reservoir fluid composition, pour point temperature, molar mass, normal boiling point, critical temperature and acentric factor of components are used [8].

Ideally, viscosity is experimentally measured in laboratory. When such direct measurements are not available, correlations from the literature are often used. Fundamentally, there are two different types of correlations in the literature. The first group of correlations is developed using randomly selected datasets [9]. Such correlations could be called generic correlations. The second group of correlations, called specialized correlations, is developed using a certain geographical area or a certain class/type of oil. Correlations using randomly selected datasets may not be suitable for certain type of oils, or certain geographical areas. Even though the authors of the generic correlations want to cover a wide range of data, specialized correlations still work better for certain types of oils. Specialized correlations represent the properties of a certain type of oil or geographical area (for which they are developed) better than the general purpose correlations[9].

2. Experimental data and analysis

In this study, PVT experimental data of three sample oils from Omani dead oils have been measured. Each sample was introduced into Viscometer tube and inserted into the heating bath, set at initial temperature of 25°C with incremental temperature of 10 °C to a final temperature of 85 °C. The viscosity of each sample was measured using Automatic Rheometer System Gemini 150/200 3X Tbar with a temperature control accuracy of ±0.1 C from Malvern company. The advantage of this instrument is particulate material tolerable, good repeatability, not so critical on misalignment and good temperature control accuracy of ±0.1 C.

Before viscosity measurement of samples, the viscometer was calibrated frequently according to the instructions using standard calibration fluids provided by the supplier company and then was checked with pure liquids with known dynamic viscosity.

The viscometer was placed in a dry place and the viscosity measurement proceeded as soon as the sample was placed on viscometer cylinder. The sample is considered to be, no more in contact with the external environment, the viscometer was operated at single shear rate 1 S^{-1} using double gap (DG 24/27)at 298.15 K to 358.15 K. The double gap measuring system consists of a hollow cylinder with diameter 27.5 mm and height 53mm that is lowered into a cylindrical grove in the outer cylinder. The sample is contained in the double annular gap between them. Application material includes mobile liquids, suspensions and emulsion. The viscosity measurements were performed ten times and the results were reported as an average. Repeating the measurement several times could help attain data as close as possible to the true value in spite of the variations that might occur in the midst of an experimental process. The following table 1 shows the results of the measurements in the laboratory for the three samples.

Plotting the data yields fig.1 in which viscosity versus temperature has been drawn for the three samples. Fig.1 shows how viscosity changes according to temperature change and how scattered the data of this reservoir are.

T (°C)	1. LEKH Incoming (cp)	2. Yibal Incoming(cp)	3. Booster Pump(cp)
25	6.0423	5.8935	34.3738
30	5.7104	5.5253	31.3382
35	5.3281	5.1092	28.6623
40	4.8819	4.6732	25.7045
45	4.6435	4.3483	23.2368
50	4.3218	3.9598	20.026
55	3.8478	3.6398	17.5592
60	3.4857	3.2408	15.55
65	3.1841	2.9537	13.7337
75	2.9678	2.5729	11.1786
85	2.6262	2.1869	8.7418

Table 1. Experimental data of three oil sample viscosities at different temperatures

Variable	Range
Oil gravity, °API	32.4 to 38.58
Pressure, atm	1
Temperature,°C	25 to 85

Number of dead oil observations = 3

Table 2. Description of data used from the samples

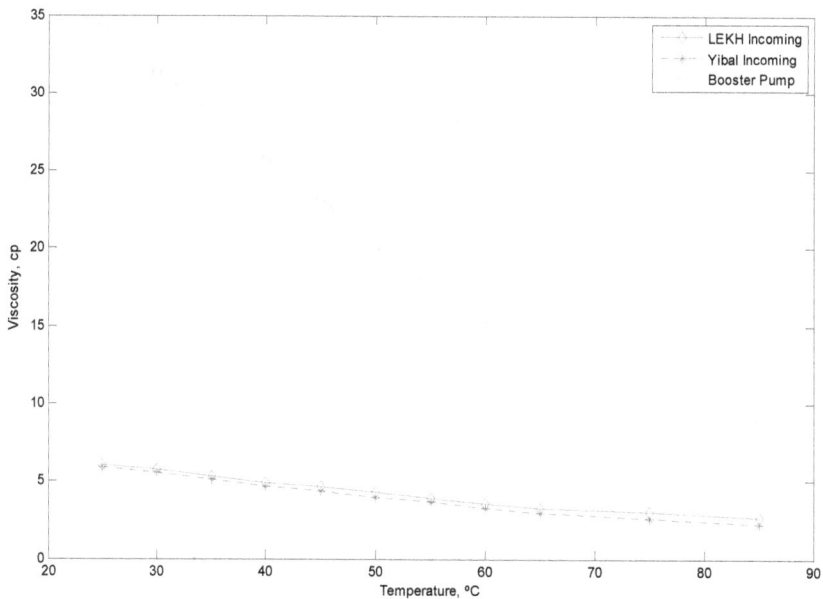

Figure 1. Experimental data of three oil sample viscosities at different temperatures

Some other properties of these oil samples (API degree and specific gravity) as reported by the laboratory are shown in table 3.

Sample Properties	1. LEKH Incoming	2. Yibal Incoming	3. Booster Pump
°API	38.58	39.34	32.4
Specific Gravity	0.832	0.8283	0.8633

Table 3. API gravity of the oil samples

3. Development of the proposed correlations

The most popular empirical models presently used in petroleum engineering calculations for predicting dead oil viscosity (gas-free crude or stock tank) are those developed by Beal [3], Beggs and Robinson [4], and Glasø [5]. Beal's model [3] was developed from crude oil data from California, Beggs and Robinson's model [4] was developed from the crude oils of an unknown location, whereas Galsø's model [5] from crude oils in the North Sea. Recently, Labedi [6], and Kartoatmodjo and Schmidt [7] presented other empirical models for estimating dead crude oil viscosity for African crudes and using data bank respectively. All of these models have expressed dead oil viscosity as a function of both oil API gravity and reservoir temperature(T), see Appendix A. When these correlations were applied to data collected considerable errors and scatter were observed. These data, therefore, were used to develop new empirical correlations for dead or gas-free crude oil as a function of API gravity and temperature. Proposed correlation is based on real data, which covers Omani oil types, given in Table 1. Best results were obtained by multiple regression analysis from the following empirical model: The correlation for dead oil viscosity was developed by plotting $\log_{10}(T)$ vs. $\log_{10}\log_{10}(\mu_{DO} + 1)$ on Cartesian coordinates. It was found that each line represented oils of a particular API gravity. The equation developed is

$$\mu_{DO} = 10^X - 1 \tag{1}$$

where

$$X = y\,T^{-0.9863} \tag{2}$$

$$y = 10^Z \tag{3}$$

$$Z = 2.9924 - 0.11027\gamma_o \tag{4}$$

μ_{DO} is the dead oil viscosity in cp and T is the temperature in °C. Table 2 suggests the acceptable range of oil °API gravity between 32.4 to 38.58 and temperature between 25 to 85 °C. Development of these correlations neglects the dependence of oil viscosity on composition, since oils of widely varying compositions can have the same gravity. Viscosity does depend on composition, and if the composition is available other correlations [5-7] exist that should be used for greater accuracy. However, the correlations presented here are easy to use and give fair accuracy and precision over an acceptable range of oil gravity and temperature. As is the case with any empirical study, extrapolation outside the range of the data used to develop the correlations should be done with caution [11,12].

4. Results and discussion

4.1. Validation of the proposed correlation

Table 4 shows the error percentage for all published dead crude oil models including the one proposed in this study for the Omani crudes. Table 4 reveals average relative error (ARE), absolute average relative error (AARE) and standard deviation (SD) for dead oil viscosity correlation respectively. ARE, AARE, and SD are defined as below.

$$ARE = \frac{1}{N}\sum_{i=1}^{N}\left(\frac{X_{experimental(i)} - X_{calculated(i)}}{X_{experimental(i)}}\right) \tag{5}$$

$$AARE = \frac{1}{N}\sum_{i=1}^{N}\left(\left|\frac{X_{experimental(i)} - X_{calculated(i)}}{X_{experimental(i)}}\right|\right) \tag{6}$$

$$SD = \sqrt{\frac{1}{N-1}\sum_{i=1}^{N}\left(\left|\frac{X_{experimental(i)} - X_{calculated(i)}}{X_{experimental(i)}}\right| - AARE\right)^2} \tag{7}$$

Where i is the sample number and N is the total number of samples which is three. The validity of the dead oil model, Eq. (1), is checked in Table 4. The proposed model shows that dead crude oil viscosity decreases as the API gravity and/or the reservoir temperature increases. Table 4 compares the behaviour of the proposed model in this study to those in previously published models. It is important to note that in this table, errors reported by the authors for their models when predicting dead crude oil viscosities are also shown. The table depicts that Kartoatmodjo and Schmidt [7] model has an average absolute percentage error as high as 40% in predicting dead oil viscosity of the crudes. It is obvious from the figure that the new correlation provides results in good agreement with experimental values.

References	Average relative error %	Average absolute error %	Standard deviation %
Beal [3]	28.5	31.6	37.3
Beggs and Robinson [3]	7.5	21.2	28.0
Glaso [5]	24.9	27.4	31.9
Labedi [6]	-16.9	29.7	42.6
Kartoatmodjo and Shmidt [7]	30.9	33.1	37.25
Present Study	-2.5	19.2	25.8

Table 4. Accuracy of dead crude oil models for estimating viscosity of Omani Crude Oils

In the following figure accuracy of the proposed model against the Omani crude oil data has been examined. As depicted in fig. 2, laboratory measured dead oil viscosity data are plotted against calculated oil viscosity data and the points fall on the line of. This obviously shows that the measured quantities for the viscosity of the samples well match with the calculated quantities from the correlation presented in section 3.

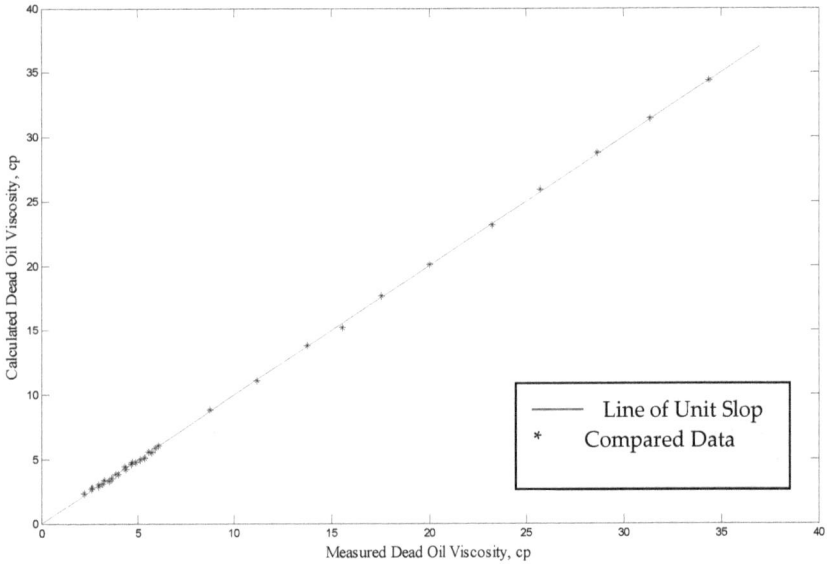

Figure 2. Crossplot of μ_{oD} versus calculated μ_{oD}(this study)

5. Conclusions

In this study, using the laboratory data of Omani Fahud-field, a new empirical viscosity correlation has been developed. The proposed correlation covers an acceptable range of validity, and is superior to other published correlations in the literature. The comparisons with previously published correlations showcased in section 4 supports the fact that the proposed correlation better predicts the viscosity of this type of crude oils.

Author details

Nabeel Al-Rawahi
Department of Mechanical & Industrial Engineering,
College of Engineering, Sultan Qaboos University, Muscat, Oman

Gholamreza Vakili-Nezhaad*, Ibrahim Ashour and Amin Fatemi
*Department of Petroleum and Chemical Engineering,
College of Engineering, Sultan Qaboos University, Muscat, Oman*

Gholamreza Vakili-Nezhaad
*Department of Chemical Engineering, Faculty of Engineering,
University of Kashan, Kashan, I. R. Iran*

Amin Fatemi
*Faculty of Civil Engineering and Geosciences, Geoscience & Engineering (G&E)
Department, Petroleum Section, Delft University of Technology, CN Delft, The Netherlands*

Acknowledgement

The research leading to these results has received funding from Petroleum Development Oman (PDO), Sultanate of Oman, through research agreement no. CTR #2009/111.

Nomenclature

T = temperature, ^{0}C
μ_{DO}= viscosity of gas-free/dead oil at T, cp
^0API = oil gravity, ^{0}API
ARE = average relative error
$AARE$ = absolute average relative error
SD = standard deviation

Appendix A. Dead oil viscosity models

A-1 Beal [1]

$$\mu_{od} = \left(0.32 + \frac{1.8(10^7)}{API^{4.53}} \right)\left(\frac{360}{T - 260} \right)a;$$

$$a = 10^{(0.43 + 8.33/API)}.$$

A-2 Beggs and Robinson [2]

$$\mu_{od} = 10^x - 1;$$
$$x = y(T - 460)^{-1.163},$$
$$y = 10^z,$$

A-3 Glasø [3]

$$\mu_{od} = [3.141(10^{10})](T - 460)^{-3.444}[\log(API)]^a;$$
$$a = 10.313[\log(T - 460)] - 36.447.$$

---

* Corresponding Author

A-4 Labedi [4]

$$\mu_{od} = \frac{10^{9.224}}{\text{API}^{4.7013}T_f^{0.6739}}.$$

A-5 Kartoatmodjo and Schmidt [5]

$$\mu_{od} = 16(10^8)T_f^{-2.8177}(\log \text{API})^x;$$
$$x = 5.7526\log(Tf) - 26.9718.$$

6. References

[1] Salem S. Al-Marri, SPE, Kuwait Institute for Scientific Research, "Optimal Transformations for Multiple Regression: Application to Oil Viscosity Correlation Above and Below Bubble Point Pressures, SPE Saudi Arabia Section Technical Symposium, 9-11 May 2009, AlKhobar, Saudi Arabia

[2] M.S. Hossain, SPE, C. Sarica, SPE, and H.-Q. Zhang, SPE, The U. of Tulsa, and L. Rhyne and K.L. Greenhill, SPE, Chevron, "Assessment and Development of Heavy Oil Viscosity Correlations ", 2005, SPE 97907-MS.

[3] Beal, C. "Viscosity of Air, Water, Natural Gas, Crude Oil and Its Associated Gases at 011-FieldTemperatures and Pressures," *Trans.,*AIME (1946) 165, 94-115.

[4] Chew, J. and Connally, C. A.: "A Viscosity Correlation for Gas- Saturated Crude Oils" Trans., AIME (1959) 216, 23-25.

[5] Lohrenz, J., Bray, B. G., and Clark, C. R.:"Calculating Viscosities of Reservoir Fluids from their Compositions" Journal of Petroleum Technology (Oct. 1964) 1171-1176 *Trans.,* AIME, 231.

[6] Glasø O.,"Generalized pressure–volume–temperature correlation for crude oil system" Journal of Petroleum Technology 1980;2:785–795.

[7] Labedi R. "Improved correlations for predicting the viscosity of light crudes", Journal of Petroleum Science Engineering (1992);8:221–234.

[8] Kartoatmodjo F, Schmidt Z. Large data bank improves crude physical property correlation. Oil and Gas Journal(1994);4:51–55.

[9] Houpemt,A. H. andLrelliez, M. B.: "Predicting the Viscosity of Hydrocarbon Liquid Phases From Their Composition,.' paper SPE 5057 presented at the SPE-AIME 49th Annual Fall Meeting, Houston, Oct. 6-9, 1974.

[10] Little. J. E. and Kennedy, H. T.: "A Correlation of the Viscosity of Hydrocarbon Systems With Pressure, Temperature and Composition," Sot. *Pet. Eng. J.* (June 1968) 157-162;*Trans.,* AIME, 243.

[11] McCain, W.D., Jr.: "Reservoir-Fluid Property Correlations – State of the Art," SPE Reservoir Engineering (May 1991) 266-272.

[12] Birol Dindoruk and Peter G. Christman, Shell Int. E & P. Inc., "PVT Properties and Viscosity Correlations for Gulf of Mexico Oils", SPE 71633

Permissions

The contributors of this book come from diverse backgrounds, making this book a truly international effort. This book will bring forth new frontiers with its revolutionizing research information and detailed analysis of the nascent developments around the world.

We would like to thank Chaoqun Liu, for lending his expertise to make the book truly unique. He has played a crucial role in the development of this book. Without his invaluable contribution this book wouldn't have been possible. He has made vital efforts to compile up to date information on the varied aspects of this subject to make this book a valuable addition to the collection of many professionals and students.

This book was conceptualized with the vision of imparting up-to-date information and advanced data in this field. To ensure the same, a matchless editorial board was set up. Every individual on the board went through rigorous rounds of assessment to prove their worth. After which they invested a large part of their time researching and compiling the most relevant data for our readers. Conferences and sessions were held from time to time between the editorial board and the contributing authors to present the data in the most comprehensible form. The editorial team has worked tirelessly to provide valuable and valid information to help people across the globe.

Every chapter published in this book has been scrutinized by our experts. Their significance has been extensively debated. The topics covered herein carry significant findings which will fuel the growth of the discipline. They may even be implemented as practical applications or may be referred to as a beginning point for another development. Chapters in this book were first published by InTech; hereby published with permission under the Creative Commons Attribution License or equivalent.

The editorial board has been involved in producing this book since its inception. They have spent rigorous hours researching and exploring the diverse topics which have resulted in the successful publishing of this book. They have passed on their knowledge of decades through this book. To expedite this challenging task, the publisher supported the team at every step. A small team of assistant editors was also appointed to further simplify the editing procedure and attain best results for the readers.

Our editorial team has been hand-picked from every corner of the world. Their multi-ethnicity adds dynamic inputs to the discussions which result in innovative outcomes. These outcomes are then further discussed with the researchers and contributors who give their valuable feedback and opinion regarding the same. The feedback is then collaborated with the researches and they are edited in a comprehensive manner to aid the understanding of the subject.

Apart from the editorial board, the designing team has also invested a significant amount of their time in understanding the subject and creating the most relevant covers. They scrutinized every image to scout for the most suitable representation of the subject and create an appropriate cover for the book.

The publishing team has been involved in this book since its early stages. They were actively engaged in every process, be it collecting the data, connecting with the contributors or procuring relevant information. The team has been an ardent support to the editorial, designing and production team. Their endless efforts to recruit the best for this project, has resulted in the accomplishment of this book. They are a veteran in the field of academics and their pool of knowledge is as vast as their experience in printing. Their expertise and guidance has proved useful at every step. Their uncompromising quality standards have made this book an exceptional effort. Their encouragement from time to time has been an inspiration for everyone.

The publisher and the editorial board hope that this book will prove to be a valuable piece of knowledge for researchers, students, practitioners and scholars across the globe.

List of Contributors

Ping Lu, Manoj Thapa and Chaoqun Liu
University of Texas at Arlington, Arlington, TX, USA

Ondrej Zavila
Vysoka Skola Banska, Technical University of Ostrava, Faculty of Safety Engineering, Ostrava, Vyskovice, Czech Republic

Hatem Kanfoudi, Hedi Lamloumi and Ridha Zgolli
Laboratory of Hydraulic and Environmental Modelling, National Engineering School of Tunis, Tunis, Tunisia

N.M.S. Hassan, M.M.K. Khan and M.G. Rasul
Power and Energy Research Group ,Institute for Resource Industries and Sustainability (IRIS), Faculty of Sciences, Engineering and Health Central Queensland University, Rockhampton, Queensland, Australia

Desmond Adair
University of Tasmania, Australia

Yiin-Kuen Fuh, Wei-Chi Huang and Jia-Cheng Ye
National Central University, Department of Mechanical Engineering & Institute of Energy, Taiwan, ROC

Antonio Vitale and Federico Corraro
Italian Aerospace Research Centre, Italy

Guido De Matteis and Nicola de Divitiis
University of Rome "Sapienza", Italy

Ladislav Halada, Peter Weisenpacher and Jan Glasa
Institute of Informatics, Slovak Academy of Sciences, Slovakia

Ralph Wai Lam Ip and Elvis Iok Cheong Wan
Department of Mechanical Engineering, The University of Hong Kong, Pokfulam, Hong Kong SAR

Emre Alpman
Department of Mechanical Engineering, Marmara University, Istanbul, Turkey

Ricardo F. Oliveira, José C. Teixeira and Luís F. Silva
Mechanical Engineering Department, University of Minho, Portugal

Senhorinha Teixeira
Production and System Department, University of Minho, Portugal

Henedina Antunes
Pediatric Department, Braga Hospital, Portugal
Life and Health Sciences Research Institute, School of Health Sciences, University of Minho, Portugal
ICVS/3B's - PT Government Associate Laboratory, Braga/Guimarães, Portugal

Nabeel Al-Rawahi
Department of Mechanical & Industrial Engineering, College of Engineering, Sultan Qaboos University, Muscat, Oman

Gholamreza Vakili-Nezhaad, Ibrahim Ashour and Amin Fatemi
Department of Petroleum and Chemical Engineering, College of Engineering, Sultan Qaboos University, Muscat, Oman

Gholamreza Vakili-Nezhaad
Department of Chemical Engineering, Faculty of Engineering, University of Kashan, Kashan, I. R. Iran

Amin Fatemi
Faculty of Civil Engineering and Geosciences, Geoscience & Engineering (G&E), Department, Petroleum Section, Delft University of Technology, CN Delft, The Netherlands